AF606465

FLOW AND TRANSPORT IN POROUS MEDIA

PROCEEDINGS OF EUROMECH 143/DELFT/2 - 4 SEPTEMBER 1981

Flow and Transport in Porous Media

Edited by

A.VERRUIJT
University of Delft

F.B.J. BARENDS
Delft Soil Mechanics Laboratory

A.A.BALKEMA/ROTTERDAM/1981

The texts of the various papers in this volume were set individually by typists under the supervision of each of the authors concerned.

ISBN 90 6191 216 4

© 1981 A.A.Balkema, P.O.Box 1675, Rotterdam, Netherlands

Distributed in USA & Canada by MBS, 99 Main Street, Salem, NH 03079

Printed in the Netherlands

Contents

Preface

The objective of Euromech Colloquia is to provide opportunities for European scientists, working in a specialized area of mechanics, to meet each other, and to discuss their current research activities. Meetings are rather small in size and informal in character.

In March 1980 the European Mechanics Committee decided to include a colloquium on Flow and Transport in Porous Media in the program for 1981, to be held at Delft, The Netherlands.

Because the response to the first announcement, which was sent to a selected number of prospective participants, was rather encouraging, with a variety of papers being offered, it was decided to print the papers submitted before the colloquium in a separate volume, published by A.A. Balkema, Rotterdam. The organizers are very happy that a large number of participants managed to complete their manuscript in the short time available.

It is our belief that the present book gives an up-to-date review of the state of the art in the field of Flow and Transport in Porous Media. This branch of science is rapidly growing, especially because of the importance of problems of pollution of groundwater, of heat transport and storage of solar energy in the ground, of hydraulics and mechanics in large granular structures, and of the simultaneous flow of different fluids in porous media. All these subjects are covered in this volume, together with a number of papers on the fundamental aspects of flow through porous media.

A. Verruijt
F.B.J. Barends

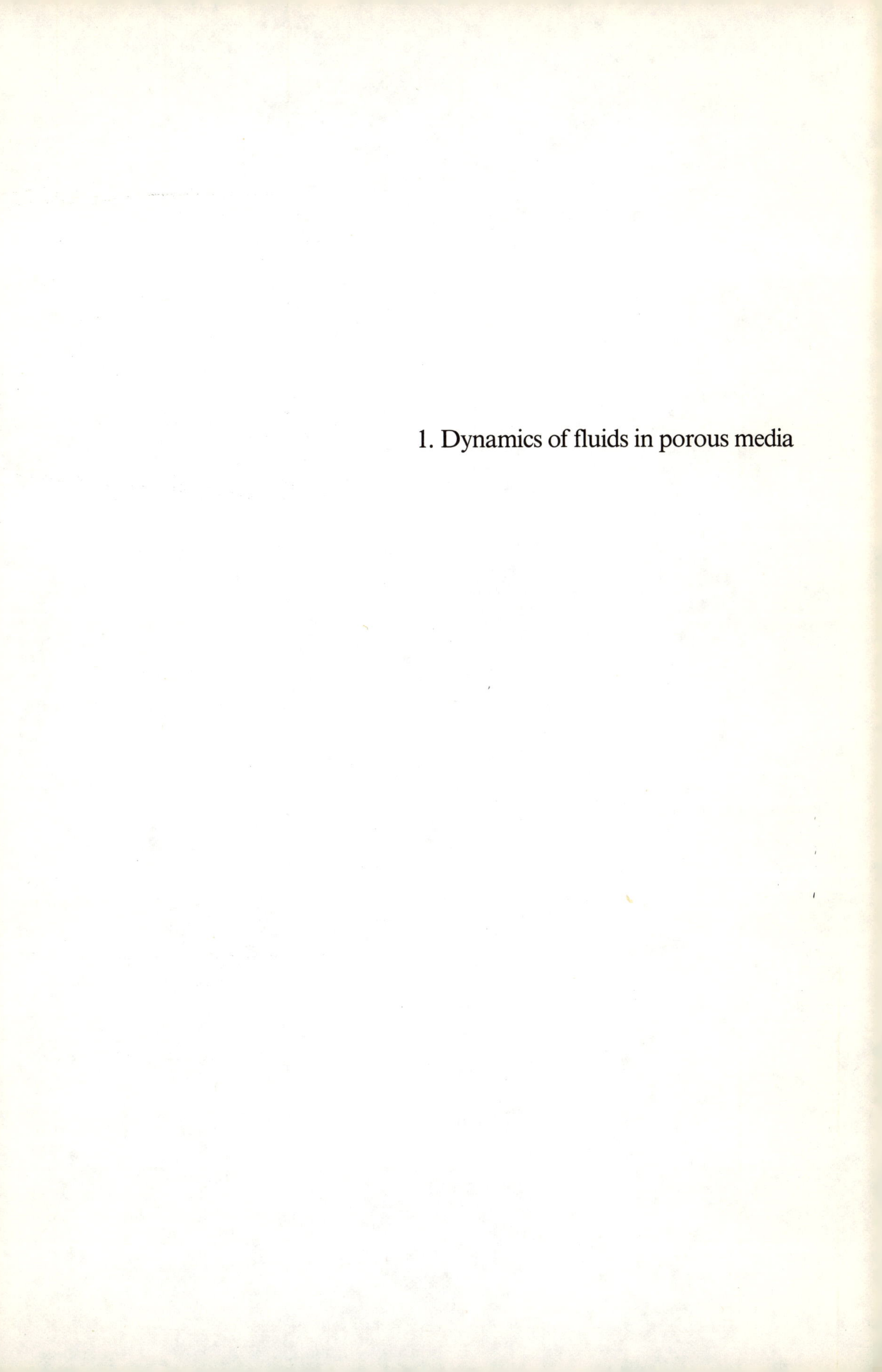

1. Dynamics of fluids in porous media

Proceedings of Euromech 143 / Delft / 2-4 September 1981

Shock-induced flow in a porous medium

B.ROGG, D.HERMANN & G.ADOMEIT
Institut für Allgemeine Mechanik, RWTH Aachen, Germany

ABSTRACT

First experiments and evaluations were made of shock-induced flow in a solid matrix. The experiments were performed in a shock tube measuring pressure in front of and within the solid matrix as a function of time and location. Qualitative information on the flow pattern was obtained from shadowgraphs. These experimental data were compared with a numerical solution obtained by solving the transport equations for two-phase flow. Tentative adjustments of the drag and heat-transfer laws of Ergun (1952) and Yoshida et al. (1962) were made to fit the numerical results to the experimental data.

1 INTRODUCTION

Flow through porous media and packed beds occur in many branches of science and technology. A number of empirical correlations about the terms which describe momentum and energy exchange between the gas phase and the matrix of solid material are available. They have however been established for low Reynolds numbers and their extrapolation to high Reynolds numbers is not permissible, as was shown by Kuo and Nydegger (1978). They also fail at high Mach numbers and steep pressure gradients. Also time dependence may become important, e.g. when the flow is induced by a shock wave, as was found by Heilig and Reichenbach (1979, 1980) who studied shock loading experimentally. They found that the drag coefficient for a single cylindrical body is strongly time dependent and its value can exceed the steady state value by a factor of up to 30, depending on Mach and Reynolds numbers. Schultz-Grunow (1972) studied the interaction of an incident shock wave with obstacles along walls. He observed increased pressure peaks depending on the details of the geometrical arrangement. Zloch (1974) investigated the decrease of shock strength in an array of axially arranged tublets experimetally and numerically. He used a shock tube to generate the incident shock. By making the area change rather large, he achieved steady-state boundary conditions at the entrance into the system of tublets. However, this is only a limiting case. For medium and large fractions of void volume the boundary conditions at the surface of the matrix change with time after shock impingement. This may be understood as an obstruction effect increasing with time.

In the present paper this more general case is studied experimentally and theoretically and first results are presented. The experiments are performed in a shock tube taking as matrix either a packed bed or an array of cylinders, with their axis perpendicular to the tube axis. Pressure profiles are measured in dependence upon time and location in front of and within the matrix. Also shadowgraphs are taken to obtain qualitative informations on the shock-induced flow pattern. These data are compared with numerical results obtained by solving the basic equations of two-phase flow using a method of characteristics. Particularly account was taken of coupling the two regions of flow in front of and within the matrix. For the drag and heat transfer the relations given by Ergun (1952) and Yoshida et al. (1962) were used. They had to be modified by a constant factor to obtain a first agreement between computational results and experimentally obtained pressure profiles. More detailed measurements of pressure and of heat transfer are in progress. It is expected that they will provide sufficient information to establish the drag and heat transfer laws valid under the conditions in more detail.

2 EXPERIMENTAL APPARATUS

The shock tube arrangement used for the experiments is represented schematically in Fig. 1. It consists of a tube of rectangular cross section 54 x 54 mm with the test section arranged at the end. The resulting

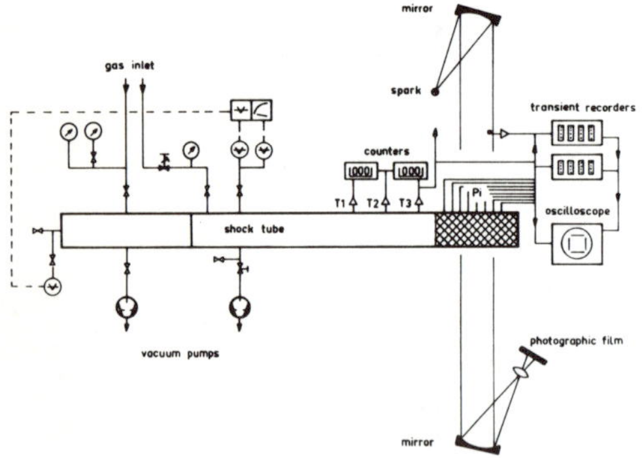

Fig. 1 Schematic diagram of the shock tube

wave system is represented schematically in Fig. 2. The Mach number of the incident shock wave was determined using film temperature gages. Pressure profiles were measured using seven gages of the type Kistler 603 B, 6031 and Piezotronics 113A21, 113A24. The signals were registered using oscilloscopes and transient recorders.

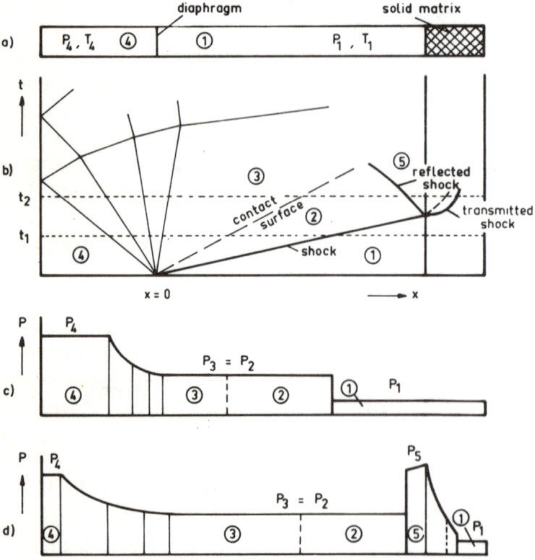

Fig. 2 Schematic representation of the wave system

Optical information on the flow pattern was obtained using a shadowgraph arrangement with single spark flashes initiated at suitable time intervals. A Cranz-Schardin multiple spark system is in preparation. As solid matrix a packed bed of spherical glas pellets of diameter 4, 5, 6 and 8 mm and also an array of cylinders 5 mm in diameter arranged at distances of 9 mm perpendicular to the tube axis was investigated.

3 EXPERIMENTAL RESULTS

Some of the experimental results are represented in Figs. 3 to 8. Figs. 3 to 6 refer to the fixed bed and Fig. 7 and 8 refer to the array of cylinders.

The shock wave incident upon the solid matrix is partially reflected and partially transmitted by the elements of the matrix. The complexity of the resulting wave and flow pattern may be gathered from the shadowgraphs of Fig. 7. In front of the array a first strong reflection is followed by various weaker shock waves which lead to an acceleration of the reflected shock and a continuous pressure increase in this region. Fig. 3 shows some typical pressure records obtained for a packed bed. The

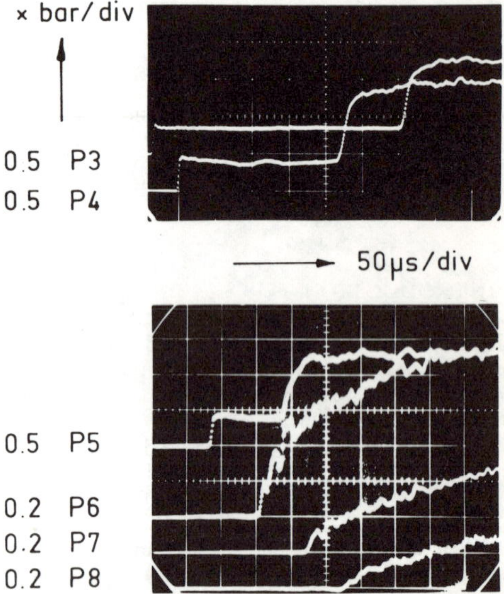

Fig. 3 Pressure records obtained for a packed bed

pressure gages No. 3 to 5 were placed in front of and No. 6 to 8 inside the bed. Probe No. 6 shows three distinctive pressure peaks which indicate shock waves, whereas probe No. 8 at a location farther inside the bed shows already an essentially continuous pressure increase, which however, as Fig. 7 shows, may still be caused by a multiplicity of weak shock waves.

Plotting the pressures measured by the

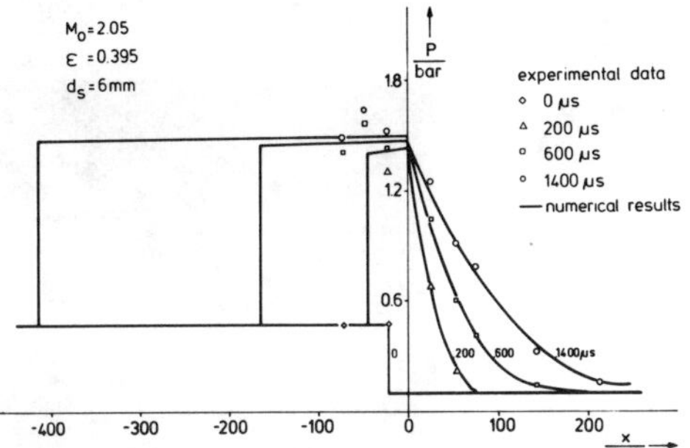

Fig. 4 Pressure p vs. location x for various times t. Points: measurements. Solid curves were obtained by numerical solution adjusting D and q.

different gages at fixed times vs. location the representations shown in Figs. 4 to 6 are obtained. In Fig. 4 values of pressure readings at various time intervals are shown. In Fig. 5 results are given for various pellet diameters and in Fig. 6 for various Mach numbers of the incident shock wave. The

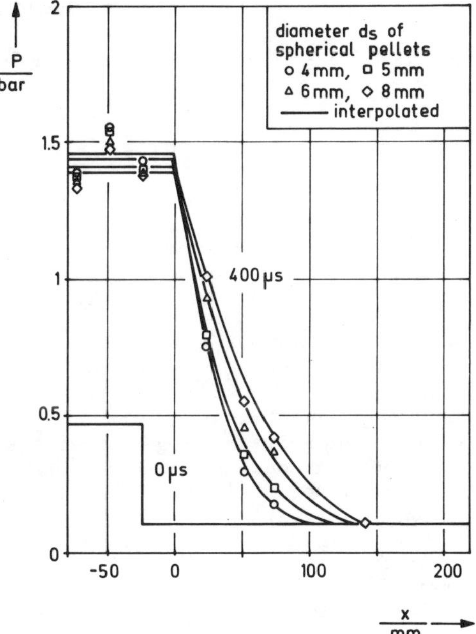

Fig. 5 Pressure p vs. location x for various diameters d_s of pellets. Points: measurements. Solid curves obtained by interpolating the measurements.

pressure level in front of the bed increases with decreasing diameter of the pellets and increasing Mach number approaching in the limit the value of a solid end wall.

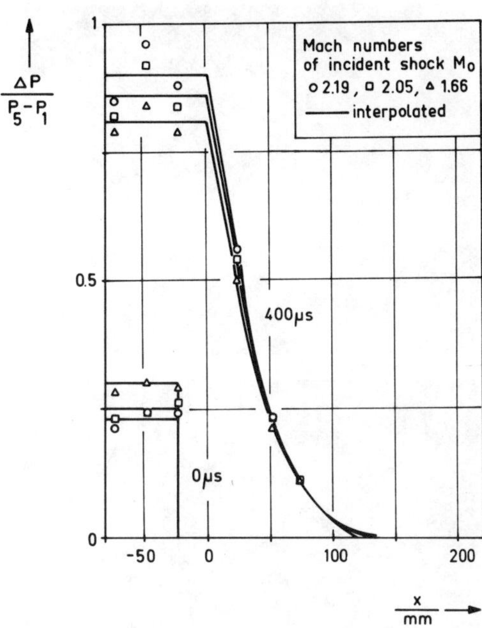

Fig. 6 Pressure p vs. location x for various Mach numbers M_o of incident shock. Points: measurements. Solid curves obtained by interpolating the measurements.

The results obtained with the array of cylinders are illustrated in Figs. 7 and 8.

Fig. 7 shows shadowgraphs of the resulting wave and flow pattern. Its various features may be explained by comparison with the flow and wave pattern observed for a single cylinder (Heilig 1980). Remarkable is the strong effect of the boundary layer along the front of the cylinders upon the light refraction and also the formation of dual vortices behind the obstacles. Between the cylinders a supersonic jet is formed which leads to a bow shock wave in front of the next cylinder. This jet and the recirculation region is separated by a shear layer well visible e.g. in the second photograph of Fig. 7. The pressure readings obtained at the locations 4 to 7 as indicated in Fig. 7 are shown in Fig. 8. The pressure curve of gage No. 4 shows quite clearly the incident shock wave and the pattern of the reflected shock waves. Also at the probes located inside the arry various waves may be clearly distinguished. The pressure values belonging to the various regions of the flow field are obtained from these measurements.

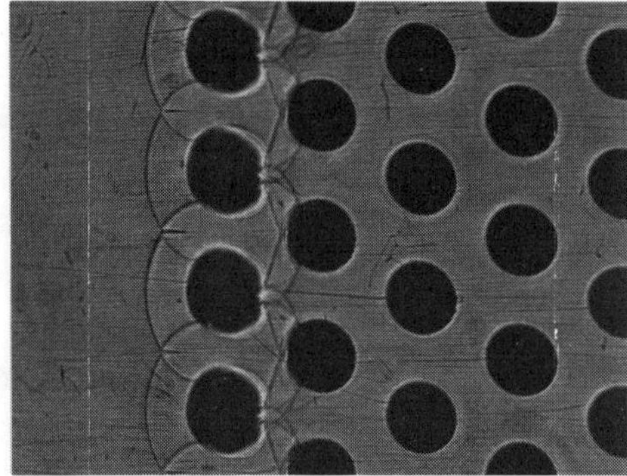

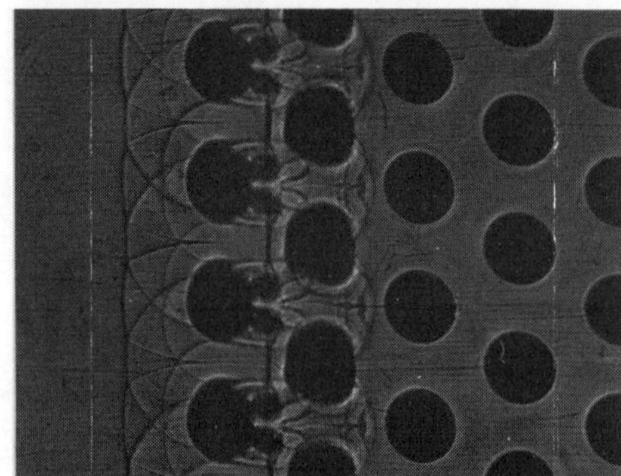

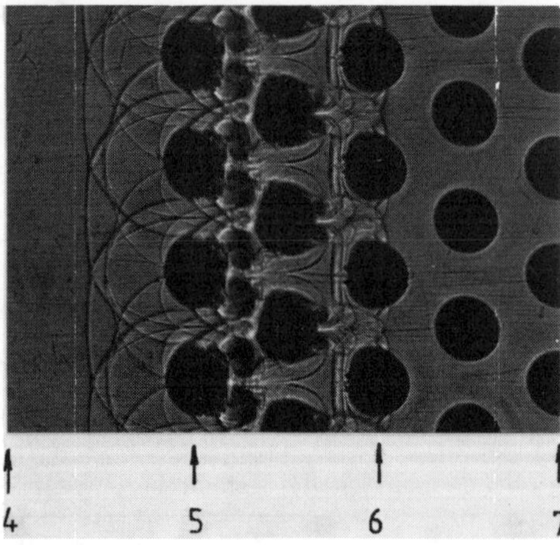

Fig. 7 Shadowgraphs of the shock induced flow through a cylindrical array taken at three distinct time intervals after shock impingement. Arrows indicate the location of pressure probes.

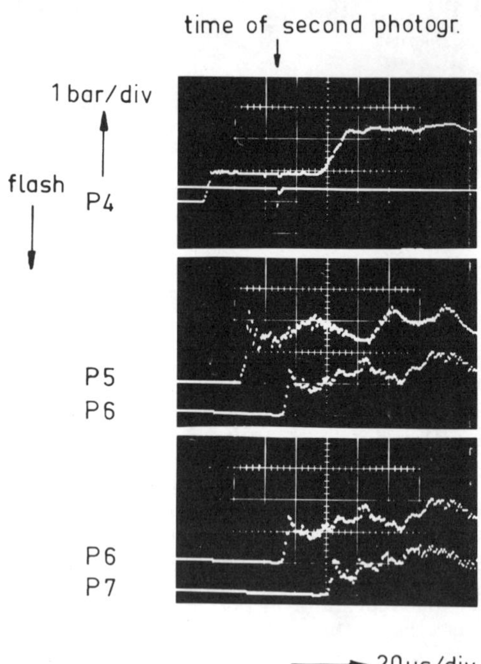

Fig. 8 Pressure readings corresponding to Fig. 7

4 THEORY: BASIC EQUATIONS

In the one-dimensional conservation equations for unsteady two-phase flow (Wallis 1969) the following terms are neglected:

a) heat loss to the tube wall
b) friction force exerted on the gas by the tube wall
c) work done due to viscous stress generated from velocity gradient in the gas phase
d) homogeneous heat dissipation
e) heat conduction in the gas phase
f) temperature dependence of constant pressure specific heat.

After rearranging various terms and inserting the perfect gas law the following set of partial differential equations is obtained (Wallis 1969, Zucrow and Hoffman 1976)

$$\frac{\partial \rho}{\partial t} + u \frac{\partial \rho}{\partial x} = -\rho \frac{\partial u}{\partial x} \tag{1}$$

$$\frac{\partial u}{\partial t} + u \frac{\partial u}{\partial x} = -\frac{1}{\rho}\left(\frac{\partial p}{\partial x} + \frac{s}{\varepsilon} D\right) \tag{2}$$

$$\frac{\partial p}{\partial t} + u \frac{\partial p}{\partial x} = -\gamma p \frac{\partial u}{\partial x} - (\gamma - 1)\frac{s}{\varepsilon}(q - Du) \tag{3}$$

The heat flux q and the drag force D are defined by (Bird et al. 1960, Krier and

Rajan 1975)

$$q = h(T - T_s) \quad (4)$$

$$D = \frac{\varepsilon^3}{3(1-\varepsilon)} \rho u^2 f \quad (5)$$

T_s denotes the surface temperature of the matrix material and f is the nondimensional friction factor. For spherical pellets the specific surface s is (Bird et al. 1960)

$$s = \frac{6(1-\varepsilon)}{d_s} \quad (6)$$

Under assumption of an uniform pellet temperature an energy balance leads to the following ordinary differential equation for T_s

$$\frac{\partial T_s}{\partial t} = \frac{s}{1-\varepsilon} \frac{q}{\rho_s c_s} \quad (7)$$

These equations are used to describe the flow inside the solid matrix and also - setting D=q=0 - the flow in the upstream region in front of the matrix. These two regions are coupled by

the energy equation for quasi-steady flow of perfect gas

$$\frac{\gamma}{\gamma-1} \frac{p}{\rho} + u^2/2 = \text{const.} \quad (8)$$

the isentropic law

$$p/\rho^\gamma = \text{const.} \quad (9)$$

and the continuity equation for quasi-steady flow

$$\rho u A = \text{const.} \quad (10)$$

The applicability of these equations as coupling conditions rests upon the fact that the extension of the transition zone between solid matrix and free flow is small compared to the characteristic length of the problem. The conditions have e.g. been discussed by Laporte (1954). They establish a dependence between porosity ε, the states in front of the incident shock wave and the Mach numbers of the incident, reflected and transmitted shock waves.

For the friction factor f the following relation (Ergun 1952, Bird et al. 1960) was used.

$$f = \frac{(1-\varepsilon)^2}{\varepsilon^3} \frac{75}{Re_s} + 0.875 \frac{(1-\varepsilon)}{\varepsilon^3} \quad (11)$$

Here the Reynolds number Re_s is defined by

$$Re_s = \varepsilon \frac{\rho u d_s}{\mu} \quad (12)$$

The heat transfer law as given by Yoshida et al., 1962 (see also Bird et al. 1960) was used

$$j_H = 0.91\, Re^{-0.51} \quad (Re < 50)$$
$$j_H = 0.61\, Re^{-0.41} \quad (Re > 50) \quad (13)$$

where the so-called Colburn factor j_H and the Reynolds number Re are defined by

$$j_H = \frac{h}{\rho u c \varepsilon} Pr^{2/3} \quad (14)$$

$$Re = \frac{\rho u \varepsilon}{s \mu \psi} \quad (15)$$

For spherical pellets the shape factor ψ is equal to one.

5 RESULTS AND DISCUSSION

In Fig. 9 some profiles obtained numerically for pressure p, temperature T and flow velocity u are plotted vs. coordinate at t = 300 μs after the shock impingement upon the solid matrix. In front of the matrix the reflected shock wave propagates to the left. A continuous increase of p and T and a decrease of u is observed between this shock wave and the surface. This may be understood as an obstruction effect of the matrix increasing with time. At the front of the matrix (location at x=0) the gas observes the balance equations formulated in Eqs. (8) to (10) resulting in an acceleration of the flow in a zone assumed here to be infinitely thin. This leads to the discontinuous changes of variables of state at x=0, causing the observed drops of pressure and temperature. The transmitted shock wave located at x ≈ 130 mm has already been attenuated considerably caused by the drag force D. At the contact surface between the gas originally seperating the matrix and the gas outside, a temperature jump and a pronounced peak are observed.

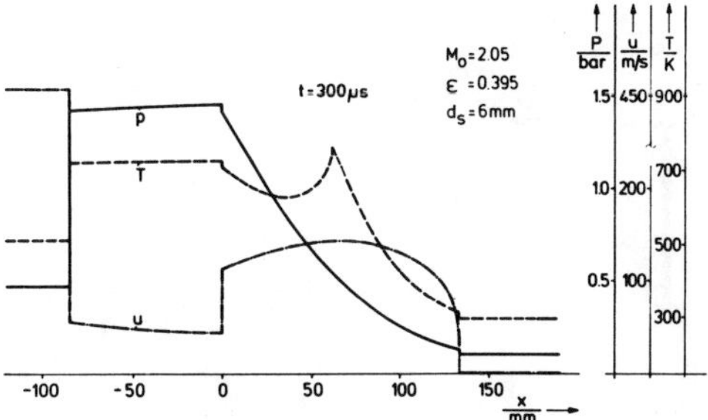

Fig. 9 Profiles of T, u and p 300 μs after shock impingement

In Fig. 10 the various influences contributing to this effect can be studied. Here the temperature profiles are plotted vs. x for a fixed time interval introducing various values for drag D and heat transfer q. A comparison between curves 1 and 2 shows that the heat flux q gives only a small contribution to the reduction of the shock strength. The comparison between curves 1 and 2 on the one side and 3 to 5 on the other side shows that this reduction is essentially due to the drag force term in the balance of momentum (Eq. (2)). Curves 3 and 5 differ in the presence of the drag force term in the energy balance (Eq. (3)), which represents the heat dissipation caused by the flow resistance of the solid matrix. It can be gathered that this term causes the strong temperature increase in the environment of the contact surface, which can be observed in comparing these two curves.

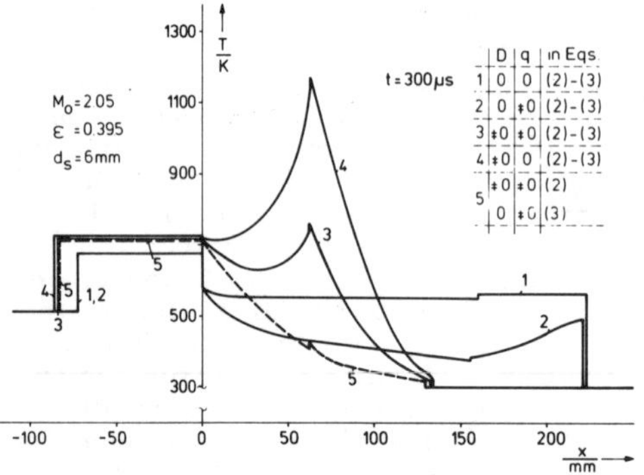

Fig. 10 Numerical solution of T vs. x for various values of D and q in Eqs. (2)-(3)

Fig. 11 shows the pressure profiles corresponding to the temperature profiles of Fig. 10 exhibiting again the strong difference in shock attenuation caused by the drag force in the momentum equation.

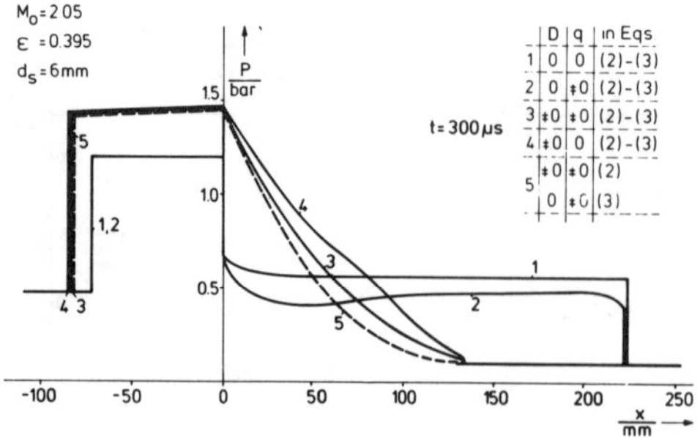

Fig. 11 Numerical solution of p vs. x for various values of D and q in Eqs. (2)-(3)

Results obtained by the numerical computation have also been plotted in Fig. 4 (solid curves). It is found that the dependences of drag force D and heat transfer q as formulated by Eqs. (11) and (13) lead to pronounced deviation between experiment and theory. Acceptable agreement for the pressure distribution inside the packed bed is achieved by multiplying the drag force term D by a factor of 1.5 and the heat-transfer term q by a factor of 4. However, this procedure does not lead to a good agreement in front of the matrix. The fact, that the original terms have to be altered may be essentially due to the unsteady behavior of the flow and to the high Reynolds and Mach numbers when the flow initially is induced by the transmitted shock wave.

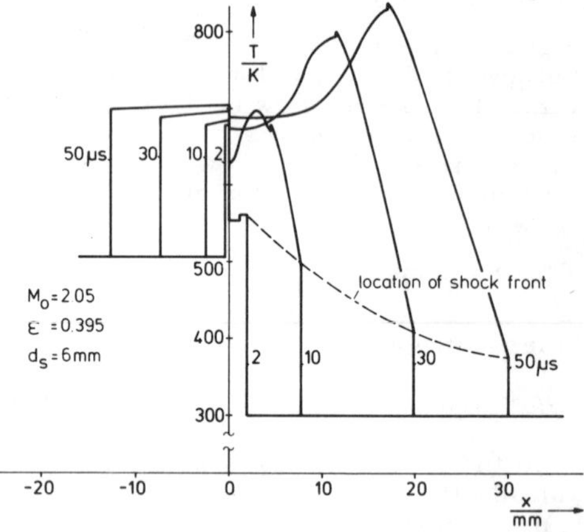

Fig. 12 Development of temperature profiles and decrease of shock strength with time obtained by numerical solution

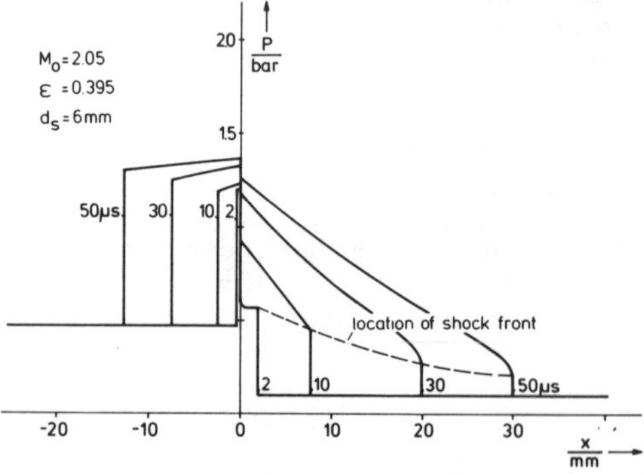

Fig. 13 Development of pressure profiles and decrease of shock strength with time obtained by numerical solution

Figures 12 and 13 show the development of temperature and pressure profiles within the solid matrix for short time intervals.

It can be seen, that the attenuation of the transmitted shock is very rapid. The reflected shock, however, is reenforced due to the obstruction of the matrix increasing with time.

6 CONCLUSIONS

Shock loading of a solid matrix consisting of a packed bed of spherical pellets and an array of cylinders perpendicular to the flow has been investigated experimentally and theoretically. The experiments were performed in a shock tube measuring the pressure distribution in dependence upon time with the Mach number of the incident shock and with pellet diameter as parameters. Also shadowgraphs were taken giving qualitativ information about the flow and wave pattern. All the essential results obtained experimentally so far (cf. Sec. 3) were confirmed by comparison with a numerical solution obtained by solving the basic balance equations of two phase flow (cf. Sec. 5). The relations used commonly for the drag force and heat transfer between the gas and the solid matrix had to be adjusted. It was found that in the average for the investigated shock induced flow they posses a higher value than under steady state conditions. For the tentative correction chosen the pressure development inside the bed could be represented sufficiently well, however not the distribution in front of the bed. Further measurements are in progress to obtain additional information about the interaction terms.

Nomenclature

- A cross-sectional area of the tube
- c specific heat
- D drag force acting on gas per unit wetted area of porous medium or of particles
- d_s diameter of pellets
- f friction factor
- h heat transfer coefficient
- j_H Colburn-factor
- M_o Mach number of incident shock
- p pressure
- Pr Prandtl number
- q heat flux per unit wetted area of porous medium or of particles
- Re Reynolds number
- s specific wetted surface area of porous medium or of packed bed
- t time
- T absolute temperature
- u velocity
- x coordinate

Greek Symbols

- γ ratio of specific heats
- ε porosity, volume fraction of void
- μ dynamic viscosity
- ρ density

Subscript

- s solid

REFERENCES

Bird, R.B., Stewart, W.E., Lightfoot, E.N. 1960, Transport Phenomena. New York, John Wiley.

Ergun, S. 1952, Fluid Flow Through Packed Columns, Chem. Engrg. Progr., 48: 89-94.

Heilig, W. 1980, Air Blast Loading of a Cylindrical Body, Int. Symp. on Flow Visualisation, Bochum: 544-549.

Heilig, W., Reichenbach, H. 1979, Measurements of the Unsteady Drag Coefficient of a Circular Cylinder, Sixième Symposium Internationale sur les Applications Militaires de la Simulation de Souffle, premiere partie, Cahors, France.

Krier, H., Rajan, S. 1975, Flame Spreading and Combustion in Packed Beds of Propellant Grains, AIAA 13th Aerospace Sciences Meeting, No. 75-240.

Kuo, K.K., Nydegger, C.C. 1978, Flow Resistance Measurement and Correlation in a Packed Bed of WC 870 Ball Propellants, Journal of Ballistics 2, 1-26.

Laporte, O. 1954, On the Interaction of a Shock with Constriction, Los Alamos Sci. Lab., Univ. of California, Rep. No. LA-1740.

Schultz-Grunow, F. 1972, Diffuse Reflexion einer Stoßwelle, EMI Freiburg, Bericht Nr. 7/72.

Wallis, G.B. 1969, One-Dimensional Two-Phase Flow. New York, McGraw-Hill.

Yoshida, F., Ramaswami, D., Hougen, O.A. 1962, Temperatures and Partial Pressures at the Surfaces of Catalyst Particles, AI. Ch. E. Journal, 8: 5-11.

Zloch, N. 1974, Abschwächung von Verdichtungsstößen infolge Reibungswirkungen und Wärmeübergang bei vollausgebildeter turbulenter Strömung, Ph.D. thesis, Technische Hochschule Darmstadt.

Zucrow, M.J., Hoffman, J.D. 1976, Gasdynamics, New York, John Wiley.

Proceedings of Euromech 143 / Delft / 2-4 September 1981

Landsubsidence due to a well in an elastic saturated subsoil

FRANS B.J.BARENDS
Delft Soil Mechanics Laboratory, Netherlands

SYNOPSIS

The landsubsidence induced by a gradual increase of groundwater recovery for domestic, industrial and agricultural use and by the production of oil and gas becomes more apparent, and the consequences for the environment become manifest. The shrinkage of compressible saturated soil layers bares a lag due to the porous flow process. Therefore, the actual observation is insufficient to merely predict subsidence. To provide correct measurements for the future a thorough comprehension of the occuring processes is required. In this article the characteristic properties of surface settlements are explained through a fundamental analysis. Comparison with measured data confirm the results obtained.

INTRODUCTION

Landsubsidence usually originates from a physical, a chemical or a mechanical process. Setting due to desiccation and erosion caused by internal or external material transport are considered as a physical process. Both can be affected by changes in the geo-hydrological equilibrium. Oxidation by bacterial activity and fire are chemical in nature. The physical and chemical processes mainly take part in the soil's upper layer(s).

To the mechanical processes belong settlement due to surface loading and shrinking or swelling caused by internal loading. The mechanical processes are not necessarily restricted to the top layer of the subsoil. Mostly, the effect decreases with depth, but this is not true where in a deep soil layer the existing situation is altered, for example by a deep well. The corresponding surface settlements will depend on the stratification and the composition of the subsoil, on the geometry of the area of recovery and on the production strategy.

As a consequence of local settlements constructions and building might be seriously damaged. This kind of damage has been observed as a result of mining activity (8 metre settlements in Limburg, the Netherlands). Large scale settlements may occur in polder reclamation. The IJ-polder near Amsterdam shows a subsidence of about 2 metre during a period of a century. The Noord-Oost polder in the Netherlands dates more recently and a settlement of some decimetres has yet been recorded. Due to oil and gas recovery well-known cases of subsidence are to be mentioned, such as in Wilmington in U.S.A. a subsidence of 9 metre and at the Bolivar Coast in Venezuela up to 4 metre (Heijnen, 1979). In the north of the Netherlands the exploration of existing gas fields has produced a maximum settlement of about 25 cm.

In general the extent of the subsidence will increase with time according to the dimension of the area of recovery. A serious measurement might be a complete stop of production (Venice, Italy).

A prediction about the development of the settlement process and a realistic estimation of the maximum possible settlements can only be provided by virtue of a good knowledge of the geological history and of the mechanical properties (Gambolati c.s., 1974). Usually this knowledge is insufficient and one is committed to apply simplified models which in the best case are calibrated with respect to measured data or are used for sensitivity studies. Such a model describes the reality poorly. Nevertheless, it may contribute substantially to a better insight and its use may lead to simple and praktically applicable results.

In this discussion the attention is drawn to the mechanical process of shrinking and swelling (consolidation) caused by oil or groundwater recovery. The fundamental case of a single well in a saturated, isolated, homogeneous porous halfmedium will be analysed following the theory of linear consolidation. A parametric formula for the subsidence due to a constant production by a single well will be derived and it is compared with some measured data.

THEORY OF LINEAR CONSOLIDATION

Consolidation refers to a deformation process, in fact the deformation of a two-phase medium consisting of a pore fluid and a deformable porous matrix. In this process the hydrodynamic behaviour of the pore fluid and the stress-strain behaviour of the porous matrix play a significant role.

The fundamentals of the consolidation theory are clearly described in literature (Biot, 1941; De Josselin de Jong, 1963; Verruijt, 1969). The essence can be summarized as follows.

"In a saturated porous medium the pore fluid prevents loading changes to be carried by the porous matrix solely. However, the pore fluid contributes as far as volume changes are concerned. As a consequence the pore pressure alters, referred to as the excess pore pressure. Due to differences in the excess pore pressure the pore fluid will flow through the pores meeting the internal resistance of the porous matrix. When the pore fluid flows out, the porous matrix will deform untill it carries the complete loading. The excess pore pressure has vanished accordingly".

In the case where a well produces a constant discharge from a deformable aquifer, a similar process occurs, but now the loading is caused by the pore pressure itself. The final state will give a stationary porous flow pattern, in which the generated internal stresses in each phase are in a state of equilibrium.

If one assumes that the deformation behaviour of the porous matrix is linear-elastic, and that Darcy's law describes the nature of the porous flow, then the mathematical formulation of the consolidation process becomes relatively simple. Under these assumptions it is called linear consolidation.

The deformation behaviour of a linear-elastic homogeneous medium can be fully described in terms of the displacement vector u_i with two independent material properties: λ and μ (Lamé's constant). The pore pressure gradient can be considered as a volumetric body force. Hence, the equilibrium of the medium is expressed by the following equation (Love, 1928):

$$(\lambda+\mu)u_{k,ki} + \mu u_{i,kk} - p_{,i} = 0. \qquad (1)$$

A second relation between u_i and p exists, to wit the storage equation, which includes Darcy's law (Verruijt, 1969):

$$\frac{K}{\gamma} p_{,kk} = n\beta\dot{p} + \dot{u}_{k,k}. \qquad (2)$$

The point on top indicates differentiation with respect to time. The parameter K represents the hydraulic permeability, γ is the specific weight of the pore fluid, n denotes the porosity and β represents the pore fluid compressibility, defined by:

$$\beta = d\gamma/\gamma dp. \qquad (3)$$

Biot (1941) introduced a scalar function ϕ and a vector function ψ according to:

$$u_i = (\phi+x_k\psi_k)_{,i} - 2\alpha\psi_i;\ \psi_{i,kk} = 0, \qquad (4)$$

to simplify equations (1) and (2). The vector x_k denotes here the spatial position. Verruijt (1968) showed that the compressibility of the pore fluid and the deformation behaviour of the porous matrix are correctly incorporated by the following expression for α:

$$\alpha = (1+n\beta(\lambda+2\mu))/(1+n\beta(\lambda+\mu)). \qquad (5)$$

Finally, equations (1) and (2) become:

$$\dot{\phi}_{,kk} = c\phi_{,kkll};\ \psi_{i,kk} = 0, \qquad (6)$$

whereas the pore pressure is related to ϕ and ψ_i according to:

$$p = (\lambda+2\mu)(\phi_{,kk} + 2\psi_{k,k}) - 2\alpha(\lambda+\mu)\psi_{k,k}. \qquad (7)$$

The constant c is the so-called consolidation coefficient:

$$c = \frac{k}{\gamma}\frac{\lambda+2\mu}{1+n\beta(\lambda+2\mu)}. \qquad (8)$$

It represents the fundamental parameter of linear consolidation.

In axially symmetrical geometries the linear consolidation process can be described by two scalar functions, E and S (McNamee & Gibson, 1960), which are correlated to ϕ and ψ_i according to:

$$\phi = -E;\ \psi_1 = \psi_2 = 0,\ \psi_3 = S. \qquad (9)$$

This renders the basic equations (6), (4) and (7) into:

$$\dot{E}_{,kk} = c\,E_{,kkll}\ ;\ S_{,kk} = 0; \qquad (10)$$

$$u_i = (-E+x_3S)_{,i} - 2\alpha S\delta_{i3}; \qquad (11)$$

$$p = (\lambda+2\mu)(-E_{,kk}+2S_{,3}) - 2\alpha(\lambda+\mu)S_{,3}. \qquad (12)$$

The spatial coordinate x_3 is measured along the axis of symmetry.

The advantage of this description is that the equations (10) are more convenient to solve provided that the boundary conditions can be expressed in terms of E and S and their derivatives and can be handled accordingly.

THE DISPLACEMENT FIELD IN A HALFMEDIUM DUE TO A WELL

Application of integral transformation techniques (Sneddon, 1951), to wit the Laplace transformation with respect to time t and the Hankel transformation with respect to the radial coordinate r, transform equations (10) for a halfmedium situated in $z \leq 0$, into:

$$\bar{E} = A_1e^{z\eta} + A_2e^{z\xi};\ \bar{S} = A_3e^{z\xi}, \qquad (13)$$

where:

$$\bar{E} = \int_0^\infty e^{-st}dt \int_0^\infty rJ_o(r\xi)E dr; \qquad (14)$$

$$\bar{S} = \int_0^\infty e^{-st}dt \int_0^\infty rJ_o(r\xi)S dr; \qquad (15)$$

$$\eta^2 = q^2 + \xi^2;\ q^2 = s/c. \qquad (16)$$

The integration constants A_i are generally functions of the transform variables s and ξ, and they are determined by the boundary conditions.

Lastly, the inverse of the transform solution will give the results required. To find the inverse usually is a complicated problem on itself, involving cumbersome mathematics.

An important fundamental solution concerns the consolidation due to a single well starting to pump a constant discharge Q from time t = 0 at depth h under the surface of a saturated porous halfmedium. If the influence of soil layers eventually situated on top are disregarded, the consolidation process of the halfmedium can be described by the previous mathematical formulation.

The problem can be analysed using the principle of superposition of part solutions, one of which is the consolidation due to a point well in an infinite porous medium.

Assuming that the surface: z = 0, is a stress-free, impervious boundary, the solution of the problem considered can be composed by the field generated by a point well at depth h under the surface and a mirror well with equal intensity acting at a distance h above the surface (see fig. I).

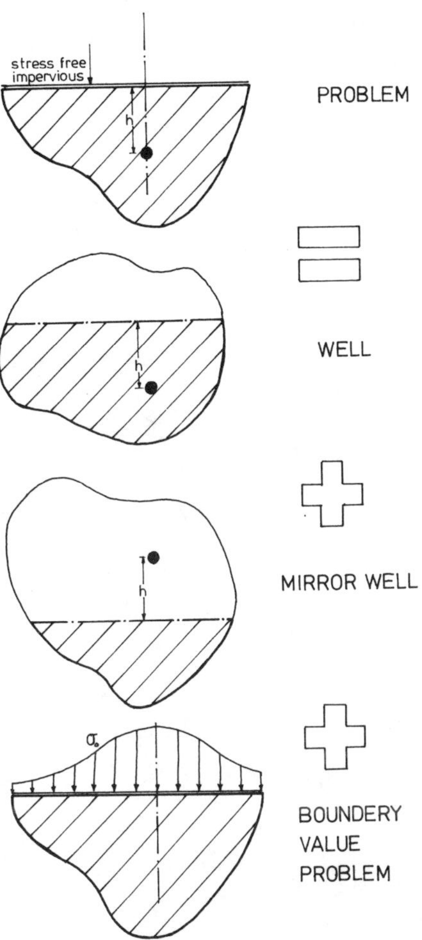

Figure I. The approach to the problem.

The induced normal stresses at the boundary have still to be elimated and this can be performed using the displacement functions E and S.

Elaboration of the sphere-symmetrical displacement field u_R in an infinite porous medium due to a point well with a radius R_o producing a constant discharge Q starting at time: t = 0, gives (De Josselin de Jong, 1963);

$$u_R = u_oF(R/R_o, 4ct/R_o^2), \qquad (17)$$

where:

$$u_o = -Q\gamma[4\pi K(\lambda+2\mu)R_o^2]^{-1}; \tag{18}$$

$$F(x,y) = \tfrac{1}{4}\int_0^y \{(x-1)e^{x-1+\frac{1}{4}y}\operatorname{erfc}(\tfrac{1}{2}\sqrt{y}+(x-1)/\sqrt{y}) + \operatorname{erfc}((x-1)/\sqrt{y})\}x^{-2}dy \tag{19}$$

Application of the so-called direct method (see appendix), which partly overcomes the complicated inverse transformation, leads to an approximate solution identical to (17), but now F satisfies the following expression:

$$F(x,y) = \frac{y}{2x^{-2}}\{\frac{x\sqrt{2}+\sqrt{y}}{\sqrt{2}+\sqrt{y}}e^{-\sqrt{2}(x-1)/\sqrt{y}} - 1\} \tag{20}$$

The exact solution and the approximate solution are presented in fig. II, and apparently the direct method gives reasonable results.

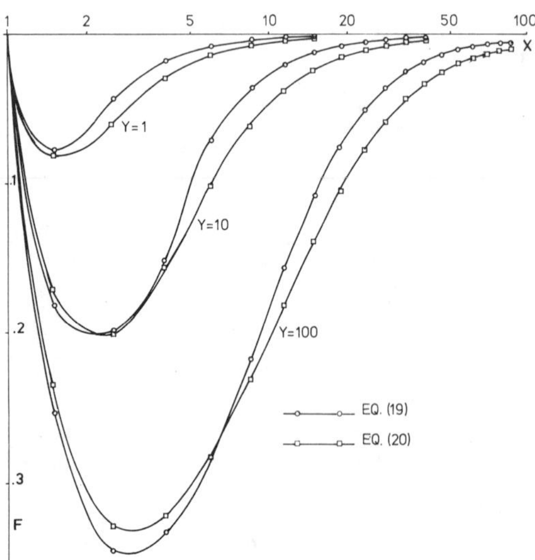

Figure II. The exact and approximate solution.

Also for the problem of a point well in a half medium the direct method is applicable.

The displacement field (17) is used to solve the problem considered here. By the introduction of a mirror well automatically two conditions at the boundary are satisfied; there are no shear stresses and the boundary is impervious. A normal stress remains and an additional solution has to be found, which elimates this stress σ_o. Application of Laplace and Hankel transformation render σ_o into:

$$\bar{\sigma}_o = \frac{-2R_o^2u_o}{\mu s\xi^3\theta(\theta-1)}\{\frac{e^{R_o\xi\sqrt{\theta^2-1}}}{1+R_o\xi\sqrt{\theta^2-1}}e^{-h\xi\theta} - \theta e^{-h\xi}\}, \tag{21}$$

where:

$$\theta = \eta/\xi; \tag{22}$$

$$\bar{\sigma}_o = \int_0^\infty e^{-st}dt\int_0^\infty rJ_o(r\xi)\sigma_o dr. \tag{23}$$

At this stage the integration constants A_i are determined. The result is:

$$A_1 = -\chi\bar{\sigma}_o/N;$$

$$A_2 = \theta\bar{\sigma}_o\{\chi+(1-\alpha)(\theta^2-1)\}/N;$$

$$A_3 = \theta\bar{\sigma}_o(\theta^2-1)/N,$$

where:

$$\chi = 2\{\alpha(\lambda+\mu)-(\lambda+2\mu)\}/(\lambda+2\mu); \tag{27}$$

$$N = (\theta-1)\{\theta(\theta+1)+\chi\}. \tag{28}$$

Substitution into equations (13) gives the transformed displacement funtion E and S. Inverse techniques will provide E and S, and by virtue of equation (11) the additional displacement field, which eliminates the surface normal stress σ_o, is found.

The solution of the problem of a point well in a halfmedium with the boundary conditions mentioned can be constructed according to the procedure presented in fig.I

THE SURFACE SETTLEMENT OF A HALFMEDIUM DUE TO A SINGLE WELL

From the solution obtained the surface settlement can be found. The contribution to the surface settlement from the well and its mirror well according to equation (17) cancel. Therefore, the generated surface settlement is completely determined by the displacement functions E and S, as they have been derived in the foregoing section. For the surface: z = 0, the vertical displacement becomes (Barends, 1971):

$$u_3(r,0,t) = -\alpha S. \tag{29}$$

The transform of S along the surface is:

$$\bar{S} = -2R_o^2u_o\{e^{-h\xi\theta} - \theta e^{-h\xi}\}/(s\xi^2N). \tag{30}$$

Elaboration for small values of time employing equations (5), (18) and (29) results into:

$$u_3(r,0,t)_{t\downarrow 0} = \frac{Qt}{2\pi(1+n\beta(\lambda+\mu))}\frac{h}{(\sqrt{r^2+h^2})^3} \tag{32}$$

For large values of time no finite expression for the settlement is found since the inverse transformation leads to improper integrals. Investigation of the settlement velocity reveals a reasonable constant value in a first period and next a gradually decreasing settlement velocity up to zero.

Using the direct method to overcome the complication of the inverse transformation results in an approximate expression for the surface settlement according to:

$$u_3(r,0,t) = \alpha R_o^2 u_o \; G(\sqrt{ct}/h), \qquad (33)$$

where:

$$G(\tau) = \int_0^\infty \{e^{-h\xi\theta} - \theta e^{-h\xi}\}\frac{d\xi}{\xi N}; \quad \theta = \sqrt{1+\tfrac{1}{2}(\tau h\xi)^{-2}}. \qquad (34)$$

It can be shown (Shapery, 1961), that in the case considered here the approximate solution converges upon the exact solution, as time tends to infinity. Inspection of the integral G shows, that it contains a logarithmic character for large values of time (Barends, 1971). Hence, no final settlement is to be expected if a continuous discharge is pumped out of a saturated, lineair-elastic halfmedium.

The largest and consequently most significant settlement occurs at the centre of the surface of the halfmedium. For different values of the parameter χ, defined in equation (27), this settlement is worked out and presented in fig. III.

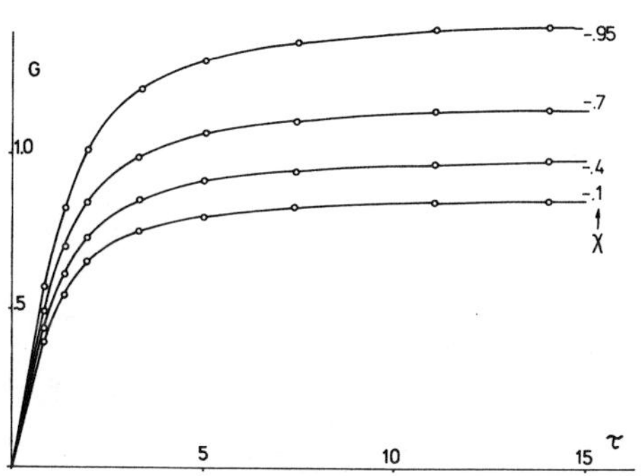

Fig. III. The centre-surface settlement.

CONCLUSIONS AND PRACTICLE IMPLICATION

Clearly one can notice from fig. III the initial period in which the surface settlement increases almost linearly with time. This period lasts about $5h^2/c$ to $10h^2/c$, where c denotes the consolidation coefficient and h represents the depth of the well. After this period the settlement decreases untill hardly any more settlement is observed. Although theoretically no absolute final value is obtained the subsidence is nearly stabilized after $100h^2/c$.

The actual settlement depends on the parameters α, u_o and χ, which have been previously defined in equations (5), (18) and (27) respectively. According to equation (33) the subsidence in the centre of the surface is proportional to α and u_o. Equation (27) can be rewritten in terms of μ and ν (Poisson's ratio):

$$\chi = \frac{1}{1-\nu}\left\{\frac{(1-2\nu) + 2(1-\nu)n\beta\mu}{(1-2\nu) + n\beta\mu}\right\}. \qquad (35)$$

From a physical point of view ν varies between 0 and ½. Since $n\beta\mu$ ought to be a positive value, the parameter χ is restricted to a value in between -1 and 0. According to fig. III the effect of the parameter χ mainly appears in the final period and reaches in the extreme case a relative effect of about 50%.

Changes in the production strategy evidently have their influence on the development of the subsidence. In the first period the subsidence is proportional to Qt, in the second period it verges to be proportional to Qln(t). This results reveal the effect of time on the development of the settlements. In the situation near Venice (see fig. IV) the initial period does not seem to have been passed.

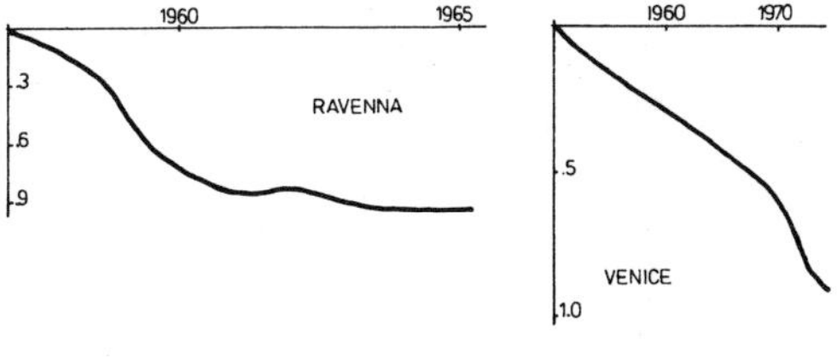

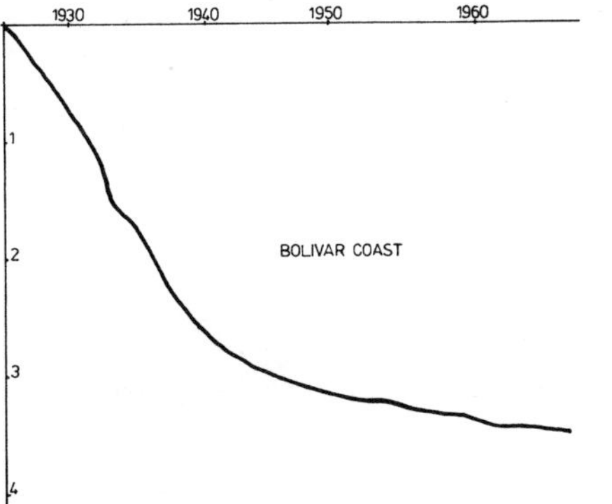

Fig. IV. Measured settlement data.

Here, a drastical decrease of production

of groundwater will have a significant effect, but it will not eliminate the effect of time on the subsidence due to the previous recovery of groundwater.

Formula (32) expresses the instanteneous settlement which in virtue of the results presented in fig. III holds for the initial period of $5h^2/c$ to $10h^2/c$. The instanteneous displacement field is identical to the surface deflection due to a nucleus of strain with an intensity of: $Qt/[2\pi h^3(1+n\beta(\lambda+\mu))]$, acting in a linear-elastic homogeneous halfspace (Mindlin & Cheng, 1942). Since at the start, at time: $t = 0$, a flow field cannot be established immediately, the instanteneous reaction of a saturated porous medium is similar to the behaviour of a homogeneous medium (see also: Gibson, 1974).

In oil and gas recovery a simple method is accepted to control the subsidence generated by the recovery. The total production is assumed to be equal to the volume released by the subsidence, i.e. the subsidence integrated over the surface area (see fig. V).

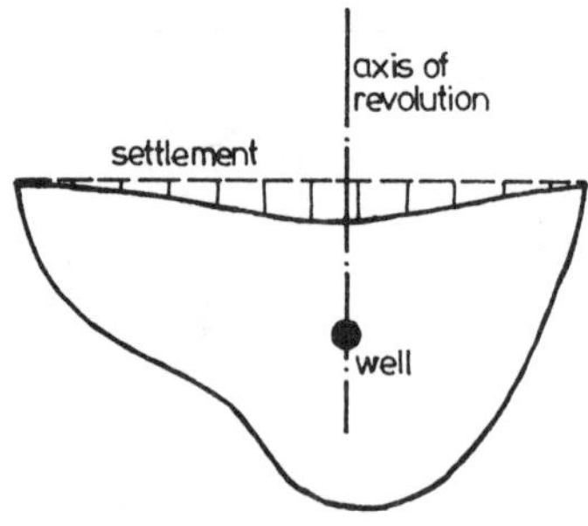

Fig. V. The volume released by subsidence.

Measurements along the Bolivar Coast oil fields confirm this idea (Escojido, 1975). Evidently, this should be valid for an incompressible pore fluid.

Evaluating the surface integral of the subsidence to wit:

$$\int_0^\infty u_3 \; 2\pi r dr, \qquad (36)$$

gives employing (32):

$$Qt/(1+n\beta(\lambda+\mu)), \qquad (37)$$

which indeed is equal to the total production in the case where the pore fluid is incompressible (β=0). It therefore confirms the method previously mentioned.

Equation (37) allows to take into account the compressibility of the pore fluid (or gas) and the stiffness of the porous matrix to some extent. In the preceding discussion the compressibility β defined in equation (3) has been considered as a constant. For gas this is an approximation and consequently equation (37) provides an estimate to evaluate the subsidence caused by gas recovery.

In reality soil is typically inhomogeneous in nature and usually groundwater recovery is established from a stratified subsoil composed of aquifers and aquitards. Nevertheless, the results presented here do have a practicle meaning in such inhomogeneous cases. Measured data presented in fig. IV fits the theoretical curves in fig. III fairly well. The initial period can be estimated and this supplies the "integral consolidation coefficient", in fact a parameter for the entire area of recovery. This value might be very different from any topically measured consolidation coefficient (oedometer). For a prediction of future subsidence the integral consolidation coefficient is very useful. In general one can state that data of the settlement history is indispensable to calibrate a (theoretical) model. Only with this information a plausible prognosis of the subsidence to be expected can be given.

REFERENCES

Barends, F.B.J. (1971), Consolidation due to a point source in a halfmedium, Delft Soil Mech. Lab., Rp. 71823-037, 48 p.

Biot, M.A. (1941), General theory of three-dimensional consolidation, J.Appl.Phys. 12:155-164.

De Josselin de Jong, G. (1963), Consolidatie in drie dimensies (Dutch), Delft Soil Mech. Lab., LGM-Mededelingen 7(3).

Escojido, D. (1975), Estimacion del hundimiento futuro a lo largo de los diques de la Costa Bolivar (Spanish), Maraven, Reservoir Eng., not REN-359, Caracas.

Gambolati, G. & Gatto, P. & Freeze, R.A. (1974), Mathematical simulation of the subsidence of Venice, Water Res. Res.10 (3):563-577.

Gibson, R.E. (1974), The analytical method in soil mechanics, The Rankine Lecture 1974, Géotechnique 24(2):115-140.

Heijnen, W.J. (1979), Analysis of surface cracking and its consequences for the safety of the Bolivar Coast dykes, Delft Soil Mech. Lab., Rp. 232390(BO).

Love, A.E.H. (1928), The stress produced in a semi-infinite solid by pressure on part of the boundary, Phil.Trans.Roy.Soc.(A)

228:377-420.
McNamee, J.M. & Gibson, R.E. (1960), Displacement functions and linear transformation applied to diffusion through porous elastic media, Qu. J. Mech. and Appl. Math.,13.
Mindlin, R.D. & Cheng, D.H. (1942), Nuclei of strain in the semi-infinite body, J. Appl. Mech.,9A:136-143.
Shapery, R.A. (1961), Approximation methods of transform inversion for visco-elastic stress analysis, Proc. IV-USA Congres Appl. Mech., 1075-1085.
Sneddon, I.N. (1951), Fourier transforms, McGraw-Hill BC, London.
Verruijt, A. (1968), The completeness of Biot's solution of the coupled thermo-elastic problem, Qu. J. Appl. Math., 26.
Verruijt, A. (1969), Elastic storage of aquifers, chap. 8 in: 'Flow through porous media' (De Wiest, ed.), Acc. Press, London.

APPENDIX

THE DIRECT METHOD

It can be shown that the inverse of a Laplace transform is well approximated by a simple substitution, whenever the derivative of the time dependent solution with respect to logarithmic time: log t, is a slowly varying function of log t. The prove is given in literature (Shapery, 1961). Here, a brief summary is presented.

The problem is to find an approximate representation of a response f(t) from the integral equation:

$$\bar{f}(s) = \int_0^\infty f(t)e^{-st}dt, \qquad (A1)$$

where $\bar{f}(s)$ is the Laplace transform of f(t) and it is a function defined at least numerically for all real nonnegative values of the transform parameter s.
It is observed that:

$$d(s\bar{f}(s))/d(\log s),$$

is a slowly varying function of log s for many visco-elastic problems, when loading is imposed as unit step functions at time: t = 0. Obviously it is then convenient to represent $s\bar{f}(s)$ as a function of log s, according to:

$$\hat{f}(u) = s\bar{f}(s) \;;\; u = \log s; \qquad (A2)$$

$$\tilde{f}(v) = f(t) \;;\; v = \log t; \qquad (A3)$$

$$w = v + u. \qquad (A4)$$

This renders (A1) into the form:

$$\hat{f}(u) = \ln 10 \int_{-\infty}^{+\infty} \tilde{f}(w-u)10^w e^{-10^w} dw \qquad (A5)$$

The weighting function $10^w e^{-10^w}$ is drawn in fig. VI, and as an approximation it can be considered as a delta function, if $\tilde{f}$ reveals sufficiently moderate changes.

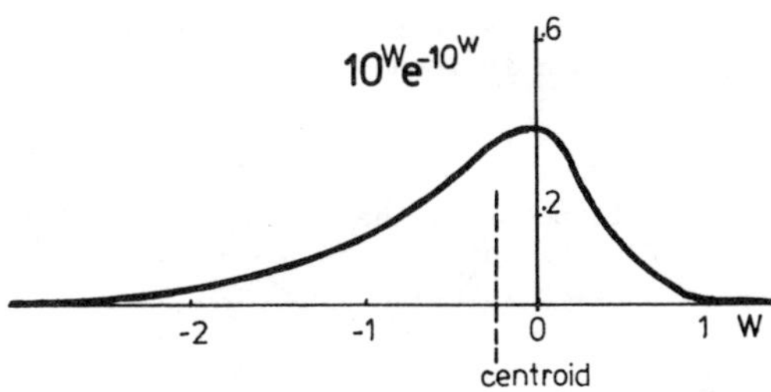

Fig. VI. The weighting function in the direct method.

Hence, replacing: $(\ln 10)10^w e^{-10^w}$ by $\delta(w-w_o)$ where:

$$\delta(w-w_o) = 0 \;,\; \text{if } w \neq w_o; \qquad (A6)$$

$$\int_{-\infty}^{+\infty} \delta(w-w_o)\, dw = 1, \qquad (A7)$$

yields an approximate inverse according to:

$$\tilde{f}(w_o-u) = \hat{f}(u). \qquad (A8)$$

Next, the value of w_o has to be found. Taylor expansion around: $v_o = w_o - u$, gives:

$$\tilde{f}(v) = \tilde{f}(v_o) + \tilde{f}'(v_o)(v-v_o) + \ldots \qquad (A9)$$

Evaluation of equation (A5) employing (A4), (A7) and (A9), as follows:

$$\begin{aligned}\hat{f}(u) &= \ln 10 \int_{-\infty}^{+\infty} \tilde{f}(w-u)10^w e^{-10^w} dw \\ &= \ln 10 \int_{-\infty}^{+\infty} [\tilde{f}(v_o) + \tilde{f}'(v_o)(v-v_o)]\, 10^w e^{-10^w} dw \\ &= \tilde{f}(v_o) + \ln 10 \int_{-\infty}^{+\infty} \tilde{f}'(v_o)(w-w_o)\, 10^w e^{-10^w} dw \\ &= \tilde{f}(w_o-u) + \{\ln 10 \int_{-\infty}^{+\infty} (w-w_o)10^w e^{-10^w} dw\}\, \tilde{f}'(v_o),\end{aligned}$$

results in a formula equal to (A8), if the integral vanishes. This provides the value w_o which after some elaboration becomes:

$$w_o = -0.58/(\ln 10). \qquad (A10)$$

When $\tilde{f}$ is closely approximated by a straight

line in the interval $|w-w_o|<1$, then from (A2), (A3) and (A4) we find:

$$w_o \approx w = \log s + \log t, \tag{A11}$$

from which follows using (A10):

$$s = s_o = 0.56/t. \tag{A12}$$

Since as a result of the approximations (w_o-u) can be expressed by v, a simple inverse formula is found:

$$[s\bar{f}(s)]_{s=s_o} = f(t). \tag{A13}$$

A slightly better value is suggested: $s_o = 1/2t$, because of the skewed form of the weighting function (see fig. VI).

Proceedings of Euromech 143 / Delft / 2-4 September 1981

Non-steady flow in confined layers

N.J.DAHL
Institute for Town and Country Planning, Copenhagen, Denmark

1 INTRODUCTORY REMARKS

In Denmark the water for domestic and industrial use is very often gained from borings into deep-seated aquifers confined between layers with a rather low permeability. The main features of the geology and the hydrology for such cases are outlined in Fig. 1.1 .

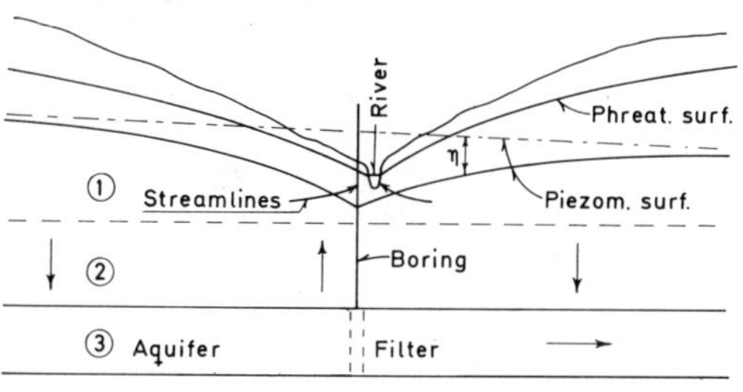

Fig. 1.1

The succession of the strata is: layer 1 is a top layer with a rather high permeability whereas layer 2, not seldom having a thickness of 50 m or more, has a (very) low permeability; layer 3 is the intrinsic aquifer resting on an impermeable layer. The infiltration of water through the unsaturated zone of layer 1 creates in the ground water body below the phreatic surface an almost horizontal flow from both sides to the river. In layer 3 the flow to a (remote) recipient is almost horizontal; to this flow belongs a piezometric surface sloping towards the recipient. The local difference in levels (above ordnance datum) between the phreatic and the piezometric surface decides whether the flow in layer 2 is - locally - going upwards or downwards.

The discharge Q through a boring, having its filter in layer 3, creates a depression η of the piezometric surface over an area around the boring. Next we conclude that this depression gives rise to an increase of the downward flow (or a decrease of the upward flow) in layer 2. The consequence of the increased downward flow in layer 2 is that the horizontal flow in layer 1 is decreased by exactly the same amount.

The aim of the following analysis is to find the changes of the flow in the three layers caused by the pumping from the boring. In this connection we have to emphasize that the discharge Q is a given function of time when the boring is part of a water supply plant. Thus we have to expect that the changes of the flow will become non-steady.

In fact we have two problems facing us. The problem described above comprises a single boring, and the change of the flow caused by the pumping will become radial and directed towards the filter. In other cases, especially when we are dealing with large water supply plants, it is necessary to install quite a few borings normally placed on a fairly straight line along a (small) river, all connected to a siphon pipe leading to the pumps. The flow problem for this case can be solved by superposition of the solutions corresponding to the individual borings. In the following we shall, however, assume that the distances between the individual borings are so small that we are entitled to interpret the compound flow as a linear flow in planes perpendicular to the line of borings. The problems attached to a single boring will be discussed in a coming paper.

2 FUNDAMENTAL EQUATIONS

The derivation and the solving of the fundamental equations demand a precise definition of the flow and its boundaries. As

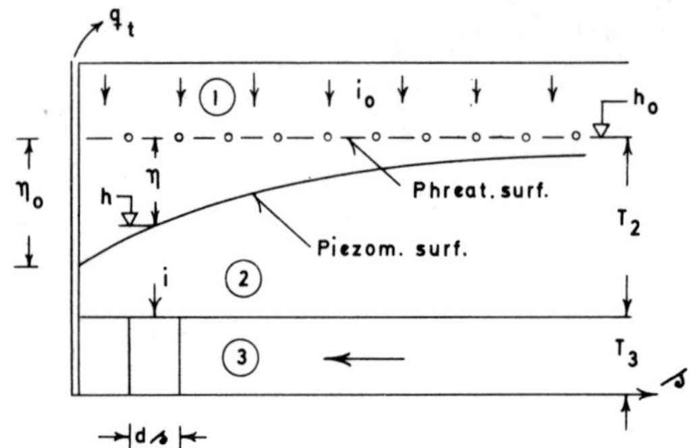

Fig. 2.1

mentioned in sect. 1 we aim at finding the changes of the natural flow caused by the pumping of water, and we define the created flow and its boundaries as shown in Fig. 2.1. Through layer 1, i.e. the unsaturated zone, a seepage i_0 originating from the precipitation is going on. The vertical seepage through layer 2, measured at the interface between layer 2 and 3, is denoted i . The surplus of water ($\sim i_0 - i$) is removed through numerous horizontal (tile-)drains; these drains are fixing the phreatic surface as a horizontal plane having the level h_0; thus the phreatic surface is not varying with the time. The seepage i through layer 2 is feeding the created flow in layer 3.

As mentioned in sect. 1 we assume that the flow is linear; this assumption involves that the filters of the boring are replaced by a (narrow) slot yielding a total discharge q_t (m^3/sec per m of the slot). A vertical line through the slot is evidently a symmetry axis which means that we have to deal with the unilateral flow shown in Fig. 2.1; for this flow we have, of course, $q_0 = \frac{1}{2}q_t$ at the slot, i.e. at $s = 0$ (m). The discharge q_t or q_0 is governed by the pumps belonging to the water supply plant and thus we must assume that q_0 is a prescribed function of time which function we shall define in the following sections.

A consequence of the flow is a lowering η of the piezometric surface η, the value of η being η_0 for $s = 0$, comp. Fig. 2.1, which lowering gives rise to the local intensity of seepage i (m^3/sec per sq.m).

The equation of energy ($\sim$ the law of Darcy) now gives for the flow in layer 2

$$(2.1) \qquad i = v = k_2 I_2 = k_2 \frac{h_0 - h}{T_2} = c_2 \eta$$

where k_2 and I_2 are the permeability and the (negative) gradient of the flow; T_2 is the thickness of layer 2, while h_0 and h are the levels (above ordnance datum) of the phreatic surface and the piezometric surface. In (2.1) we have introduced the definition

$$(2.2) \qquad c_2 = \frac{k_2}{T_2}$$

and we now assume that c_2 is a constant independent of the distance s from the slot. It is observed that η is defined by

$$(2.3) \qquad \eta = h_0 - h$$

which implies that the phreatic surface and the piezometric surfaces are coinciding when no pumping is going on.

The local discharge q in layer 3 is also found from the equation of energy; we get

$$(2.4) \qquad q = -k_3 T_3 I_3 = S\frac{\partial h}{\partial s} = -S\frac{\partial \eta}{\partial s}$$

where k_3 and I_3 are the permeability and the (negative) gradient in layer 3; the negative sign arises from the fact that the direction of the flow is opposite to the positive direction of the s-axis. In (2.4) we have introduced the transmissivity S defined by

$$(2.5) \qquad S = k_3 T_3 \qquad (m^2/sec)$$

and we shall in the following assume that S is independent of the distance s .

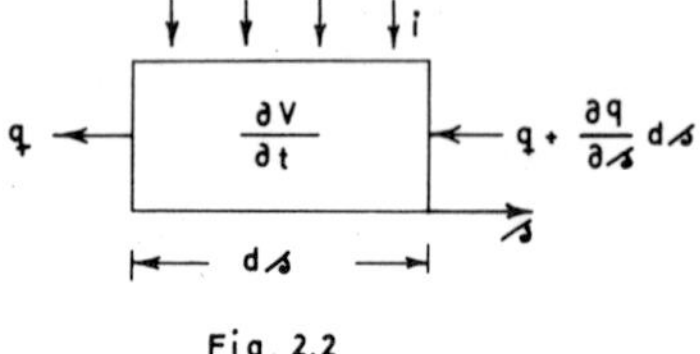

Fig. 2.2

Finally, we have to derive the equation of continuity for the non-steady flow in layer 3, comp. Fig. 2.2. For the interval of time dt we get

$$(q + \frac{\partial q}{\partial s} ds)dt + i \cdot ds \cdot dt = q dt + \frac{\partial V}{\partial t} \cdot ds \cdot dt$$

or

$$(2.6) \qquad \frac{\partial q}{\partial s} - \frac{\partial V}{\partial t} = -i$$

where $\partial V/\partial t$ is the increase of pore volume per unit of time in a vertical column having the sectional area of 1 sq.metre.

The change of pore volume is due to the change of pore water pressure caused by the lowering of the piezometric surface. Consequently we put

$$(2.7) \qquad \frac{\partial V}{\partial t} = -M\frac{\partial \eta}{\partial t}$$

where M is a dimensionless constant of (almost) the same nature as the coefficient of elasticity (Young's modulus); the minus sign arises from the fact that an increase of η gives a decrease of V. Thus we have

$$(2.8) \qquad \frac{\partial q}{\partial s} + M\frac{\partial \eta}{\partial \tau} = -i$$

Differentiation of (2.4) gives

$$(2.9) \qquad \frac{\partial q}{\partial s} = -S\frac{\partial^2 \eta}{\partial s^2}$$

and, using (2.1) and (2.9), q and i are eliminated from (2.8); we get

$$(2.10) \qquad S\frac{\partial^2 \eta}{\partial s^2} - M\frac{\partial \eta}{\partial t} = c_2\eta$$

where η is a function of s and t.

We now introduce the dimensionless function $y(s,t)$ defined by

$$(2.11) \qquad \eta(s,t) = \eta_0 y(s,t)$$

where η_0 is a constant to be discussed later on.

Insertion of (2.11) in (2.10) gives

$$(2.12) \qquad \frac{S}{c_2}\frac{\partial^2 y}{\partial s^2} - \frac{M}{c_2}\frac{\partial y}{\partial t} = y$$

which equation makes it natural to introduce the substitutions

$$(2.13) \qquad x = \sqrt{\frac{c_2}{S}}\, s \quad \text{and} \quad \tau = \frac{c_2}{M}\, t$$

where x and τ are dimensionless coordinates of position and time. By insertion of (2.13) in (2.12) we get the final form of the equation

$$(2.14) \qquad \frac{\partial^2 y}{\partial x^2} - \frac{\partial y}{\partial \tau} = y$$

in which we recognize the equation of transport of heat in solids.

Insertion of (2.11) and (2.13) in (2.4) now gives

$$(2.15) \qquad q = -S\eta_0\sqrt{\frac{c_2}{S}} \cdot \frac{\partial y}{\partial x} = q_0 \cdot u(x,\tau)$$

where we have introduced the function $u(x,\tau)$ defined by

$$(2.16) \qquad u(x,\tau) = -\frac{\partial y}{\partial x}$$

and the definition

$$(2.17) \qquad q_0 = \eta_0\sqrt{Sc_2}$$

which definition chains together the discharge q_0 and the lowering η_0 of the piezometric surface, both calculated at the slot, i.e. for $x = 0$; we shall revert to this fact later on.

Thus we conclude that we have to solve the system of equations

$$(2.18) \qquad \frac{\partial^2 y}{\partial x^2} - \frac{\partial y}{\partial \tau} = y$$

$$(2.19) \qquad u = -\frac{\partial y}{\partial x}$$

combined with the necessary and actual boundary and initial conditions. It is obvious that equation (2.19) causes no trouble and consequently we concentrate our attention on equation (2.18).

3 BOUNDARY AND INITIAL CONDITIONS

The boundary condition at $s = 0$ ($\sim x = 0$) and for $s \to \infty$ ($x \to \infty$) as well as the initial condition for $t = 0$ ($\tau = 0$) have to be derived from the technical conditions prescribed for the actual problem. The equations (2.18) and (2.19) are, however, linear and this fact entitles us to make use of the law of superposition. For this reason we therefore intend to solve the system of equations corresponding to rather simple conditions and then - using such solutions as toy-bricks - to build up solutions corresponding to more complex conditions.

At $x = 0$, i.e. at the slot, the standard condition is that the yield of water is decided by the capacity of the pumps. Thus we conclude that $q(0,t)$ is a prescribed function of time and, comparing (2.4) and (2.11), we realize that this condition can be exchanged for the condition

$$(3.1) \qquad \left.\frac{\partial y}{\partial x}\right|_{x=0} = f(\tau)$$

where $f(\tau)$ is a prescribed function. We choose this function as simple as possible, viz.

$$(3.2)\quad -f(\tau) = \begin{cases} 1 & \text{for } \tau \geq 0 \\ 0 & \text{"} \ \tau < 0 \end{cases}$$

which is, in fact, nothing but the well known unit function.

By experience we know that the lowering of the piezometric surface caused by the pumping of water is imperceptible far away from the slot. This leads to the condition

$$(3.3)\quad y(x,\tau) \to 0 \qquad \text{for } x \to \infty$$

Finally, the initial condition for $\tau = 0$ is in principle a free function. Since it is our aim to find the changes of the flow caused by the pumping, starting at the point of time $t = 0$ ($\sim\tau = 0$), the initial condition must be

$$(3.4)\quad y(x,0) \doteq 0 \qquad \text{for } \tau = 0$$

and the conditions (3.2) - (3.4) now form the necessary conditions.

4 THE TOY-BRICK SOLUTION AND ITS USE

We now introduce the substitution

$$(4.1)\quad y(x,\tau) = e^{-\tau} \cdot z(x,\tau)$$

defining the function $z(x,\tau)$. By differentiation we get

$$(4.2)\quad \frac{\partial^2 y}{\partial x^2} = e^{-\tau} \cdot \frac{\partial^2 z}{\partial x^2}$$

$$(4.3)\quad \frac{\partial y}{\partial \tau} = -e^{-\tau} \cdot z + e^{-\tau}\frac{\partial z}{\partial \tau}$$

and insertion in (2.18) then gives after a simple reduction

$$(4.4)\quad \frac{\partial^2 z}{\partial x^2} - \frac{\partial z}{\partial \tau} = 0$$

We now have to solve the homogenous equation (4.4) instead of the inhomogenous equation (2.18) with, of course, corresponding conditions. From (3.1), (3.2) and (4.1) we get

$$\left.\begin{matrix} -1 \\ 0 \end{matrix}\right\} = e^{-\tau}\frac{\partial z}{\partial x}$$

or

$$(4.5)\quad \frac{\partial z}{\partial \tau} = \begin{cases} -e^{\tau} & \text{for } \tau \geq 0 \\ 0 & \text{"} \ \tau < 0 \end{cases}$$

when $x = 0$. For $x \to \infty$ we get from (3.3) and (4.1)

$$(4.6)\quad z(x,\tau) \to 0 \qquad \text{for } x \to \infty$$

and, finally, we get from (3.4) and (4.1)

$$(4.7)\quad z(x,0) = 0 \qquad \text{for } \tau = 0$$

as the initial condition.

Using the theory of Laplace-transforms we get the subsidiary equation to (4.4)

$$\frac{d^2 Z}{dx^2} - pZ + z(x,0) = 0$$

which gives, cf. (4.7),

$$(4.8)\quad \frac{d^2 Z}{dx^2} - pZ = 0$$

The complete solution to (4.8) is

$$Z(x,p) = Ae^{\sqrt{p}x} + Be^{-\sqrt{p}x}$$

where A and B are constants of integration. The condition (4.6) shows that we have to put $A = 0$ and thus we get

$$(4.9)\quad Z(x,p) = Be^{-\sqrt{p}x}$$

from which we get

$$\frac{dZ}{dx} = -B\sqrt{p}\,e^{-\sqrt{p}x}$$

or, for $x = 0$,

$$(4.10)\quad \frac{dZ}{dx} = -B\sqrt{p}$$

This expression we have to equalize with the Laplace transform of (4.5); we have

$$(4.11)\quad L\{e^{\tau}\} = \int_0^{\infty} e^{-p\tau} \cdot e^{\tau}\, d\tau = \frac{1}{p-1}$$

and (4.10) and (4.11) now give

$$-\frac{1}{p-1} = -B\sqrt{p}$$

or

$$B = \frac{1}{(p-1)\sqrt{p}}$$

to be inserted in (4.9). We get

$$(4.12)\quad Z(x,p) = \frac{1}{(p-1)\sqrt{p}}\, e^{-\sqrt{p}x}$$

which is the final solution of the subsidiary equation and, consequently, the Laplace transform of the solution to equation (4.4) taking into account the conditions (4.5), (4.6), and (4.7)

The inverse of (4.12) is found in tables of Laplace transforms, e.g. Carslaw and Jaeger, app. V (1978); we find

$$(4.13)\quad \begin{cases} z(x,\tau) = \frac{1}{2} e^{\tau} \{e^{-x} \cdot \text{erfc}[-] - \\ \qquad\qquad - e^{x} \cdot \text{erfc}\,[+]\} \\ z(x,\text{neg}) = 0 \end{cases}$$

where erfc denotes the complementary error function and where we, for the sake of brevity, have used the symbols

$$(4.14)\quad [-] = \frac{x}{2\sqrt{\tau}} - \sqrt{\tau}$$

$$(4.15)\quad [+] = \frac{x}{2\sqrt{\tau}} + \sqrt{\tau}$$

Combining (4.1) and (4.13) we get the demanded y-solution which becomes

$$(4.16)\quad \begin{cases} y(x,\tau) = \frac{1}{2}\{e^{-x} \cdot \text{erfc}[-] - e^{x} \cdot \text{erfc}\,[+]\} \\ y(x,\text{neg}) = 0 \end{cases}$$

but we want to make sure that the conditions (3.2), (3.3), and (3.4) are fulfilled by (4.16). Tables of error functions, Carslaw and Jaeger, app. II (1978), show that

$$\left. \begin{array}{l} e^{-x}\,\text{erfc}[\frac{x}{2\sqrt{\tau}} - \sqrt{\tau}] \to 0 \\ e^{x}\,\text{erfc}[\frac{x}{2\sqrt{\tau}} + \sqrt{\tau}] \to 0 \end{array} \right\} \text{for } x \to \infty$$

which shows that the boundary condition (3.3) is fulfilled. For $\tau \to 0$ we have

$$\left. \begin{array}{l} \text{erfc}\,[\frac{x}{2\sqrt{\tau}} - \sqrt{\tau}] \\ \text{erfc}\,[\frac{x}{2\sqrt{\tau}} + \sqrt{\tau}] \end{array} \right\} \to \text{erfc}(\infty) = 0$$

which shows that the initial condition (3.4) is fulfilled.

It is a bit more difficult to show that the boundary condition (3.2) is fulfilled. By differentiation of (4.16) we get, cf. (4.14) and (4.15),

$$\frac{\partial y}{\partial x} = \frac{1}{2}\Big\{ - e^{-x} \cdot \text{erfc}[-] - \frac{2}{\sqrt{\pi}} \cdot \frac{1}{2\sqrt{\tau}}\, e^{-x} \cdot e^{-[-]^2} - e^{x} \cdot \text{erfc}[+] + \frac{2}{\sqrt{\pi}} \cdot \frac{1}{2\sqrt{\tau}} \cdot e^{x} \cdot e^{-[+]^2}\Big\}$$

$$(4.17)\quad \frac{\partial y}{\partial x} = \frac{1}{2}\{-e^{-x} \cdot \text{erfc}[-] - e^{x} \cdot \text{erfc}[+]\}$$

since we have

$$\frac{2}{\sqrt{\pi}} \cdot \frac{1}{2\sqrt{\tau}} \cdot e^{-x} \cdot e^{-[-]^2} = \frac{2}{\sqrt{\pi}} \cdot \frac{1}{2\sqrt{\tau}} \cdot e^{-x} \cdot e^{-[+]^2}$$

which is shown as follows; evidently, it suffices to show that we have

$$e^{-x} \cdot e^{-[-]^2} = e^{x} \cdot e^{-[+]^2}$$

or that, cf.(4.14) and (4.15),

$$-x - [\frac{x}{2\sqrt{\tau}} - \sqrt{\tau}]^2 = x - [\frac{x}{2\sqrt{\tau}} + \sqrt{\tau}]^2$$

$$-x - \frac{x^2}{4\pi} + x - \tau = x - \frac{x^2}{4\pi} - x - \tau$$

$$-\frac{x^2}{4\tau} - \tau = -\frac{x^2}{4\tau} - \tau$$

q.e.d.

From (4.17) we now get for $x = 0$

$$(4.18)\quad \left.\frac{\partial y}{\partial x}\right|_{x=0} = \frac{1}{2}\{-\text{erfc}(-\sqrt{\tau}) - \text{erfc}\sqrt{\tau}\}$$

$$= \frac{1}{2}\{-(1 + \text{erf}\sqrt{\tau}) - (1 - \text{erf}\sqrt{\tau})\}$$

$$= -1$$

which shows that the boundary condition (3.2) is fulfilled.

We thus conclude that the prescribed conditions attached to the problem are fulfilled by (4.16). It is, however, also interesting to analyse (4.16) for the case $\tau \to \infty$; from (4.16) we get, cf. (4.14) and (4.15),

$$(4.19)\quad y(x,\tau) \to \frac{1}{2}\{e^{-x} \cdot \text{erfc}(-\sqrt{\tau}) - e^{x} \cdot \text{erfc}\sqrt{\tau}\}$$

$$\to \frac{1}{2}\{e^{-x} \cdot 2 - 0\} \to e^{-x} \quad \text{for } \tau \to \infty$$

In the same way we get from (4.17)

(4.20) $\partial y/\partial x \to \frac{1}{2}\{e^{-x} \cdot \operatorname{erfc}(-\sqrt{\tau}) - e^{x} \cdot \operatorname{erfc}\sqrt{\tau}\} \to -e^{-x}$ for $\tau \to \infty$

and we note that (4.19) and (4.20) are nothing but the solutions to (2.18) and (2.19) in the case of a steady state, i.e. if $\partial y/\partial \tau = 0$.

The solution (4.16) has to be supplemented with the solution to (2.19); comparing with (4.17) we immediately get

$$(4.21)\quad \begin{cases} u(x,\tau) = \frac{1}{2}\{e^{-x} \cdot \operatorname{erfc}[-] + e^{x} \cdot \operatorname{erfc}[+]\} \\ u(x, \text{neg}) = 0 \end{cases}$$

and from (4.18) we conclude that $u(x,\tau) \to 1$ for $x \to 0$.

Remembering that $y(x,\tau)$ is related to the piezometric surface, cf. (2.11), and to the seepage through layer 2, cf. (2.1), whereas $u(x,\tau)$ is related to the local discharge in layer 3, cf. (2.15), we interpret these solutions as follows. Before $t = 0$ $(\sim \tau = 0)$ we have a steady state of ground water flow through the layers 1, 2, and 3. After the start, the pumps are yielding a constant gain of water, and if this gain is maintained over a very long time, the flow will approach a new steady state, cf. (4.19) and (4.20). Thus the solutions (4.16) and (4.21) are giving the transition from the former steady state of flow to the latter. We shall make use of this result later on.

For the solutions (4.16) and (4.21) the conditions $y(x,\text{neg}) = 0$, resp. $u(x,\text{neg}) = 0$, are attached. These conditions have their origin in the unit function (3.2); if instead of (3.2) we had used the function

$$(4.22)\quad -f(\tau) = \begin{cases} 1 & \text{for } \tau \geqq \tau_a \\ 0 & \text{"} \quad \tau < 0 \end{cases}$$

where τ_a is an arbitrary (positive) constant we would have achieved exactly the same solutions except for two things, viz. we would have had to write $\tau - \tau_a$ instead of τ everywhere in both solutions and the conditions $y(x,\text{neg}) = 0$ resp. $u(x,\text{neg}) = 0$ would furthermore have to be read as

$$(4.23)\quad y(x,\tau) = u(x,\tau) = 0$$

if $\tau - \tau_a < 0$.

We are now able to set up the toy-brick solutions mentioned in the heading of this section. We do this by defining the functions, comp. (4.16) and (4.21),

$$(4.24)\quad \begin{cases} \phi(x,\tau-\tau_a) = \frac{1}{2}\{e^{-x} \cdot \operatorname{erfc}[-] - e^{x} \cdot \operatorname{erfc}[+]\} \\ \phi(x,\text{neg}) = 0 \end{cases}$$

and

$$(4.25)\quad \begin{cases} \psi(x,\tau-\tau_a) = \frac{1}{2}\{e^{-x} \cdot \operatorname{erfc}[-] + e^{x} \cdot \operatorname{erfc}[+]\} \\ \psi(x,\text{neg}) = 0 \end{cases}$$

where the symbols [-] and [+] now are defined by

$$(4.26)\quad [-] = \frac{x}{2\sqrt{\tau - \tau_a}} - \sqrt{\tau - \tau_a}$$

$$(4.27)\quad [+] = \frac{x}{2\sqrt{\tau - \tau_a}} + \sqrt{\tau + \tau_a}$$

in which τ_a will act as a free parameter.

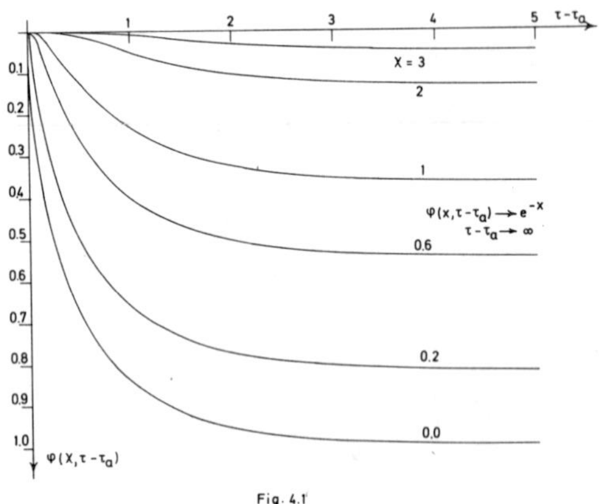

Fig. 4.1

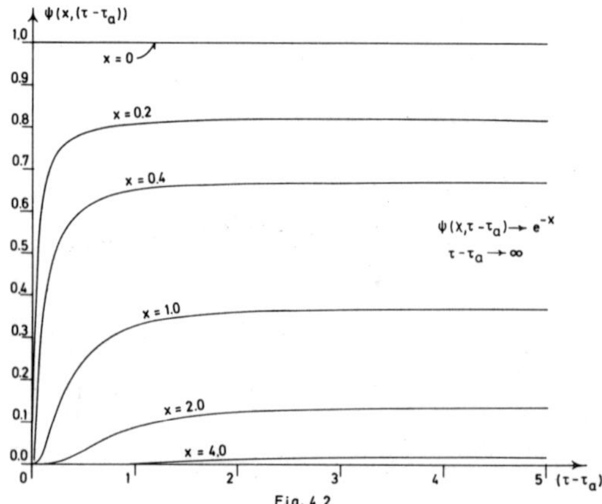

Fig. 4.2

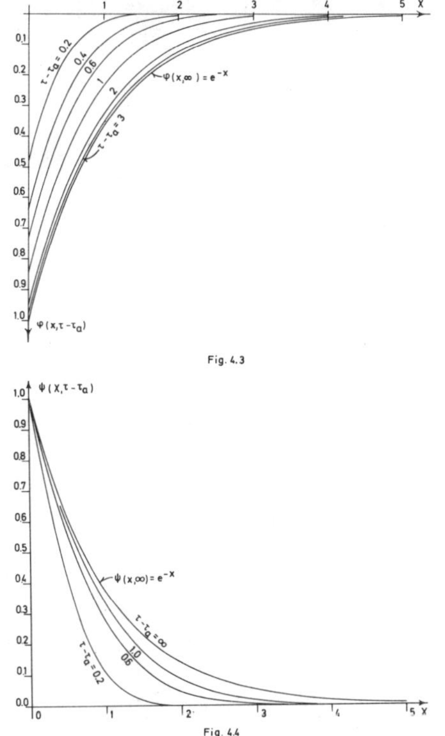

Fig. 4.3

Fig. 4.4

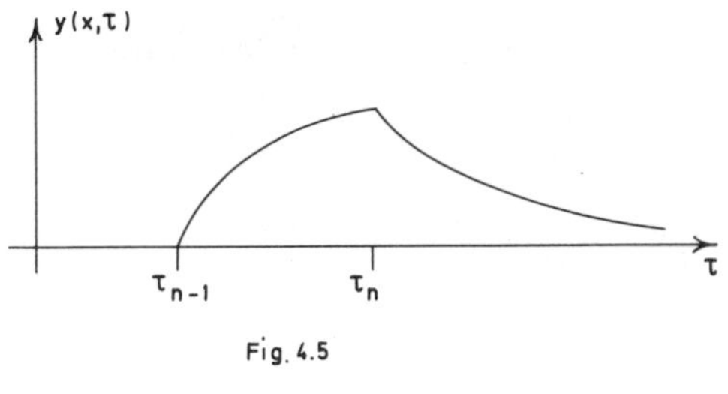

Fig. 4.5

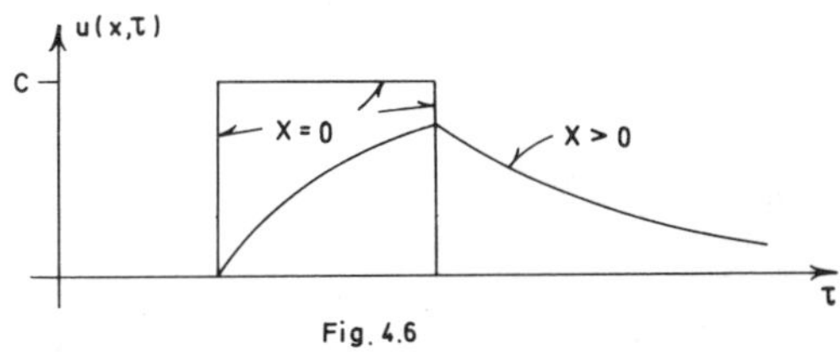

Fig. 4.6

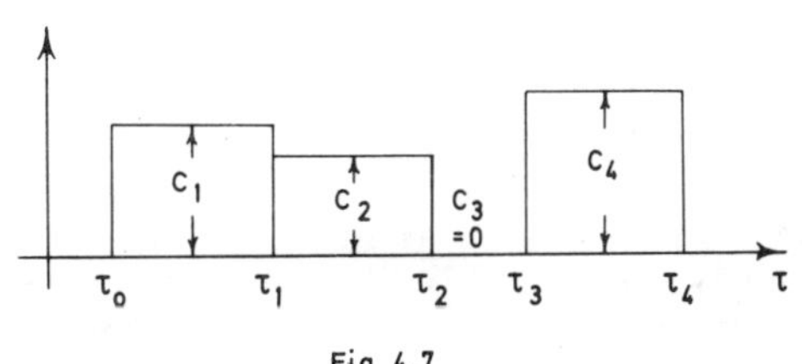

Fig. 4.7

Fig. 4.1 and 4.2 show the courses of $\phi(\;)$ and $\psi(\;)$ as functions of $\tau - \tau_a$ with x as a parameter whereas Fig. 4.3 and 4.4 show their courses as functions of x with $\tau-\tau_a$ as a parameter; from the figures we conclude that both $\phi(\;)$ and $\psi(\;)$ have reached their asymptotical limits with a fairly good approximation if one of the conditions

(4.28) $\quad \tau - \tau_a > 5 \qquad$ or $\qquad x > 5$

is fulfilled.

We now establish the forms

(4.29) $\quad y(x,\tau) = c\{\phi(x,\tau-\tau_{n-1})-\phi(x,\tau-\tau_n)\}$

(4.30) $\quad u(x,\tau) = c\{\psi(x,\tau-\tau_{n-1})-\psi(x,\tau-\tau_n)\}$

where c is an arbitrary constant and where τ_{n-1} and τ_n are two arbitrary points of time for which we have $\tau_n > \tau_{n-1}$. We now realize that (4.29) and (4.30) are particular solutions to (2.18) and (2.19) owing to the linearity of these equations.

The solution (4.29) satisfies the following conditions. For a fixed value of x we have $y(x,\tau) = 0$ if $\tau<\tau_{n-1}<\tau_n$ because of the condition $\phi(x,neg) = 0$. For $\tau_{n-1} \leqq \tau < \tau_n$ the second term in the brackets disappears because of the same condition. For $\tau \to \infty$ we get $y(x,\tau) \to 0$ since the two terms in the brackets have the common limit e^{-x}, cf. (4.20). Thus the course of $y(x,\tau)$ will be as sketched in Fig. 4.5; it is observed that the curve shown in Fig. 4.5 is nothing but a graphical representation of the difference between two identical curves of the type shown in Fig. 4.1, the curves being displaced mutually with the distance (in time) $\tau_n - \tau_{n-1}$.

What has been said about the course of $y(x,\tau)$ can in principle be repeated word by word for $u(x,\tau)$. The course of $u(x,\tau)$ is sketched in Fig. 4.6; however, we have to emphasize the fundamental difference between the curve for $x=0$ and the curve for $x > 0$.

The solutions (4.29) and (4.30) are based on the law of superposition. Nothing prevents us, however, to persist in the use of this law as follows. The τ-axis is divided into intervals by the constant values τ_0, τ_1, τ_2, ... , cf. Fig. 4.7; to each of the intervals a constant value c_1, c_2, c_3, ... is ascribed. The law of superposition now gives the solutions

(4.31) $\quad y(x,\tau) = \sum c_n\{\phi(x,\tau-\tau_{n-1})-\phi(x,\tau-\tau_n)\}$

(4.32) $\quad u(x,\tau) = \sum c_n\{\psi(x,\tau-\tau_{n-1})-\psi(x,\tau-\tau_n)\}$

where it is very important to take into

account the conditions $\phi(x,neg) = 0$, resp. $\psi(x,neg) = 0$ attached to the individual terms of the types $\phi(\ )$, resp. $\psi(\ )$. We shall in coming numerical examples demonstrate the use of the solutions (4.31) and (4.32).

It is realized that the solutions (4.31) and (4.32) are nothing but a systematic use of the law of superposition and the toy-brick solutions (4.23) and (4.24) together with the parameters $c_1, c_2, c_3, \dots$ and $\tau_0, \tau_1, \tau_2, \tau_3, \dots$. It is the mere presence of these parameters which makes the solution so flexible.

5 CONSTANT GAIN OF WATER

In this and the next section we shall apply the solutions found in sect. 4 to some technical problems. Here we assume that the total gain of water is given by

$$(5.1) \qquad q_t = \begin{cases} 2q_0 & \text{for } t \geq 0 \\ 0 & \text{"} \quad t < 0 \end{cases}$$

where q_0 is the unilateral discharge to the slot, cf. Fig. 5.1. This situation corresponds to the starting up of a new water

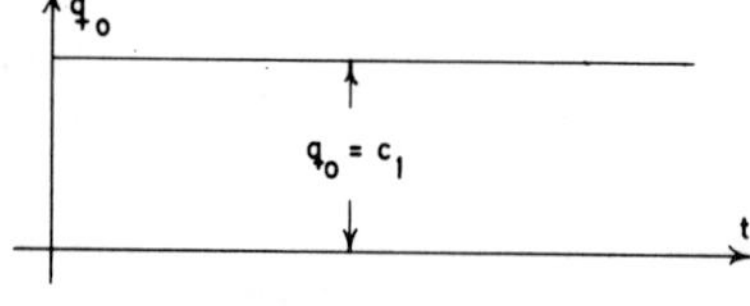

Fig. 5.1

supply plant running with a constant yield for a very long time.

Comparing (2.15) and (4.31) we now get

$$(5.2) \qquad q(s,t) = \tfrac{1}{2} q_t \cdot u(x,\tau) = \tfrac{1}{2} q_t \cdot \psi(x,\tau)$$

since in this case we must have $c_1 = \frac{1}{2} q_t$, $t_0 = 0$, and $t_1 = \infty$. For the special case $t \to \infty$ we get, cf. (4.20),

$$(5.3) \qquad q(s,\infty) = \tfrac{1}{2} q_t \cdot e^{-x}$$

From (2.17) we now get

$$(5.4) \qquad \eta_0 = \frac{q_0}{\sqrt{Sc_2}} = \frac{1}{2\sqrt{Sc_2}} \cdot q_t$$

and comparing with (2.11) and (4.29) we find

$$(5.5) \qquad \eta(s,t) = \frac{1}{2\sqrt{Sc_2}} q_t \cdot \phi(x,\tau)$$

where we have to use the same values of c_1, t_0, and t_1. For the special case $t \to \infty$ we get, cf. (4.19),

$$(5.6) \qquad \eta(s,\infty) = \frac{1}{2\sqrt{Sc_2}} q_t \cdot e^{-x}$$

Finally we get from (2.1) and (5.5)

$$(5.7) \qquad i(s,t) = \tfrac{1}{2}\sqrt{\frac{c_2}{S}} \cdot q_t \cdot \phi(x,\tau)$$

having the limit for $t \to \infty$

$$(5.8) \qquad i(s,\infty) = \tfrac{1}{2}\sqrt{\frac{c_2}{S}}\, q_t \cdot e^{-x}$$

The solutions (5.2), (5.5), and (5.7) are giving the (unilateral) local values of the discharge $q(s,t)$, the lowering $\eta(s,t)$ of the piezometric surface, and of the seepage $i(s,t)$ at the interface between layer 2 and 3 caused by the starting up of the water supply plant. It is observed that the local values of $q(s,t)$, $\eta(s,t)$, and $i(s,t)$ are proportional to the total gain of water q_t. Furthermore the formulas reveal how the geo-hydraulic parameters S and c_2 are entering into this proportionality.

The formulas (5.3), (5.6), and (5.8) show that the influence of the gaining of water is insensible if we have $x > 5$; reverting to (2.13) we then get

$$(5.9) \qquad s_{max} = 5\sqrt{\frac{S}{c_2}}$$

as the largest distance for which the influence is perceptible. The lines $s = s_{max}$ to both sides of the slot might be called the watersheds belonging to the water supply plant.

Moreover, we get from (2.13) and (4.28) that the transition, defined by (5.2), (5.5), and (5.7), from the original steady state to the new steady state, defined by (5.3), (5.6), and (5.8), is passed after a time t_{max} defined by

$$(5.10) \qquad t_{max} = 5 \cdot \frac{M}{c_2}$$

where t_{max} defines the phrase "a very long time".

Finally, we have to add that the symmetry around the slot (∿ the river), used in the foregoing (cf. 5.1), is by no means a sine qua non. Instead of (5.1) we are entitled to write

$$(5.11) \qquad q_t = q_{\ell,0} + q_{r,0}$$

where $q_{\ell,0}$ and $q_{r,0}$ denote the unknown

discharges at the slot from the left and the right, and where q_t is given function of time. Moreover, it is not necessary to ascribe the same numerical values to the geo-hydraulic parameters S, c_2, and M for the left and the right side.

The formulas (5.2) - (5.8) still hold true if we everywhere exchange $\frac{1}{2}q_t$ for $q_{\ell,0}$, resp. $q_{r,0}$, but we have to supplement (5.11) with the condition

$$(5.12) \qquad \eta_\ell(0,t) = \eta_r(0,t)$$

since the lowering of the piezometric surface must be the same at the slot for the left and the right flow. Moreover it should be emphasized that we cannot expect the ratio $q_{\ell,0} / q_{r,0}$ to be independent of time; for this reason the necessary calculation must be based on a trial-and-error method.

A NUMERICAL EXAMPLE

In this example we choose the following values of the individual parameters:

Thickness of layer 2	T_2=50 m
" " " 3	T_3=10 m
Permeability of layer 2	$k_2=2.10^{-7}$ m/sec
" " " 3	$k_3=10^{-3}$ "
Storage coefficient	$M=10^{-3}$

from which we calculate the conversional factors, cf. (2.13),

$$\sqrt{\frac{c_2}{S}} = \sqrt{\frac{k_2}{T_2 \cdot k_3 \cdot T_3}} = 6.325 \cdot 10^{-4} \quad m^{-1}$$

$$\frac{c_2}{M} = \frac{k_2}{T_2 M} = 4 \cdot 10^{-6} \quad sec^{-1}$$

Insertion of these factors in (5.9) and (5.10) then gives

$$s_{max} = \frac{5}{6.325} \cdot 10^4 \text{ m} \qquad \sim 7.9 \quad \text{km}$$

$$t_{max} = \frac{5}{4} \cdot 10^6 \text{ sec} \qquad \sim 350 \quad \text{hours}$$

showing that the width of the catchment area is around 16 km and that the transition takes around 350 hours. It should be emphasized that these figures correspond to $e^{-x} \sim 0.007$, cf. (5.3), (5.6), and (5.8); if we had chosen $e^{-x} \sim 0.05$ we would have found $s_{max} \sim 4.7$ km and $t_{max} \sim 210$ hours.

For the gain of water we choose $q_t = 0.2$ m^3/m per hour which gives around 1750 m^3/m per year. Moreover we get $\eta_0 = 4.392$ m and i_0 = 555 mm/year at the slot.

Fig. 5.2 shows graphically the transition of q(s,t) as a function of time and with s as a parameter. For s = 0 q(0,t) gets its final value immediately after the start of the pumps, comp. the "unit"function, Fig. 5.1; for s different from zero the effect of the pumping is delayed and the time lag is increasing with the distance from the slot.

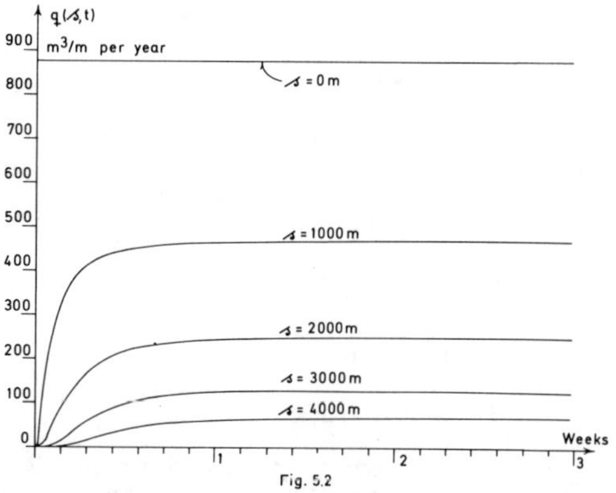

Fig. 5.2

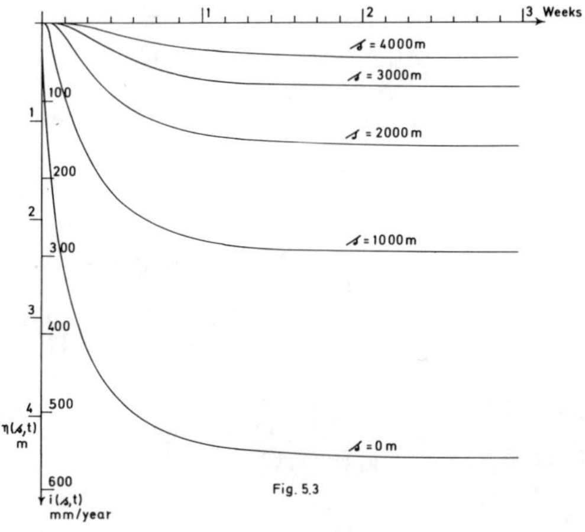

Fig. 5.3

Fig. 5.3 shows in the same way the transition of η(s,t) as a function of time and with s as a parameter. Moreover, Fig. 5.3 is provided with a scale for i(s,t), since i(s,t) is proportional with η(s,t), cf. (2.1). For s = 0 we find a very rapid growth of η(0,t) and i(0,t); for s different from zero we once more get a delay with a time lag increasing with the distance from the slot.

From both figures we observe that the transition for q(s,t), η(s,t), and i(s,t) is passed within, say, two weeks.

6 NON-STEADY GAIN OF WATER

The pumping of water must in all essentials correspond with the load on the water supply plant. Normally this means that q_t is varying as a staircase function, comp. Fig. 6.1,

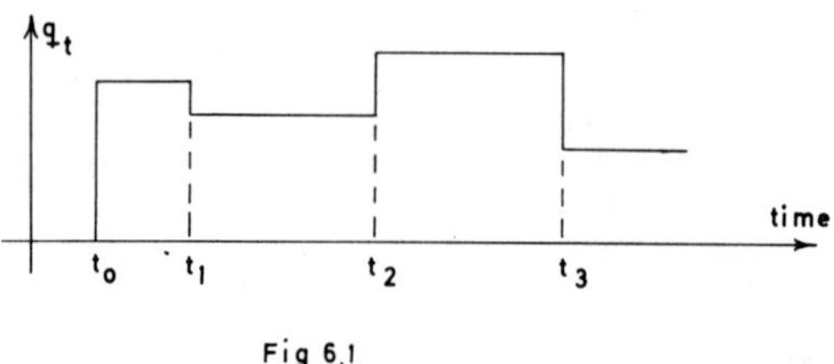

Fig. 6.1

where the size of q_t over the individual intervals is determined from the activated pumps; in the same way t_0, t_1, t_2, ... represent the time for the starting and stopping of the pumps.

We now have to use the toy-brick solutions as follows. Comparing (2.15) and (4.32) we get for the local discharge

$$(6.1) \quad q(s,t) = \tfrac{1}{2} q_t \cdot u(x,\tau) = \tfrac{1}{2} \sum q_{t,n} \{\psi(x,\tau-\tau_{n-1}) - \psi(x,\tau-\tau_n)\}$$

and correspondingly we get for the local lowering of the piezometric surface, cf. (2.11), (2.17), and (4.31),

$$(6.2) \quad \eta(s,t) = \eta_0 \cdot y(x,\tau) = \frac{1}{2\sqrt{Sc_2}} \sum q_{t,n} \{\phi(x,\tau-\tau_{n-1}) - \phi(x,\tau-\tau_n)\}$$

and for the local seepage

$$(6.3) \quad i(s,t) = c_2 \cdot \eta(s,t) = \tfrac{1}{2}\sqrt{\frac{S}{c_2}} \cdot \sum q_{t,n} \{\phi(x,\tau-\tau_{n-1}) - \phi(x,\tau-\tau_n)\}$$

In these formulas we have to insert the $q_{t,n}$ as the value of q_t for the individual intervals. In the same way, the values of τ_0, τ_1, τ_2, ... correspond to the points of time t_0, t_1, t_2, ... It is realized that each addend in the summations corresponds to a certain interval on the time-axis and that the summation means that we have to add - graphically or numerically - a number of figures of the type shown in Fig. 4.5 and 4.6, the figures being displaced in time according to the individual intervals.

We have to emphasize that the correctness of the solution implies that we have $q_t = 0$ if $t < t_0$; on the other hand, formula (5.10) shows that the effect from interval No 1 has disappeared if we have $t - t_1 > t_{max}$. This means that the effect of possible events before $t = t_0$ has disappeared after the lapse of a certain time.

A NUMERIC EXAMPLE

In this example we shall demonstrate the use of the formulas (6.1) - (6.3) in a case where the gaining of water is a rhythmic function over a period equal to one week.

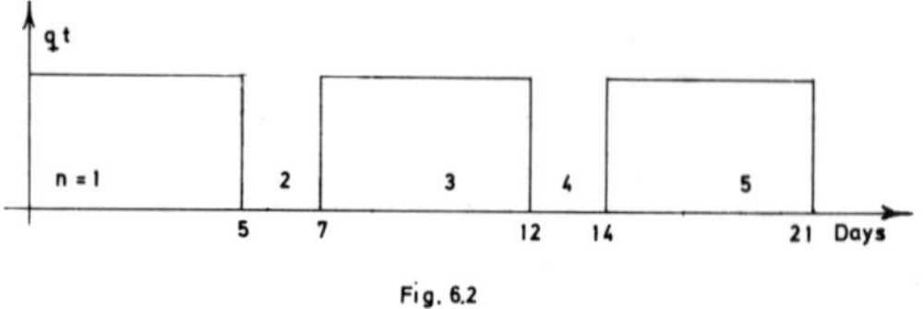

Fig. 6.2

As shown in Fig. 6.2 we assume that the plant is working 5 days per week with intervening stopping intervals of 2 days. Thus we get for q_t, comp. Fig. 6.2,

$q_{t,n}$ = 0.2 m^3/m per hour for n odd

$q_{t,n}$ = 0 " " n even

where $q_{t,n} = 0.2\ m^3$/m per hour is the same value as used in sect. 5. The values of τ_0, τ_1, τ_2, ... must of course correspond to $t = 0, 5, 7, 12, \ldots$ days. For the geo-hydraulic parameters we take - for the sake of comparison - the same values as used in the numerical example treated in sect. 5.

Fig. 6.3 shows the variation of q(s,t) as a function of time with s as a parameter; Fig. 6.4 shows in the same way the variation of η(s,t) and i(s,t). It is evident from the figures that the expected rhythmic state is established during 3 weeks. It is also evident that the amplitudes of the local oscillations of q(s,t), η(s,t), and i(s,t) are decreasing with increasing values of the distance from the slot, and, due to the direct proportionality of q_t to q(s,t), η(s,t), and i(s,t), the oscillations tend to take place around ordinates being 5/7 of the asymptotic ordinates found in Fig. 5.2 and 5.3. We finally note that the time lag between the oscillations is - also in this case - increasing with the distance from the slot.

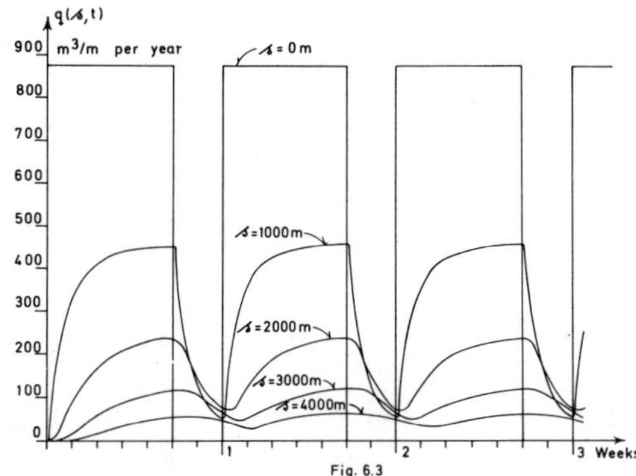

Fig. 6.3

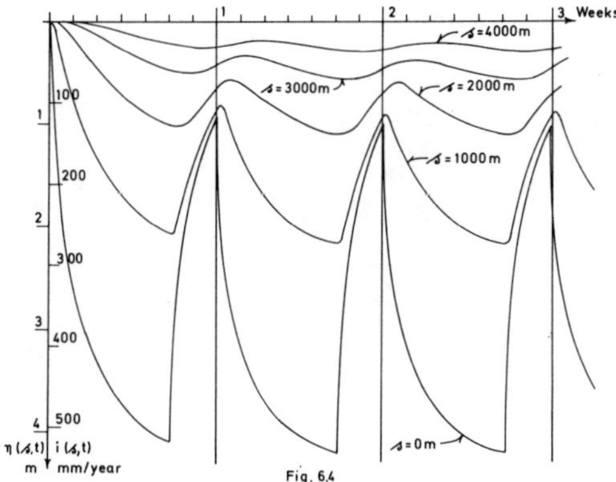

Fig. 6.4

7 CLOSING REMARKS

The solution found in the foregoing must be interpreted as a mathematical model describing the changes of the natural flow of ground water caused by the gaining of water. The fundamental partial differential equations have been linearized only in order to obtain a solvable system of equations; this proces has made it necessary to assume that the fundamental geo-hydraulic parameters S, c_2, and M are constants independent of time and with the same value over the whole catchment area. If these assumptions are not fulfilled, the solution must be accepted only as an approximative one, of course.

It is, however, necessary to discuss an assumption introduced in relation to the deduction of the fundamental equations. We have, in fact, assumed that the position of the phreatic surface was unaffected by the gaining of water, cf. Fig. 2.1. Whether this surface is affected or not demands, however, an analysis of the water balance not only for the catchment area as defined in sect. 5, but also the catchment areas belonging to the river as well as to the aquifer, cf. Fig. 1.1.

It is the author's intention to treat this problem in a coming report. To avoid misunderstandings we shall, however, mention a problem related to a numerical result found in sect. 5. In this example we found (a change of) the seepage i larger than 500 mm per year in the vicinity of the borings (the slot). This figure exeeds by far the net-precipitation (precipitation minus evapotranspiration) in Denmark, which is 200 - 300 mm per year. The normal practice (in Denmark) is, however, that the borings for water supply are placed in the river valleys, and a closer examination has in several actual cases shown that the natural phreatic surface is lying lower than the natural piezometric surface belonging to the aquifer, cf. Fig. 1.1. In such cases a natural upward flow exists in layer 2, and nothing then prevents the establishing of an (artificial) downward flow of the size found in the numerical examples. It should be added that the total gain of water q_t = 1750 m^3/m per year distributed evenly over the catchment area having a width of 15.8 km, leads to a mean seepage of i_m = 110 mm per year. A mean seepage of this size is very often found for water supply plants in Denmark.

Proceedings of Euromech 143 / Delft / 2-4 September 1981

Unsteady groundwater flow due to flood routing in open channels

A. DIACON, I.SETEANU, M.ERHAN & R.POPA
Polytechnical Institute of Bucharest, Romania

1. INTRODUCTION

The paper is concerned with the analysis of one-dimensional unsteady flow in open channels when the seepage discharge to the neighbouring aquifer through the river banks is significant.

By choosing as a coupling parameter the seepage discharge, the Saint Venant system and Boussinesq equation are numerically solved with an implicite scheme (double sweep algorithm).

The boundary conditions for flood routing consist of the known upstream hydrograph and stationary downstream water level at the junction with a major reservoir. A steady water table level at far distance of river reach is imposed as boundary condition for groundwater flow. As initial conditions, a gradually-varied flow in open channel is assumed together with the corresponding steady water table in aquifer.

Finally, some concise numerical results are presented for a given case study.

2 UNSTEADY FLOW IN OPEN CHANNELS

The one-dimensional, unsteady flows in open channels and rivers are often of nearly-horizontal type (long-wave motions). A hydrostatic pressure distribution and uniform velocity over any cross section are assumed. For small bed slopes, the turbulent exchange of energy and momentum within the water body may be reasonably modelled as a "resistance" to flow, defined by the shear stress at the limit of bed boundary layer, and acting opposite to flow velocity. In such circumstances, the Chézy law is valid

$$V = C\sqrt{R.S_f} \qquad (1)$$

where V = average cross-sectional velocity, C = Chézy coefficient, R = hydraulic radius ($R = A/\chi$), A = cross-sectional area, χ = wetted perimeter, and S_f = frictional slope. The Chézy coefficient is, usually, given in terms of the bed roughness coefficient, n, by Manning's formula

$$C = \frac{1}{n}.R^{1/6} \qquad (2)$$

2.1 Governing equations

Considering an infinitesimal element of water body bounded by two cross sections (Fig.1), the following general equations for one-dimensional, unsteady flow in open channels (the Saint Venant system) may be written:

- momentum equation

$$\frac{\partial Q}{\partial t} + \frac{\partial}{\partial x}\left(\frac{\alpha Q^2}{A}\right) + g.A.\frac{\partial h}{\partial x} + \frac{g.Q.|Q|}{C^2.A.R} = 0 \qquad (3)$$

in which Q = A.V = flow discharge, h = water surface elevation above a datum, g = gravity acceleration, α = coefficient of cross-sectional nonuniformity of flow velocity (Coriolis), x = space coordinate in longitudinal direction, and t = time;

- continuity equation

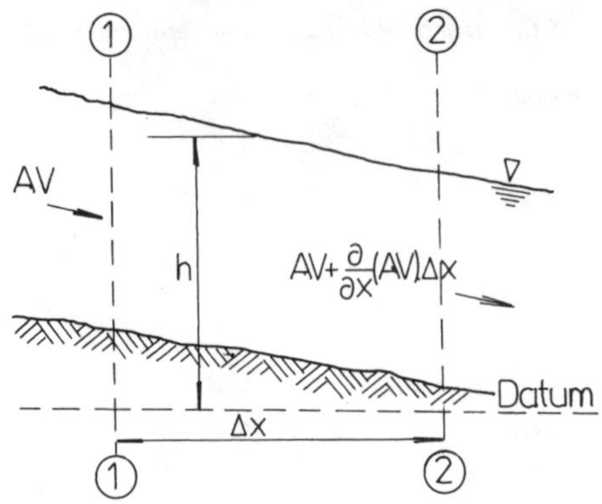

Fig. 1. Sketch of water element

$$\frac{\partial A}{\partial t} + \frac{\partial Q}{\partial x} - q = o \qquad (4)$$

where q = specific (per unit length) lateral inflow/outflow.

The numerical evaluation of the derivatives of A has to be done with some caution (Diacon 1976) as:

$$\left(\frac{\partial A}{\partial t}\right)_{x=ct} = \frac{\partial A}{\partial h}\cdot\frac{\partial h}{\partial t} = B\cdot\frac{\partial h}{\partial t} \qquad (5)$$

$$\left(\frac{\partial A}{\partial x}\right)_{t=ct} = \left(\frac{\partial A}{\partial x}\right)_{\substack{h=ct\\t=ct}} + \left(\frac{\partial A}{\partial h}\right)_{\substack{x=ct\\t=ct}}\cdot\frac{\partial h}{\partial x}$$

$$= \lim_{x_1 \to x_2} \frac{A(x_2,h)-A(x_1,h)}{x_2 - x_1}$$

in which B = width at water surface

If Coriolis coefficient, α, is accepted as constant, one obtains:

$$\frac{\partial}{\partial x}\left(\frac{\alpha Q^2}{A}\right) = \frac{2.\alpha.Q}{A}\cdot\frac{\partial Q}{\partial x} - \frac{\alpha Q^2}{A^2}\left(\frac{\partial A}{\partial x}\right)_{t=ct} \qquad (6)$$

$$= \frac{2.\alpha.Q}{A}\cdot\frac{\partial Q}{\partial x} - \frac{\alpha Q^2}{A^2}\cdot\frac{\delta A}{\delta x} - \frac{\alpha Q^2}{A^2}.B.\frac{\partial h}{\partial x}$$

where $\frac{\delta A}{\delta x} = \left(\frac{\partial A}{\partial x}\right)_{\substack{h=ct\\t=ct}}$, and the Saint Venant system becomes:

$$\frac{\partial Q}{\partial t} + \frac{2.\alpha.Q}{A}\cdot\frac{\partial Q}{\partial x} - \frac{\alpha.Q^2}{A^2}\cdot\frac{\delta A}{\delta x} - \qquad (7)$$

$$\frac{\alpha Q^2}{A^2}.B.\frac{\partial h}{\partial x} + g.A.\frac{\partial h}{\partial x} + \frac{g.Q.|Q|}{C^2.A.R} = o$$

$$B.\frac{\partial h}{\partial t} + \frac{\partial Q}{\partial x} - q = o \qquad (8)$$

The above system consists of two nonlinear, hyperbolic type, partial differential equations with flow discharge, Q, and water level, h, as unknown functions, while the x - coordinate and time t are the independent variables.

The analytical treatment of the problem involves great difficulties due to non-linearity and complexity of coefficients. Consequently, a numerical procedure is required. In order to develop such a scheme, a computational grid is defined by discrete points where Q and h values are to be alternatively calculated (Fig. 2).

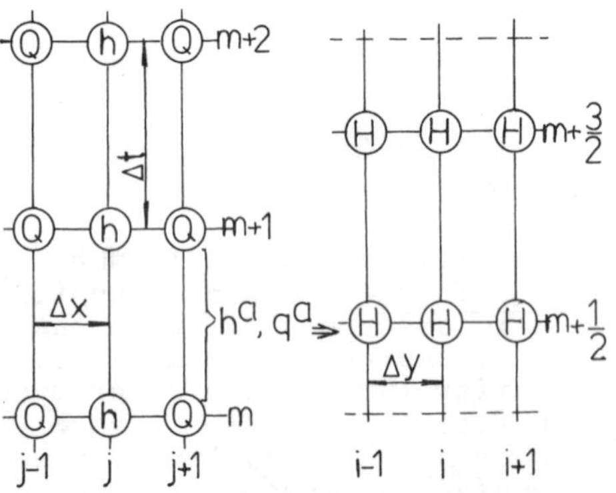

Fig. 2. Computational networks for open channel and groundwater flows.

Note. For a trapezoidal cross section (Fig. 3) the following relations are used:

$$B = b + 2.h'.ctg\beta$$
$$A = h'.(b + h'.ctg\beta) \qquad (9)$$
$$R = A/(b + 2.h'.csc\beta)$$

in which b = bottom width, h' = water depth, and β = angle between the side walls and datum.

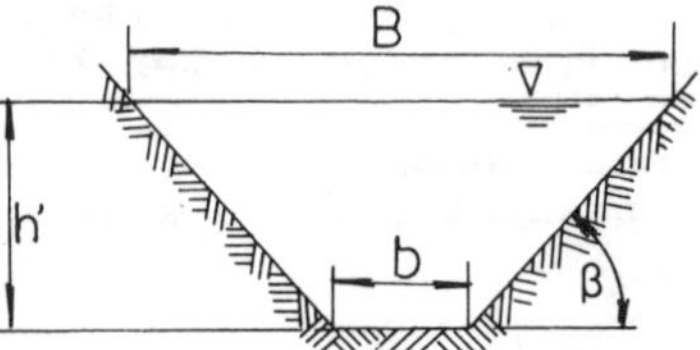

Fig. 3. Trapezoidal cross section

2.2 Difference scheme

Accepting a constant space-step, Δx, the Saint Venant system (7)-(8) may be approximated by the following implicit difference equations:

$$\frac{1}{4.\Delta x}.L_j^{(m)}.(h_{j+1}^{m+1}-h_{j-1}^{m+1}+h_{j+1}^{m}-h_{j-1}^{m}) + \frac{1}{\Delta t}.(Q_j^{m+1}-Q_j^{m}) - \frac{1}{2.\Delta t}.M_j^{(m)}.(h_{j+1}^{m+1}- h_{j+1}^{m}+h_{j-1}^{m+1}-h_{j-1}^{m}) + \frac{g.Q_j^{m+1}.|Q_j^m|}{(C^2.A.R)^{(m)}} +$$

$$(\frac{\alpha.q}{A})_j^{(m)}.(Q_j^{m+1}+Q_j^m) - (\frac{\alpha.Q^2}{A^2})_j^{(m)}.\frac{A_{j+1}^{(m)a}-A_{j-1}^{(m)a}}{\Delta x} = o \quad (1o)$$

and

$$\frac{1}{4.\Delta x}.(Q_{j+1}^{m+1}-Q_{j-1}^{m+1}+Q_{j+1}^{m}-Q_{j-1}^{m}) + B_j^{(m)}.\frac{h_j^{m+1}-h_j^m}{\Delta t} - q_j^{(m)} = o \quad (11)$$

in which:

$$L = g.A - \frac{\alpha.Q^2}{A^2}.B \; ; \quad M = \frac{2.\alpha.Q}{A}.B$$

A notation like Q_j^m specifies the value of discharge, Q, at time m.Δt for grid-point x_j. The superscript (m) denotes the time-moment (m+1/2) .Δt, whereas $A_{j+1}^{(m)a}$ represents the cross-sectional area for grid-point x_{j+1} (Q-point), at time (m+1/2).Δt, corresponding to average water elevation between the adjacent h-grid-point values.

The resistance term is expressed as a combination including the known Q^m, and unknown Q^{m+1} values, in order to avoid the computational instability mentioned by Abbott, 1967.

The relations (1o) and (11) may be rearranged in the forms:

$$A_j^1.Q_{j+1}^{m+1}+B_j^1.h_j^{m+1}+C_j^1.Q_{j-1}^{m+1}=D_j^1 \quad (12)$$

$$A_j^2.h_{j+1}^{m+1}+B_j^2.Q_j^{m+1}+C_j^2.h_{j-1}^{m+1}=D_j^2 \quad (13)$$

where:

$$A_j^1 = \frac{1}{4.\Delta x}; \quad A_j^2 = A_j^1.L_j^{(m)} - \frac{M_j^{(m)}}{2.\Delta t}$$

$$B_j^1 = \frac{B_j^{(m)}}{\Delta t}; \quad B_j^2 = \frac{1}{\Delta t} + \frac{g.|Q_j^m|}{(C^2.A.R)_j^{(m)}} + (\frac{\alpha.q}{A})_j^{(m)}$$

$$C_j^1 = -A_j^1; \quad C_j^2 = C_j^1.L_j^{(m)} - \frac{M_j^{(m)}}{2.\Delta t}$$

$$D_j^1 = A_j^1.Q_{j-1}^m+B_j^1.h_j^m+C_j^1.Q_{j+1}^m+q_j^{(m)}$$

and

$$D_j^2 = A_j^2.h_{j-1}^m + \frac{1}{\Delta t} - (\frac{\alpha\,q}{A})_j^{(m)}.Q_j^m + C_j^2.h_{j+1}^m + \frac{1}{\Delta x}.(\frac{\alpha\,Q^2}{A^2})_j^{(m)}.(A_{j+1}^{(m)a}-A_{j-1}^{(m)a})$$

The coefficients of equations (12) and (13) are evaluated for h-, and Q- grid-points, respectively, excepting the terms with a-superscript that are calculated for h-points with $h_j^a = (h_{j+1}+h_{j-1})/2$.

The system (12)-(13) is to be solved by two simultaneous tri-diagonal algorithms (double sweep algorithm (Abbott 1973) . Having Q_{j+1}^{m+1} and h_j^{m+1}, together h_{j+1}^{m+1} and Q_j^{m+1} interrelated by the following equations:

$$Q_{j+1}^{m+1} = E_j^1.h_j^{m+1} + F_j^1 \quad (14)$$

$$h_{j+1}^{m+1} = E_j^2.Q_j^{m+1} + F_j^2 \quad (15)$$

one may obtains a set of recurrence relations:

$$E_{j-1}^2 = \frac{-C_j^1}{A_j^1.E_j^1+B_j^1}; \quad F_{j-1}^2 = \frac{D_j^1-A_j^1.F_j^1}{A_j^1.E_j^1+B_j^1} \quad (16)$$

$$E_{j-1}^1 = \frac{-C_j^2}{A_j^2.E_j^2+B_j^2}; \quad F_{j-1}^1 = \frac{D_j^2-A_j^2.F_j^2}{A_j^2.E_j^2+B_j^2}$$

Assuming as initial conditions the known discharges, Q_j, and water surface elevations, h_j for a gradually-varied flow, the computational procedure starts from the downstream boundary condition (lake junction) where the water level h_L is maintained constant. This condition may be identically satisfied providing that

$$E_{J-1}^2 = o \text{ and } F_{J-1}^2 = h_L$$

J being the total number of x-grid-points. From here, with relation such as (16), E_{J-2}^1, F_{J-2}^1, E_{J-1}^2, F_{J-1}^2, and so on, and finally E_1^2, F_1^2 may be found succesively (the first sweep). On the other hand, since Q_1^{m+1} is given for the upstream end (the known flood hydrograph), h_2^{m+1} may be derived from relation (15), and so on, up to the downstream end (the second sweep).

It may be noted that, since the coefficients A^1, A^2, and others, have some terms defined at the intermediate time-moment $(m+1/2).\Delta t$, a succesive approximations procedure is required starting with their values at time-moment $m.\Delta t$ and iterating two or three times for the same computational step.

To achieve the coupling of the unsteady flow in river with the unsteady groundwater flow through the adjacent aquifer, the water table curves for each cross section along the channel are to be determined, assuming that the time-grid-points of seepage analysis correspond to the centers of time-steps of previous grid (Fig. 2). For this, the corresponding hydraulic parameters have to be time-averaged, and the seepage time-grid-points staggered by $\Delta t/2$ with respect to the open-channel flow within the x-t plane.

3 UNSTEADY LATERAL SEEPAGE

The unsteady groundwater flow through the aquifer adjacent to the channel may be described with a quite good approximation by the partial differential equation of Boussinesq that, for a plane-parallel flow appears as

$$\frac{\partial}{\partial y}\left(H.\frac{\partial H}{\partial y}\right) = \frac{p}{k}.\frac{\partial H}{\partial t} \qquad (17)$$

where H = water table elevation above the impervious layer, p = porosity, k = infiltration Darcy coefficient, and y = horizontal coordinate in any cross section along the river (Fig.4).

Pinder & Sauer, 1971 analyzed such a problem with a two-dimensional unsteady model, while Zitta & Wiggert, 1971 utilized the Boussinesq scheme for a rectangular cross section channel.

To simplify the analysis, due to greater scale of groundwater flow, vertical river banks are assumed whereas the specific seepage discharge, q, is to be computed.

The above relation may be written as follows:

$$\frac{\partial}{\partial t}(H^2) = 2.\lambda.H.\frac{\partial^2}{\partial y^2}(H^2) \qquad (18)$$

where $\lambda = k/(2.p)$.

With finite differences one obtains:

$$\frac{1}{2.H_i^m}.\frac{(H^2)_i^{m+1}-(H^2)_i^m}{\Delta t} = \lambda.\frac{(H^2)_{i+1}^{m+1}-2.(H^2)_i^{m+1}+(H^2)_{i-1}^{m+1}}{(\Delta y)^2} \qquad (19)$$

After some manipulations, this equation becomes

$$A_i^3.(H^2)_{i-1}^{m+1}+B_i^3.(H^2)_i^{m+1}+C_i^3.(H^2)_{i+1}^{m+1} = D_i^3 \qquad (2o)$$

in which

$$A_i^3 = -C_i^3 = -1;\; B_i^3 = 2 + \lambda_e/H_i^m;$$

$$D_i^3 = \lambda_e.H_i^m, \text{ and } \lambda_e = \frac{(\Delta y)^2}{2.\lambda.\Delta t}$$

The initial water table curves correspond to the river gradually-varied flow levels, being different from one cross section to an other. The boundary conditions for the unsteady seepage are:

- the constant water table level far of the channel, H_b;
- the same water table level as for river surface at bank junction

The set of simultaneous relations (2o) for i = 2,3,...,I-1 (I being the total number of lateral space-grid-points) forms, once again, a tri-diagonal, linear system of implicit, finite difference equations

The corresponding recurrence relations are

$$E_{i-1}^3 = \frac{-A_i^3}{B_i^3+C_i^3.E_i^3}\;;\; F_{i-1}^3 = \frac{D_i^3-C_i^3.F_i^3}{B_i^3+C_i^3.E_i^3} \qquad (21)$$

Using value of water table level far of the river, the term $(H^2)_I^{m+1}$ may be computed. One chooses

$$E_{I-1}^3 = o \quad\text{and}\quad F_{I-1}^3 = (H^2)_I^{m+1} = H_b^2$$

after that the quantities E_{I-2}^3, $F_{I-2}^3,\ldots,E_1^3$, F_1^3 are succesively evaluated by relations (21). In the second sweep, one starts with $(H^2)_1^{m+1}$ value (corresponding to actual water level in river), and water table elevations at all grid-points of any cross section are derived by recurrence relation

$$(H^2)_{i+1}^{m+1} = E_i^3.(H^2)_i^{m+1} + F_i^3 \qquad (22)$$

The specific (per unit length) seepage discharge is computed as

$$q = -p.\int_0^{\infty} \frac{\partial H}{\partial t}.dy \qquad (23)$$

The relation (23) represents a result of mass conservation.

In discret form, one obtains

$$q = -p.\sum_{i=1}^{I} (H_i^{m+1} - H_i^m).\frac{\Delta y}{\Delta t} + q_I =$$

$$q_I + \frac{p.\Delta y}{2.\Delta t}.\Big[(H_1^{m+1} + 2.H_2^{m+1} + \ldots. + 2.H_{I-1}^{m+1} + H_I^{m+1}) - (H_1^m + 2.H_2^m + \ldots + 2.H_{I-1}^m + H_I^m)\Big] \qquad (24)$$

As soon as the q_i values at actual time-moment are derived for all sections along the river reach under study, the procedure advances by a new computation of open channel regime at the next time-moment. The new water surface levels are derived, after that the seepage analysis is resumed. This computational algorithm is succesively applied during the whole time period of interest.

4 NUMERICAL RESULTS

The above computational scheme was applied to a case study with a channel reach extended over 19 km upstream of a lake. A longitudinal bed slope of o.oooo1, a Manning roughness coefficient n = o.o167 over the bottom of the channel and n = o.o45 over the lateral banks were choosen. The prismatic channel with trapezoidal cross section had the bottom width b = 15 m and slope of lateral banks 1:15. With 2o grid-points (Δx = 1ooo m), the computer program has been run for more 5o time-steps (Δt = 72oo s). A constant water surface elevation at lake junction h_L = 8 m was imposed. The input natural flood hydrograph was

$$Q = Q_o.(1 + \sin(3.636.10^{-5}.t))$$

for $t \leq 86400$ s, and $Q = Q_o$ for $t > 86400$ s. This analytical form of input hydrograph corresponds to a sinusoidal impuls extended over a half period of 432oo s. The Coriolis coefficient has choosen to be $\alpha = 1.1$ while the average river discharge was Q_o = 24o mc/s. The aquifer porosity was p = 42 %, and the Darcy coefficient of infiltration k =

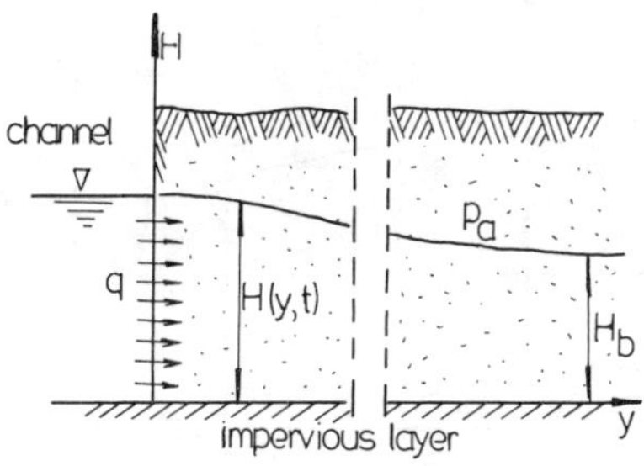

Fig. 4. Sketch of groundwater flow

o.oo338 m^2/s. The cross sectional grid consisted of 2o points with step Δy = 25 m.

In Figs. 5 and 6 the variation of discharge and water level curves with time t as parameter and space grid points along the abscissa are represented. The lateral specific seepage discharge was an average value of q = o.oo992 m^2/s.

This computational procedure can be used to predict the water table variation in aquifer, as result of the flood routing through the analysed channel reach, together with the evaluation of water table changes produced by an intensive irrigation system.

5 REFERENCES

Abbott, M.B. & Ionescu, F. 1967, On the numerical computation of nearly-horizontal flows, J.Hydr. Res. 5; 2

Abbott, M.B. & Verwey, A. 1973, Some aspects of the design system SIVA, Proc. XV-th Congress of IAHR, Istambul, 5.

Diacon, A., Seteanu, I., Popa, R. & Roman, P. 1976, Mathematical modelling of thermal river pollution, Int. Symposium on Unsteady Flow in Open Channels, BHRA, New castle-upon-Tyne, England.

Pinder, S. & Sauer, S.P. 1971, Numerical simulation of flood wave modification due to bank storage effects, Water Resour.Res., 7; 1.

Zitta, V.L. & Wiggart, J.M. 1971 Flood routing in channels with bank seepage, Water Resour.Res., 7; 5.

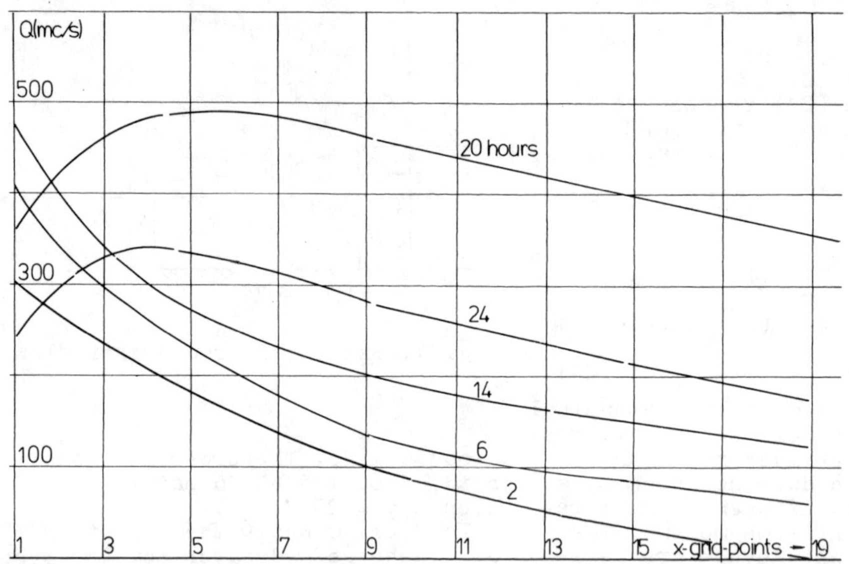

Fig. 5. Time-variations of discharge along the channel

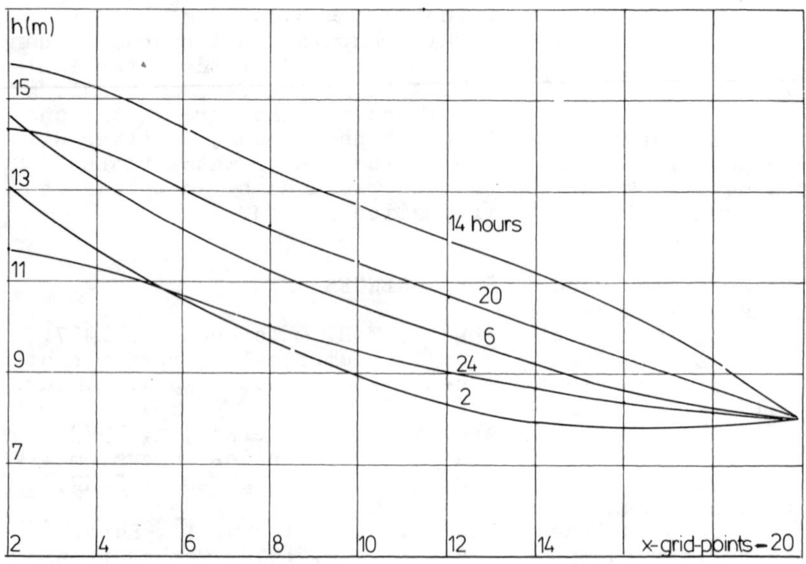

Fig. 6. Time-variations of water level along the channel

Proceedings of Euromech 143 / Delft / 2-4 September 1981

Non-darcy flow; A state of the art

A'ALIM A.HANNOURA
University of New Orleans, Louisiana, USA

FRANS B.J.BARENDS
Delft Soil Mechanics Laboratory, Netherlands

1. INTRODUCTION

The probably earliest mentioned nonlinear form describing flow in a porous medium is due to Kröber (1884), quoted by Bear c.s. (1965). However, Forchheimer's publication in 1901 is commonly marked as the start of a general understanding that Darcy's law is not universally valid for porous flow. Since then non-Darcy flow has been in continuous investigation.

Recent developments in oil and gas production, in groundwater recovery and management, rockfill dam and rubble mound structure design and lately the use of granular beds for the storage of solar energy, have demanded a more accurate determination of the hydraulic conductivity of porous media. Much excellent work has been done and most of it has been extensively reported in various textbooks. This paper reviews some of the prominent contributions to the study of the hydrodynamic aspects of this so-called non-Darcy flow. Attention is given to practical application using numerical methods.

2. THE PHYSICAL INTERPRETATION OF DARCY'S LAW

The transport process of water through a porous medium involves two substances: water and the porous matrix, and therefore, it will be characterised by specific properties of these two substances. The process can be described in terms of an equilibrium of forces (De Josselin de Jong, 1969). The driving force necessary to move a specific volume of pore water at a certain speed through a porous medium is in equilibrium with the resistance force generated by internal friction between the pore water and the pore structure.

In most granular media the driving action can be produced by a force F resulting from a pore pressure gradient $p_{,i}$ and the gravity acting per unit volume ρg.
The driving force per unit volume becomes:

$$F = -p_{,i} - \rho g z_{,i} \,, \tag{1}$$

where z is measured vertically upwards, ρ is the fluid density and g the gravity acceleration. The coordinate z is only significant as to fix the direction of the gravity action and it does not matter whether the density ρ is a constant or a variable in space, since the unit volume considered can be conceived as a physical point.

It is sometimes possible to introduce a potential ϕ to express the volumetric driving force F, according to:

$$F = -\rho g \phi_{,i} = -p_{,i} - \rho g z_{,i} = \rho g I. \tag{2}$$

In this form ϕ can be identified with a vertical head. It is generally referred to as the piezometric head. Formula (2) is valid for irrotational flow fields.

The resistance force R generated by internal friction is characterized by Darcy's law. This law provides a relation between the filter velocity q and the corresponding volumetric resistance force R, according to:

$$R = \rho g W q = \rho g q / K. \tag{3}$$

W is called the coefficient of resistance of the flow, K is called the hydraulic conductivity of the porous medium. These coefficients (W=1/K) are related to the substances being considered and they contain information regarding the properties concerning porous flow. Equation (3) reflects the internal constitution of the substances in the form of a volumetric flux and a generated volumetric force. In this respect it represents a constitutive relation for porous flow. Such a relation can only be stated by observation and Darcy (1856) was the first to verify this relation by experiments.

Beside the term 'filter velocity' or the 'Darcy velocity' also definitions like 'specific discharge' or 'volumetric flux' or 'bulk velocity' are frequently used for the same quantity q.

The equilibrium condition is satisfied by:

$$F = R. \tag{4}$$

Introduction of equation (2) and (3) yields:

$$\phi_{,i} = -Wq \quad \text{or} \quad q = -K\phi_{,i} = KI, \tag{5}$$ ®

which represents a linear relation between the potential gradient I and the filter velocity q, provided that W or K does not depent on ϕ or q. Expression (5) is usually identified with Darcy's law and in this form it comprises two fundamental concepts: equilibrium between driving and resistance forces and a constitutive relation for porous flow.

In general Darcy's law describes a certain regime of the porous flow process. A different behaviour is noticed at high velocities in coarse beds and at low velocities in fine porous media, referred to as the post-linear regime (see fig. 1). Actually, at all stages the internal equilibrium, equation (4), is satiesfied. The correct description of the nature of the flow process is embedded in the constitutive law, equation (3). Non-Darcy flow implies a nonlinear constitutive law formulated as a relation between a flux (for example the filter velocity q) and the generated resistance force R.

® Note: I denotes a "negative" gradient.

3. THE NATURE OF FLOW IN POROUS MEDIA

In non-Darcy flows the main problem is to estiablish a relation between two measurable parameters: the pressure gradient and the velocity of the pore fluid. Porous flow was initially conceived as being analogous to pipe flow, which was rather well understood at that time. The porous matrix was represented by a system of capillary tubes. By this concept the irregularity in natural porous media, referred to as tortuosity ,a quantity introduced by Carman (1937) and extensively expounded by Whitaker (1966) and Bear (1972) is obviously not considered. Nevertheless, the similarity of porous flow with pipe flow clearly reveals the physical background of the different regimes of porous flow commonly recognised in literature. Lateron it was shown that the absence of abrupt transition in flow type observed in pipe flow but not in porous flow reveals more similarity with flow around a sphere, where no critical transition value has been found (Oseen flow). Therefore, in this section a short outline of the experience with pipe flow and Oseen flow is given projected on the observed behaviour of porous flow. A similar approach is found in many papers (Forchheimer,1901, 1930; Muskat,1937; Missbach,1937; Goldstein,1938; De Lara,1955; Dudgeon,1964; Rumer,1969; Bear,1972).

Stokes's law which dates from 1851 indicates that the drag on a sphere moving through a fluid is proportional to the first power of its velocity provided its Reynolds number, based on the diameter as a characteristic length, is less than about 0.10. As the Reynoldsnumber increases the linear relationship between drag and velocity breaks down The exponent on the velocity increases with increasing Reynoldsnumber approaching a maximum value of 2.0 for pure turbulent flow (Rouse, 1961). Some useful analogies between flow in porous media and flow passing a single particle can be drawn. Obviously in a porous medium there will be an interference of flow patterns which will affect the drag on the particles of the medium. Therefore, it will not be possible to obtain quantitive estimates of drag forces (i.e. internal friction) in a porous medium from the drag characteristics of the individual particles of the medium. However,there is a clear correspondence between the flow regimes in a porous medium and the flow regimes for flow around a single particle.

In both cases a linear laminar or viscous regime exists at low Reynoldsnumber: the drag is proportional to the velocity. In a porous medium this type of flow is called Darcy flow (Darcy, 1856); in the case of a single particle it is referred to as Stokes' flow (Stokes, 1851). Viscous forces predominate over inertial forces and the drag is primarily due to skin friction and viscosity. The streamlines are steady (and stable) with a strong tendency towards symmetry for and after the particle.

As the inertial forces, i.e.,the local acceleration due to convective and centrifugal effects, increase relatively to the viscous forces the streamlines become more distorted and the drag increases more rapidly than linearly with the velocity. However, the streamlines remain stable, they do not fluctuate. Some steady secondary flow may appear at the higher velocities of this regime. This flow regime is sometimes referred to as Oseen flow (Oseen, 1927) in the case where flow is considered passing a particle. In porous media it is called nonlinear laminar or steady inertial flow (Lindquist, 1933; Karadi, 1961).

Close inspection of the flow pattern shows that the symmetry mentioned for linear laminar flow is lost and inertia becomes important. A similar phenomenon has been noticed in porous media (Chauveteau c.s., 1967). For this regime the hydraulic gradient I in the porous medium can be approximated by:

$$I = (a_1 + b_1 q)q, \tag{6}$$

in which q is the filter velocity, a_1 and b_1 are constants for a particular medium.

Increasing the Reynoldsnumber still further, in fact increasing the inertial forces with respect to the viscous forces, may cause the streamlines to become unstable and separate from the particle. Thus the flow pattern in a certain region becomes unsteady and fluctuates in a periodic or random manner (Dudgeon,1964; Wright,1969). The flow is locally turbulent.

The drag in this regime depends on both the viscous and the inertial forces. Increasing the Reynoldsnumber will increase the relative importance of turbulence. Thus in this regime Oseen's solution is not applicable for the sphere and equation (6) does not describe the flow correctly (Dudgeon, 1964). The hydraulic gradient for the porous medium could now be represented by:

$$I = (a_2 + b_2 q)q, \tag{7}$$

in which, a_2 and b_2 are constants for a particular medium, most probably different form a_1 and b_1. It should be noted that the drag on a single particle is affected by the location of the point of flow separation. This may change with Reynoldsnumber and is influenced by the particle shape and the surface roughness.

In a fully turbulent regime at very high Reynoldsnumbers turbulence exists everywhere in the flow region and flow separation is stabilized. The hydraulic gradient can be taken as:

$$I = bq^2. \tag{8}$$

A striking example of the quadratic flow law is measured by Hajdin, who investigated porous flow in karstic media (Boreli, 1978). It is still a point of discussion whether in a porous medium a fully turbulent regime can be generated, since the interference of neighbouring particles represent a limitation to the development of an undisturbed turbulent flow (Goldstein,1938; Rumer,1969). Actual turbulence starts at the larger pores while in the smaller ones flow is still laminar (Bear, 1972).

There is another non-Darcy flow regime which is associated with extremely small velocities (Swartszendruber,1962; Slepicka, 1961; Habid,1969; Kovacs,1969; Kutilec,1969; Benis,1968; Bear, c.s.; 1965, Bear,1972; Barends,1980). A stationary resistance might exist while the actual flow is yet zero. This limit can be identified as a so-called initial gradient or threshold gradient below which there is no flow. It occurs in clayey soils and it is attributed to rheological non-Newtonian behaviour, in particular as a result of molecular linkage or blocking air bubbles. It affects the flow and can be considered as a velocity damping factor, adjusting the flow of affinite phases in very fine pores. The opposite is observed as well. For gas flow in a porous medium an increase in the hydraulic conductivity at low velocities exist due to the mean free path of the gas molecule motion (Bear, 1972). It is referred to as the slip phenomenon (Scheidegger, 1960).

4. DARCY'S LAW, A COUPLING IN A TWO-PHASE MECHANICAL SYSTEM

The motion of a pore fluid in a granular bed is controlled by an action force resulting from pore pressure gradients, gravity, capillarity and other effects, and by a reaction force resulting from inertia and internal friction in the fluid itself (viscosity) and along the interface between the fluid and the particles. The equilibrium of these forces constitute the equation of motion with two independent variables.
Using the principle of conservation of mass permits to solve the equation of motion.

As an example the one-dimensional case is considered, where a volume water HBdx is exposed to the action force:

$$\frac{\partial F}{\partial x}dx, \tag{9}$$

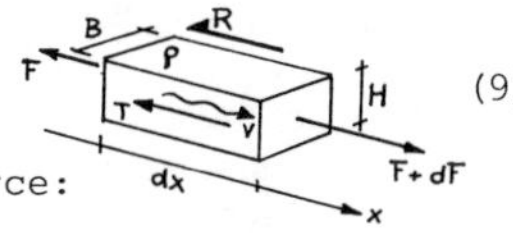

and the inertia force:

$$T = \rho HBdx\frac{Dv}{Dt}. \tag{10}$$

Here, D/Dt denotes the water-substantial derivative:

$$\frac{Dv}{Dt} = c_m \frac{\partial v}{\partial t} + \beta v \cdot \frac{\partial v}{\partial x}, \tag{11}$$

in which c_m denotes a coefficient related to the virtual mass effect arising in unsteady turbulent flow, and β is the momentum distribution coefficient related to the velocity distribution.
On the volume of water acts a friction force R, which embeds the constitutive relation for porous flow:

$$R = \rho gHB\,dx\,(v-u)/K. \tag{12}$$

Here, u represents the local velocity of the

granular bed.
The equilibrium states:

$$\frac{\partial F}{\partial x} = \rho HB\left[\frac{Dv}{Dt}+\frac{g}{K}(v-u)\right]. \quad (13)$$

In general the parameter K is a function of v (nonlinear flow) and of x (inhomogeneity). Since the action force F can be expressed in terms of the pore pressure p according to:

$$F = -pHB, \quad (14)$$

the equilibrium equation (13) becomes:

$$-\frac{\partial p}{\partial x} = \rho\frac{Dv}{Dt}+\frac{\rho g}{K}(v-u). \quad (15)$$

With respect to soils and granular beds the volume of water saturating the pores equals n (volumetric porosity) times the bulk volume of soil (particles plus water). Following the previous analysis the equilibrium of the pore water is governed by:

$$-n'p_{,i}-n\rho gz_{,i} = n\rho\frac{Dv_i}{Dt}+n^2\frac{\rho g}{K}(v_i-u_i) \quad (17)$$

valid for more dimensions and including the gravity, measured along the negative z-axis. n' represents the cross-sectional porosity.

Since p and u_i are related to the granular skeleton behaviour, the equilibrium of the granular medium must be considered as well. This equilibrium is composed of an action force resulting from intergranular contact forces (effective stresses), pore pressure and internal friction forces generated by the pore water motion, and of a reaction force due to inertia of the granular particles (having density $\bar{\rho}$). The intergranular contact forces are represented by a fictitious stress averaged with respect to the bulk volumetric cross-sectional surface. This stress: σ'_{ij}, is referred to as effective stress or granular stress.

The individual grain is considered as incompressible. This condition is not strictly necessary, but it simplifies the equations. The pore pressure surrounding the grains acts in the grain's inside resulting in a term: (1-n")p, where n" represents the areal free surface porosity.

The equilibrium of the granular skeleton is governed by:

$$-(1-n'')p_{,i} - (1-n)\bar{\rho}gz_{,i} + \sigma'_{ij,i} + n^2\frac{\rho g}{K}(v_i-u_i) = (1-n)\bar{\rho}\frac{\mathbb{D}u_i}{\mathbb{D}t} \quad (18)$$

Here, $\mathbb{D}/\mathbb{D}t$ denotes the granular-substantial derivative.

Although the various porosity factors represent different geometrical parameters depending on the stress situation each one in its own manner, the difficulty of precise measurement provokes the trivial assumption:

$$n = n' = n''. \quad (19)$$

Adding equation (17), which describes the equilibrium of the pore water, and equation (18), which describes the equilibrium of the granular skeleton, eliminates the coupling term:

$$-p_{,i} - (n\rho+(1-n)\bar{\rho})gz_{,i} + \sigma'_{ij,i} = (1-n)\bar{\rho}\frac{\mathbb{D}u_i}{\mathbb{D}t} + n\rho\frac{Dv_i}{Dt}. \quad (20)$$

It is not possible to combine the inertia terms to one composite effect preserving the physical nature, since the velocity of the pore water u_i and the velocity of the granular medium v_i are basically independent variables. One can consider special cases, where the motion is constraint, for example an equal motion ($u_i=v_i$) or a rigid granular skeleton ($u_i=0$). Then it is possible to derive onesingle equation, which covers the corresponding rheological behaviour.

Equilibrium in soils is governed by total stresses σ_{ij}. For a saturated soil with volumetric weight γ the state of equilibrium can be expressed by:

$$-\gamma z_{,i} + \sigma_{ij,i} = \frac{\gamma}{g}\frac{\mathbb{D}u_i}{\mathbb{D}t}, \quad (21)$$

in which the motion of the pore water and the grains is assumed equal ($u_i=v_i$). Comparison with (20) yields:

$$\sigma_{ij,i} = \sigma'_{ij,i} - p_{,i}\ ; \quad (22)$$

$$\gamma = n\rho + (1-n)\bar{\rho}, \quad (23)$$

and this may lead to Terzaghi's concept (Barends, 1980):

$$\sigma_{ij} = \sigma'_{ij} - p\delta_{ij}. \quad (24)$$

Groundwater flow problems are usually solved while disregarding inertia and assuming a rigid skeleton. Thus any effective stress σ'_{ij} can be adapted. The governing equations are for this situation:

$$-p_{,i} - \rho gz_{,i} = n\frac{\rho g}{K}v_i\ ; \quad (25)$$

$$-(1-n)p_{,i} - (1-n)\bar{\rho}gz_{,i} + \sigma'_{ij,i} + n^2\frac{\rho g}{K}v_i = 0\ . \quad (26)$$

Conservation of mass, for example: $v_{i,i}=0$,

allows to solve equation (25) for p and v_i for given boundary conditions, and allows to find the corresponding stress σ'_{ij} from equation (26).

If the soil is not rigid the initial pore pressure distribution has to match equilibrium of the granular skeleton according to equation (26). It will give rise to deflections and deformations of the skeleton. Equations (25) and (26) have to be solved simultaneously. In general the induced deformations involve volumetric strain leading to a mass conservation equation, which includes "storativity". In this respect the following mass conservation equation holds for the pore water:

$$-(\rho v_i)_{,i} = \partial(n\rho)/\partial t. \qquad (27)$$

Introduction of Darcy's law, i.e., K is not dependent on the velocity v_i, and assuming a logarithmic compressibility law for the pore water, according to:

$$\beta_p = d\rho/\rho dp, \qquad (28)$$

renders equations (25) and (27) into the so-called storage equation (Barends, 1980):

$$\frac{K}{\rho g}\{p_{,ii} + \beta_p \rho g p_{,i} z_{,i}\} = n\beta_p \frac{\mathbb{D}p}{\mathbb{D}t} + \frac{\mathbb{D}\varepsilon}{\mathbb{D}t}, \qquad (29)$$

in which ε is the volumetric strain of the soil skeleton.

Equations (21), (24) and (29) describe the consolidation process in deformable soils, which can be solved if the constitutive behaviour of the soil deformation process is known, i.e. a stress-strain law: $\varepsilon(\sigma'_{ij})$.

In a dynamic environment the inertia terms are not negligible, and equations (17), (18) and (27) are to be solved together with a stress-strain law and a porosity-strain relation (Van der Kogel,1977) determining the unknown variables: σ'_{ij}, p, v, u and n (note: $\partial\varepsilon/\partial t$ is related to $u_{i,i}$).

The interaction term in equations (17) and (18):

$$n\frac{\rho g}{K}(v-u), \qquad (30)$$

actually the internal friction force is usually supposed to be linear.

Introduction of the so-called filter velocity q, defined according to:

$$q = n(v-u), \qquad (31)$$

clearly shows the similarity with Darcy's law, provided K is a constant. In coarse gravel beds and rubble mound structures the porous flow behaviour is nonlinear and turbulent. The development of viscous boundary layers and added mass effects should be considered as well. A non-Darcy law will provide a better modelling of the real physical process, and as such it can be substituted in the equations (17) and (18).

In a pure dynamic environment of propagating waves inside the media the interaction between the two phases considered will probably be of a very different nature. The convective motion of pore water in a diffusion type process is quite different from the induced motion due to propagating density waves. The common assumption for the interaction term in the equations of motion for each phase is a linear behaviour, i.e., Darcy's law (Van der Kogel,1979).

A complete theory for the net internal resistance due to unsteady porous flow in a dynamical environment is not available. Several contributions have to be considered: the unsteady hydrodynamic drag in a porous medium, the force required to accelerate the actual fluid mass, the force required to establish a viscous field, entrained air effects and the influence of deformation of the granular skeleton. In recent studies (Hannoura, 1980) some experimental evidence is shown to state the existence of several of these effects. Research papers on the subject of deformation in embankments have been presented (Eisenstein,1979; Barends, 1980), but they usually concern sandy and clayey material.

At this moment these effects are only accounted for in an approximate way by using experimental overall turbulent permeability values. The actual flow is calculated using linearized interaction and incorporating the nonlinear flow effects in an iterative procedure (Hannoura,1978; Barends,1978; Kalliontzis,1981).

Although for more than a century since Darcy's law, the attention of many researchers has been directed to the phenomenon of pore fluid flow and much of the fundamental characteristics have been elucidated, the interaction of pore fluid and porous matrix is not yet completely understood. Beside the effects previously mentioned it is stressed that the major part of experimental research covers the one-dimensional case. Extending the concepts found to more dimensional flow situations unfolds new difficulties to coop with, such as non-coaxiality of pressure gradients and fluxes in the case of turbulent porous flow (induced anisotropy) and the eventuation of rotational effects due to inhomogeneity in the porous matrix, which renders the concept of a potential describing the flow process as incomplete. For coarse granular media the macroscopic approach, in fact dealing with a filter velocity and a hydraulic conductivity, might be an approximation too rough to describe the intrinsic properties. In granular filter constructions it is not seldom that a layer thickness is not more than a few particles (for example the armor layer of a rubble mound breakwater). A particulate approach simulating the microscopic processes nowadays possible due to the enormous development in fast computers, is promising.

In conclusion one might state that simplified approaches must be adopted with the obvious consequence that the results obtained can be regarded as acceptable under the restrictions specified. In this respect the next section will deal with the great variety of approximate, generalised porous flow equations, in which the nonlinearity is considered as a deviation from the linear flow; Darcy's law.

5. CLASSIFICATION OF NON-DARCY POROUS FLOW EQUATIONS

Distinction is made between the following formulations of non-Darcy flow:

- The exponential form:

$$I = aq^m \tag{32}$$

- The series form:

$$I = I_o + aq + bq^2 \tag{33}$$

- Graphical representations
- Statistical models.

5.1. The exponential form of non-Darcy flow

Groundwater flow was initially conceived as being analogous to pipe flow, while the porous medium was represented into a system of fine tubes. This has lead to the exponential form:

$$I = aq^m, \tag{32}$$

in which the coefficient m reprents a value somewhere in between 1 and 2. The earliest description of porous flow with equation (32) is probaly due to Kröber (1884) and Smreker, mentioned in Bear c.s. (1965). According to Jaeger (1956) this form is attributed to Prony. Missbach (1937) and Izbash (1931) proposed similar equations. The coefficients a and m are constants determined experimentally. White (1935) obtained a value for m equals to 1.8 based on experiments with gas flow. Muskat (1937) attempted to find expressions for a and m in terms of fundamental material properties using the method of dimensional analysis and similitude. Following the method suggested by Muskat two physically relevant numbers are used:

- Reynoldsnumber: $Re = qD/\nu$; (34)
- Galileo number: $Ga = \nu^2/D^3g$; (35)

in which D is some relevant length and ν denotes the fluid kinematic viscosity. This yields for equation (32) the proportionality (Barends, 1980):

$$a (:) \ \nu^{2-m} \ D^{m-3} \ \rho. \tag{36}$$

Considering the exponent of the viscosity ν in this formula, it becomes clear that, since for Newtonian fluids the one with a larger viscosity will generate a larger resistance at a similar filter velocity, values for m exceeding 2 will be physically unreasonable. This is an important conclusion, merely a consequence of dimensional analysis taking into account density versus friction (Reynoldsnumber) and friction versus gravity (Galileo number).

Esconde (1953), Wilkins (1955), Parkin (1963) and Curtis (1965) are some of the experimenters who have used the exponential form. A wide variation of the value of the exponent m has been quoted. De Lara (1955) realised that both a and m would not be constant in the transition from laminar to turbulent flow. This variation in a and m has been recognised by investigators who have carried out tests over a sufficiently wide range of velocities. From experimental results Anandakrishnan c.s. (1963); Dudgeon (1964); Rumer c.s. (1966) and Basak (1977) proposed the existence of several post linear flow regimes, each of which could be described by an exponential nonlinear relation. Equations of the form:

$$I = fq^2/2gD, \tag{39}$$

have been proposed to generalise the flow in these regimes. The friction factor f, is usually related to the Reynoldsnumber. The existence of a wide range of friction factor-Reynoldsnumber relationships as determined by experiments (Lindquist, 1933; Bakhmeteff c.s., 1937; Yalin c.s., 1961; Anderson, 1963; Dudgeon, 1964; Ward, 1964; Wright, 1968) indicates the difficulty of this form to yield one general relation for different flow regimes in various types of porous media.

Burke c.s. (1928) (quoted by Scheidegger, 1960) suggested:

$$f = C(1-n)/Dn^3, \tag{40}$$

where C is a constant and n denotes porosity.

Bakhmeteff c.s. (1937) suggested:

$$f = C/(Re^{0.2}\, n^{1.67}). \qquad (41)$$

De Lara (1955) suggested:

$$f = C_D/n^5, \qquad (42)$$

where C_D represents the drag coefficient measured in a uniform flow for one single sphere related to Reynolds number. Dudgeon (1964) suggested a similar form:

$$f = Re^m/n^5. \qquad (43)$$

Rumer c.s (1966) suggested:

$$f = \left(\frac{\alpha D}{2n^2}\right)\left(\frac{1-n}{\lambda\beta k}\right)^{\frac{1}{2}} C_D, \qquad (44)$$

in which αD^2 is related to the effective particle's cross section, βD^3 is related to the relevant particle's volume, k is the intrinsic permeability and λ is a coefficient which accounts for the effect of neighbouring particles.
Barends c.s. (1978) suggested:

$$f = C_D\, \alpha(1-n)/\beta n^5, \qquad (45)$$

showing that this formulation corresponds to the fundamental analysis of Rumer (1969) and fits the experimental result of De Lara (1955) by taking into account the interference of neighbouring particles in a different manner. Relation (45) has been experimentally verified and has been applied in the design process of the granular foundation of the Eastern-Scheldt storm surge barrier in the Netherlands. Oliver (1967) used an exponential equation similar in form to the Chezy equation to compute seepage through a rockfill dam:

$$q = C\left(\frac{n}{1-n}\,\frac{V}{A}\right)^{m_1} I^{m_2}, \qquad (44)$$

in which, V represents the rock volume, A the surface area of the rock, C is a constant and m_1 and m_2 are exponents to be measured experimentally.

The main disadvantage of the exponential form is that the exponent is usually a function of the local Reynolds number, thus creating a difficulty to treat two-dimensional flow if large variations occur. Barends (1978) suggested, that it is preferred to define the nonlinearity embedded in a linearized coefficient in an iterative numerical procedure for more-dimensional flow in terms of the gradient rather than the filter velocity, since the gradient is less sensitive to successive adjustments in the flow pattern during iterative computations. He applied the form:

$$q = -KI \;;\; K = \sqrt{2gD/f|I|}, \qquad (46)$$

where f is given in expression (45). This form was suitable for large variations in the Reynolds number.

To match the ante-linear flow behaviour the exponential form can be used. For this case of very slow velocities in a porous media with fine pores a relevant number, i.e.

- Weber number: $We = \rho g D^2/\sigma$, (47)

in which σ represents the surface tension (dim.: kg/s^2), is applied to characterize the nature of flow. It can be verified by dimensional analysis that using the Re, Ga and We number in the exponential form (32) will yield (Barends, 1980):

$$I = (\rho\nu/\sigma)^{m-1}\, q^m/K. \qquad (48)$$

A similar result has been mentioned by Slepicka (1961). It clearly reveals the threshold gradient at m tending to zero:

$$I_o = (\rho\nu/\sigma)^{-1}/K. \qquad (49)$$

The factor $(\rho\nu/\sigma)$, referred to as the coefficient of molucular linkage, has the dimension of s/m and therefore, it can be interpreted as a velocity damping factor.

5.2. The series form of non-Darcy flow

Forchheimer (1901) was probably the first to suggest a series form to describe the nonlinear relationship between the gradient I and the filter velocity q at large Reynolds numbers:

$$I = (a+bq)q, \qquad (50)$$

in which a and b are constants. This equation better reflects the characteristics of flow around a sphere (Oseen flow) than the flow through rough pipes, since in the latter case the transition from linear to turbulent behaviour is sharp at a critical value of Reynolds number, which is much higher than the value corresponding with turbulent flow in porous media. Moreover, the transition from laminar to turbulent flow in porous media is a very gradual process. A large number of proposed non-linear equations, resulting from various theoretical and/or experimental investigations are reviewed by Scheidegger (1960), Kirkham (1967), Bear (1972) and McCorquodale c.s. (1978). Polubarinova Kotchina (1952) added an unsteady term:

$$I = (a+bq)q + c\partial q/\partial t, \qquad (51)$$

which can account for a virtual mass effect

in non-steady turbulent flow.
To formulate the ante-linear porous flow behaviour is usually attempted by extending Forchheimers equation:

$$I = I_o + (a+bq)q, \tag{52}$$

where I_o is called the initial or threshold gradient. An extensive review of this effect is presented by Kutilec (1969).

On the basis of experiments several researchers suggested a flow equation in the form of:

$$I = fq^2/2gD, \tag{53}$$

in which the friction coefficient is a function of the flow properties.
Lindquist (1933) arrived at:

$$f = 2500 + 4Re. \tag{54}$$

Chaveteau c.s. (1967) have obtained a similar expression.
Goldstein (1938) defined a friction coefficient according to:

$$f = \nu(1+C\,Re)/q\sqrt{k}, \tag{55}$$

which yields after some elaboration the results of Sunada (1965).

Rose (1945, 1949, 1950) used the principles of dimensional analysis in the study of non-Darcy flow and his experiments are best fitted with a friction coefficient according to:

$$f = \{C_1 Re^{-1} + C_2 Re^{-\frac{1}{2}} + C_3\} F(D/d,n), \tag{56}$$

in which C_1, C_2 and C_3 are constants, and d is the diameter of a relevant pore. This equation can be written as (McCorquodale, 1970):

$$I = aq + a_1 q^{1.5} + bq^2, \tag{57}$$

which shows similarity with observation of flow around a sphere (Allen, 1900).

In some cases experiments are fitted with a power series:

$$I = aq + bq^2 + b_1 q^3 + \ldots \tag{58}$$

In respect of the discussion in the previous section the use of this series representation must be dissuaded on pure physical grounds, at least for Newtonian fluids.

Many researchers have attempted to parameterize equation (50) and proved reliability of their formulas with experiments.
Carman (1937) suggested:

$$a = 180\,\frac{(1-n)^2\nu}{n^3 gD^2}; \tag{59a}$$

$$b = 2.87\,\frac{(1-n)^{1.1}\nu^{0.1}}{n^3 gD^{1.1}}\,q^{-0.1} \tag{59b}$$

obtained from experiments.
Ergun (1952) suggested:

$$a = 150\,\frac{(1-n)^2}{n^3 gD^2}; \tag{60a}$$

$$b = 1.75\,\frac{1-n}{n^3 gD}. \tag{60b}$$

Muskat (1946) gives:

$$a = \alpha\,\frac{(1-n)^3\mu}{n^2 D^2}; \tag{61a}$$

$$b = \beta\,\frac{(1-n)\rho}{n^3 D}, \tag{61b}$$

where α and β are shape factors.
Schneebeli (1955) suggested:

$$a = 1100\,\frac{\nu}{gD^2}; \tag{62a}$$

$$b = \frac{12}{gD}, \tag{62b}$$

for a granular bed of spheres at a Reynolds number beyond 2.
Irmay (1958) gives:

$$a = \alpha\,\frac{\nu(1-n)^2}{gD^2(n-n_o)^3}; \tag{63a}$$

$$b = \frac{\beta(1-n)}{gD(n-n_o)^3}; \tag{63b}$$

$$c = \frac{1}{g(n-n_o)}, \tag{63c}$$

for equation (51). Here n_o denotes the stagnant porosity.
Sheidegger (1960) suggested:

$$a = C_1\frac{\nu T^2}{gn}; \tag{64a}$$

$$b = C_2\frac{T^3}{gn^2}, \tag{64b}$$

in which C_1 and C_2 are constant, and T represents the medium's tortuosity.
Ward (1964) obtained from experiments:

$$a = \frac{\nu}{gk}; \tag{65a}$$

$$b = 0.55/g\sqrt{k}, \tag{65b}$$

where k represents the intrinsic permeability.
Sunada (1965) gives:

$$a = \frac{\nu}{gk}; \tag{66a}$$

$$b = CD/gk, \tag{66b}$$

where C is a constant.
Blick (1966) applied a model of capillary tubes with varying diameter and obtained for air flow:

$$a = \frac{32\nu}{gnD^2}; \tag{67a}$$

$$b = \frac{C_D}{2Dgn^2}, \tag{67b}$$

with a special interpretation of D (relevant length) and C_D corresponding to his model.

Watson (1963) and Stark (1968) obtained Forchheimer's equation (50) by integrating the Navier-Stokes equations at microscopic level. Irmay (1964) and Ahmed (1967) did the same including the inertia terms. Ahmed (1967) suggested:

$$a = \frac{\nu}{gk}; \tag{68a}$$

$$b = (g\sqrt{ck})^{-1}, \tag{68b}$$

in which c is a constant.

Bachmat (1965) integrates the Navier-Stokes equations for a specific model including the convective term. Stark (1968) showed how a computer analysis will help to find expressions for the factors a and b.

Ward(1964) unlike Todd(1959) used a characteristic length of $\sqrt{k}$ (k is the intrinsic permeability) instead of a relevant diameter D. It is a point of discussion whether a diameter of a granular bed of various particles is representative, since the actual flow takes part through the pores. Based on experiments Barends (1978) suggested that D_{20} fits best to model non-Darcy flow. Collin (1961) suggested $\sqrt{k/n}$ as a relevant length. Ng (1969) has made studies on crushed rock in an attempt to adapt the Ward equation (65). His results indicate that the turbulent term does not correctly represents his data.

McCorquodale (1970) modified Ward's equation in order to obtain a non-dimensional Forchheimer equation for crushed rock. He gives for crushed rock and river gravel based on experiments by Ng (1969), McCorquodale (1969), Nasser (1970) and Hannoura (1978) and based on evaluation of literature:

- Low Reynolds numbers:

$$\frac{Ignd^2}{\nu q} = 4.6 + 0.79\,\frac{dq\sqrt{n}}{\nu}; \tag{69}$$

- High Reynolds numbers:

$$\frac{Ign\delta^2}{\nu q} = 70.0 + 0.54(0.5+0.5\,\frac{f_e}{f_o})\frac{\delta q\sqrt{n}}{\nu}, \tag{70}$$

where d is a relevant pore size and δ is the effective pore hydraulic radius, f_e is the Darcy friction for rough surface and f_o for smooth surface. Kovacs (1969) presented his own work and reviewed other research on Newtonian and non-Newtonian non-Darcy flow in porous media. He suggested a non-dimensional flow equation based on the Kozeny-Carman hydraulic radius model:

$$\left[\frac{Ign\delta^2}{4.6\nu q}\right]^{\frac{3}{4}} = 1 + \left[\frac{(0.5+0.5f_e/f_o)\delta q}{21.8\,\nu n}\right]^{\frac{3}{4}}. \tag{71}$$

Cox (1977) gave a short review of different forms of the non-Darcy flow equation, and compared the different forms using experimental data emphasizing the percentage of deviation in correspondence to the experimental error. According to him Forchheimer's equation gives the best results. To overcome numerical difficulties in application Cox reformulated the flow equation as:

$$I = Cq;$$

$$C = 1/(\tfrac{1}{2}a+\sqrt{(\tfrac{1}{2}a)^2+b|I|}). \tag{72}$$

Nonlinearity is incorporated by the coefficient C, which is related to the parameters a and b in the Forchheimer equation (50).

5.3. Graphical representations of non-Darcy flow

In establishing a relation for porous flow, i.e., an equation between the gradient I and the filter velocity q, a number of researchers have presented their data in graphical form. A schematic representation of the gradient-filter velocity relation is shown in Fig. 1, reflecting the different regimes of flow.

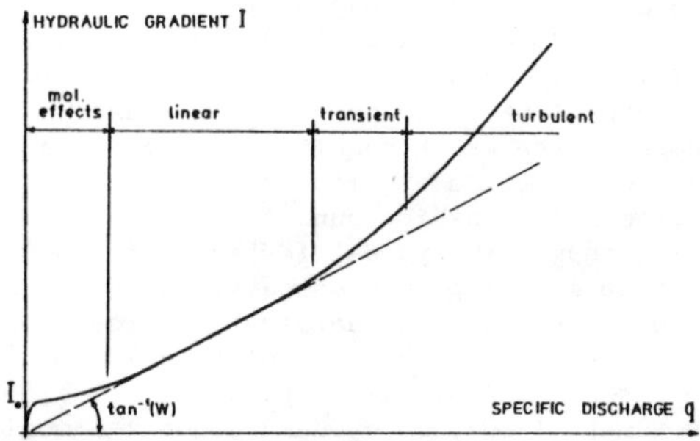

Fig. 1. Idealized porous flow law.

Prinz (1930) made a dimensional plot of I versus q for various materials. He placed the limit on Darcy's law at the point on each curve where it deviates from a straight line. Mott (1950) presented a paper on the movement of particles through a fluid and the movement of a fluid through a bed of particles; similarly wall effects and resistance curves were obtained.

The experimental results of Bakhmeteff c.s. (1937), Saunders c.s. (1940) and others were used by Mott. Dudgeon (1964) also plotted I versus q, but on log-log scale. He attempted to define various flow regimes by the straight line segments of his curves. Dudgeon's work included angular and rounded granular material with filter velocities in the range of 10^{-6} to 0.4 m/s. Dudgeon has presented his results in a dimensionless friction factor versus Reynoldsnumber using a relevant length of D_{50}.

Barends (1980) gave a similar representation on log-log scale for a material varying from sand, gravel, slags, crushed stones up to a size: D_{50} = 8 cm. The velocities varied from 10^{-6} to 0.1 m/s. The data were used to fit the exponential form of non-Darcy flow. Comparing the grain distribution curves with the I versus q plots made clear that it is preferred to use D_{20} as a relevant length in the formulas.

De Lara (1955) and Le Mehauté (1957) plotted data projected on the well-measured relation between the drag coefficient C_D and the Reynoldsnumber for a single sphere in a uniform flow and suggested form factors to account for the tortuosity and the interference of neighbouring particles. Le Mehauté studied scale effects of the non-Darcy flow equation.

Lane (1964) gave some experimental data for crushed rock and rounded glacial gravels. Wright (1969) used a friction factor versus Reynoldsnumber plot. Zampaglione (1969) used a friction factor and the Reynoldsnumber to present his experimental results on a Fanning type diagram. He used an adjusted filter velocity: $q_n = q\,n^{1.5}$, and a reduced relevant length similar to the characteristic length $\sqrt{k}$ used by Harleman c.s. (1963), Ward (1964), Massarani (1967) and Ahmed c.s. (1969).

Kovacs (1969) has presented his data and those of Zunker, Lindquist and Nagy in a log-log plot. He studied the applicability of the relationship found for various particle shapes and porosities and concluded that spherical particles packed at various porosities could adequately be represented but that disc shaped particles could not be represented. A similar conclusion was made by Rumer (1969). This would suggest a modification of the shape factor to account for the geometrical form and the angularity of the materials (see Kovacs' equation (71) in section 5.2).

5.4. Statistical models for non-Darcy flow

Ferrandon (1948) used a statistical approach to investigate the mathematical representation of the conductivity. Already Versluijs (1915), quoted by Vreedenburgh (1936) showed that the conductivity can be represented by a second-rank symmetrical tensor modelling anisotropical linear porous flow. De Josselin de Jong (1969) introduced an instructive method of scatterning mechanics, which proves the tensor character. Scheidegger (1956, 1960) has extended his own statistical treatment of Darcy's flow which was based on Gibbs' work to the case of non-Darcy flow. He introduces a probability-density function, P(x,t), for a particle to be at a point x at a time t:

$$P(x,t) = \frac{1}{(4\pi\bar{D}t)^{3/2}}\, e^{-(x-\bar{x})^2/4\bar{D}t}, \qquad (73)$$

in which $\bar{D}$ is a dispersivity factor. Expressing for the purely turbulent regime the particle displacement during a small time step, he obtained the pore water velocity q from which: $\bar{x} = qt$ can be determined. Apperently the factor $\bar{D}$ can be found by observing the dispersion of a tracer in the medium.

Wittman (1980) studied the statistical representation of pore geometries and their effect on the conductivity in the linear and nonlinear regimes of flow, taking into account the inertia effect due to secondary flow on microscopic scale.

6. OTHER ASPECTS OF FLOW IN POROUS MEDIA

The previous sections of this paper have, mainly, considered the steady saturated flow in porous media. In this section, a brief presentation of unsteady, unsaturated, and hydrothermal convection flow aspects in porous is made.

The effect of the unsteadyness of the flow in porous media on the non-Darcy hydraulic conductivity has been, often, considered to be negligible. This assumption may be a valid one in case of low Reynolds number and highly compacted fine porous media. Recently, in an experimental investigation of this phenomenon, Hannoura (1978, 1878a) has shown that the virtual mass coefficient could be of a significant value for non-Darcy flow in coarse granular media of the type encountered in breakwaters and rockfill structure. He assumed that the steady flow coefficients a and b in Forchheimer's equation to be valid for unsteady flow too. Then by comparing the experimental steady and unsteady flow results, he was able to estimate the virtual mass coefficient.

On the unsaturated non-Darcy flow in porous media, the authors have found that this phenomenon has attracted the attention of hydrologists and soil scientists for a considerable period of time. Most of these efforts have been concentrated toward the characterization of a certain function to describe the hydraulic conductivity of the unsaturated media. A good review of these efforts was summarized in the work of Morel-Seytoux (1973), and Philip (1969).

Zeller (1961) presented measurements on the effect of air in unsteady phreatic flow cases. Irmay (1958) included a stagnant porosity term to account for entrapped air pockets (see section 5.2). Verruijt (1969) extended the theory of flow in saturated soils to a semi-saturated porous medium, in which air bubbles are present.

The existence of air in the pore fluid causes two fundamental modifications, one with respect to the compressibility of the mixture and one with respect to the flux. It is the question whether the gas moves independently, is blocked in the pores or is conveyed by the pore fluid. Barends (1979, 1980) derived different equations of motion for the mixture in each case and found on physical grounds an expression for the composite compressibility β_p, which contains a discontinuity due to collapsing bubbles. A serious reduction of the overall conductivity due to presence of air is to be expected. A factor of $(1-b)^3$ is suggested, where b denotes the volume of stagnant air. Teunissen (1981) extensively discussed the influence of air bubbles.

The application of two-phase flow models to air-water flow in porous media was found in the publications of Hannoura and McCorquodale (1978b). Other research on air/gas bubble flow in porous media has been, often, devoted towards industrial applications, e.g., O'Neill (1972).

Another important aspect of flow in porous media is due to hydrothermal convection. This phenomenon is encountered in oil reservoirs, solar energy storage, and thermal insulation purposes. Combarnous and Bories (1975) has reported the results of research done in the "Groupe d'Etude I.F.P. -I.M.F. sur les Milieux Poreux". The non-Darcy permeability term was estimated either by the Kozeny-Carmen relationship or through in situe measurements. An extensive survey of the hydrothermal convection in porous media literature is found in their report.

7. NUMERICAL SOLUTIONS OF NON-DARCY FLOW PROBLEMS

Since the governing equations for non-Darcy flow are nonlinear, analytical solutions for linear equations are not applicable. However, in some instances the nonlinear system can be transformed into a linear system and the available analytic techniques are applied. In this section a summary of different solutions for non-Darcy flow in porous media is presented.

Methods of Transformations: Kristianovich (1940) introduced a method of solving steady non-Darcy flow. His method involved a number of transformations and an analytic function. Engelund (1953) transformed the nonlinear hydraulic conductivity term in the field equation by replacing it with a function of the flow velocity to solve for groundwater movement in aquifers. A similar approach was employed by Sollitt and Cross (1972) to solve wave transmission through rectangular permeable breakwaters.

Finite Difference Solution: Curtis and Lawson (1967) and Fenton (1968) have applied the finite difference method to non-Darcy flow using an exponential form for the hydraulic conductivity. Their solutions were applied to rockfill dams. A finite difference solution for wave dissipation in rectangular rubble-mound breakwater was developed by Nasser (1974). Forchheimer's equation was used to represent the non-Darcy flow resistance. Trollope (1971) used a finite difference solution to solve for complex flow through porous media.

Method of Characteristics: Ali (1968) applied the one-dimensional method of characteristics to one and two well problems. Forchheimer's equation was used to represent the frictional forces. Later Peters (1974) and Edgell (1974) extended Ali's approach to two space dimensions.

Finite Element Solution: Engelund (1953) referred to the possibility of applying variational calculus to steady non-Darcy flow problems in porous media. The unsteady non-Darcy flow solution was later developed by McCorquodale (1970) and Volker (1969). Their solutions have been developed for rockfill embankments. Cox (1977) and Dudgeon (1978) have applied the finite element method to groundwater flow and well design. Barends (1976) developed a user-oriented computer program, the SEEP code, based on a finite element formulation. Different types of problems can be treated, including transient flow in inhomogeneous three dimensional porous media having a flexible geometry under general time-dependent boundary conditions. Turbulent porous flow is accounted for by using a linearized conductivity and updating the nonlinearity in an iterative manner applying the exponential form of non-Darcy flow. Fast convergence is found in this procedure; usually a few cycles are sufficient to correctly adjust the initial conductivity values. Time dependent phreatic flow problems are treated using an explicit finite difference method to solve the moving boundary condition. Convective terms and infiltration effects are taken into account. Recently ecological effects (marine growth) on the turbulent hydraulic conductivity have been considered for a coarse granular structure by solving a convective nutrient flow problem simultaneously and adjusting the conductivity values iteratively in the corresponding flow in the granular bed (Barends c.s., 1981).

A hybrid numerical model was developed by Hannoura in 1978. The model was developed for wave motion in rubbel-mound and rockfill structures. A finite difference scheme governed by the characteristics directions was used to integrate in time (one-dimensional) and a finite element analysis was used to integrate in space (two-dimensional). An iterative procedure was used to linearize and update the non-Darcy hydraulic conductivity which was described by a Forchheimer type equation.

A survey of free boundary value problems in porous media is published by Bruch (1980). Other modeling techniques have been reviewed by Prickett (1975).

8. CONCLUSIVE REMARKS

An examination of the extensive research on the subject of non-Darcy flow in porous media reveals that there is no universal equation which is valid over a wide range of Reynolds number.

Recently, McCorquodale, Hannoura and Nasser (1978) published a paper in which their research results and the published results of others were combined to produce a generalized equation for the non-Darcy flow in coarse porous media. They discussed the influence of the following parameters on the statistically correllated equation: (1) choice of the characteristic length for the porous media; (2) effect of grain friction and (3) effect of the permeameter's wall on the experimental results. Dudgeon (1966), Mott (1950), Saunders and Ford (1940), and Rose (1949) have noted the importance of the wall effect in permeater tests.

The graphical representation of the non-Darcy flow as published by Arbhabhirama and Dinoy (1973) shows the possibility of developing such a plot (friction coefficient versus Reynolds number). However, when it was examined by Hannoura (1978) it was found to be limited to fine sand.

The experimental results reported by Hannoura and McCorquodale (1978) on air-water flow in porous media and virtual mass for granular material have to be extended to a wider range of experimental parameters before using them in praktical applications.

In closing our article, we note that the amount of non-Darcy flow literature extends far beyond the few papers we cited here. We must also note that, whereas the general knowledge of the non-Darcy flow in porous media may be considered fairly sufficient to handle practical problems, further research is needed.

9. REFERENCES

Ahmed, N. & Sunada, D.K. (1969), Nonlinear flow in porous media, J.Hydr.Div., ASCE, HY6(Nov.).

Ahmed, N. (1967), Physical properties of porous medium affecting laminar and turbulent flow of water. Ph.D.Thesis, Colorado State Univ., Fort Collius, Colo.

Ali, I. (1968), Unsteady unconfined flow towards gravity wells, Ph.D.Thesis, Univ. of Leeds.

Allen, H.S. (1900), The motion of sphere in a viscous fluid, Phil.Mag.,vol.30.

Anandakrishnan, M. & Varadarajulu, G.H. (1930), Laminar and turbulent flow through sand, J.Soil Mech. and Found.Eng.Div., ASCE,vol.89,SM5, Sept.

Anderson, K.E.(1963) Water Well handbook, Missouri Water Well Drillers Assoc., 2nd. ed.

Bachmat, Y. (1965), Basic transport coefficients as aquifer characteristics, IASH Symp.Hydr. of Fract.Rocks, Dubrovinik, vol.1.

Bakhmeteff, B.A. & Feodoroff, N.V. (1937), Flow through granular Media, J. of Appl. Mech.,4.A.

Barends, F.B.J. (1976), SEEP, a finite element code for groundwater flow, Delft Soil Mech.lab., Rp.74-680215.

Barends, F.B.J. & Thabet, R. (1978), Groundwater flow and dynamic gradients. Symp. Found. Aspects of Coastal Struct., Delft, vol.6.

Barends, F.B.J. (1978), Advanced methods in groundwater flow computation, Delft Soil Mech.lab., LGM-Mededelingen 19.

Barends, F.B.J. (1979), The compressibility of an air-water mixture in porous media, Delft Soil Mech.lab., LGM-Mededelingen 20.

Barends, F.B.J. (1980), Nonlinearity in groundwater flow, Ph.D.Thesis, Delft Univ.

of Technology, Delft.

Barends, F.B.J. (1980), Slope stability of river and sea dykes, 2nd. Int.Conf.Inland Waterways & Harb.Eng., Basrah.

Barends,F.B.J & Van Driel,P.J. & De Quellerij,L. (1981) Ecological aspects of the mechanical stability of granular structures, to appear SEME, Bruges, 1982

Basak, P. (1977), Non-Darcy flow and its implications to seepage problems, I.Irrigation and Drainage Div.,ASCE, vol.103, IRA (Dec.).

Bear, J. & Zaslavsky, D. & Irmay, S. (1965) Physical principles of water percolating and seepage, Unesco Arid Zone Research XXIX, Paris.

Bear, J. (1972), Dynamics of flow in porous media, American Elsevier, New York,

Benis, A.M. (1968), Theory of non-Newtonian flow through porous media in narrow three-dimensional channels, Int.J. Nonlinear Mech.,vol.3.

Blick, E.F. (1966), Capillary orifice model for high speed flow through porous media, Proc.Design and Devel., No.1-vol.5.

Boreli, M. (1978), Hétérogénéité des milieux poreux (rocheux) et l'effect d'échelle, IAHR Symp.Effect d'Echelle en Milieux Poreux, Thessaloniki.

Bruch, Jr., J.C. (1980), A survey of free boundary value problems in the theory of fluid flow through porous media, variational inequality approach-Part II, Advances in Hydrose.,Vol.3, Academic Press, New York, N.Y.

Carman, P.C. (1937), Fluid through granular beds, Frans.Inst. of Chem.Eng.,vol.15, London.

Chauveteau, G. & Thirriot, C.I. (1967), Régimes d'écoulement en milieux poreux et limite de la loi de Darcy, la Houille Blanche, Assoc. pour la Diff. de la Doc. Hyd., No.2, Grenoble, France.

Collins, R.E. (1961), Flow of fluids through porous material, Reinhold, New York, N.Y.

Combarnous, M.A. & Bories, S.A. (1975), Hydrothermal convection in saturated porous media, Advances in Hydrosc., Vol.10, Academic Press, New York, N.Y.

Cox, R.J. (1977), A study of near well groundwater flow and the implications in well design. The Univ. of New South Wales.

Curtis, R.P. (1965), Flow over and through rockfill banks, Univ. of Melbourne, Dept. of Civ.Eng.

Curtis, R.P. & Lawson, J.D. (1967), Flow over and through rockfill banks, J. of Hyd.Div.,ASCE, Vol.93, HY5(Sept.).

Darcy, H. (1856), Les fontaines publiques de la ville de Dijon, Dalmont, Paris.

De Josselin de Jong, G. (1969), Generating functions in the theory of flow through porous media, Chap.8 in 'Flow through porous media', De Wiest (ed.), Acad.Press, London.

De Josselin de Jong, G. (1969), The tensor character of the dispersion coefficient in anisotropic porous media, IAHR Symp. Fund.Transp.Phen. Porous Media, Haifa, vol.4.

De Lara, G.C. (1955), Coefficient de perte de charge en milieux poreux basé à l'équilibre hydrodynamique d'un massif, La Houille Blanche, vol.2.

Dudgeon, C.R. (1964), Flow of water through coarse granular materials, M.Eng.Thesis, Univ. of New South Wales.

Dudgeon, C.R. (1966), An experimental study of flow of water through coarse granular materials, La Houille Blanche, vol.21.

Dudgeon, C.R. (1976), Wall effect in permeameters, J.of Hyd. Div.,ASCE,vol.93, HY5(Sept.).

Dudgeon, C.R. & Cox, R.J. (1978), Flow to large diameter bottom entry wells, Proc. of 2nd. Int.Conf. on FE in Water Res., Pentech Press, London, U.K.

Edgell, G.J. (1974), Unsteady Flow to Gravity Wells, Ph.D.Thesis, Univ.of Leeds.

Eisenstein, Z. & Law, S.T.C. (1980), The role of constitutive laws in analyzis of embankments, 3rd. Int.Conf. Numerical Meth.in Geomechanics, vol.4, Balkema, Rotterdam.

Engelund, F. (1953), On the laminar and Turbulent flow of groundwater through homogeneous sand, Trans. Danish Acad. Tech.Sce.,No.3.

Ergun, S. & Orning, A.A. (1949), Fluid flow through randomly packed columns and fluidized beds, J. of Ind.Eng.Chem.,vol.41, No.6.

Escande, L. (1953), Experiments concerning the infiltration of water through a rock mass, Proc.of IAHR, Minnesota, (Sept.).

Fenton, J.D. (1968), Hydraulic and stability analysis of rockfill dams, Civ.Eng.Report, Univ. of Melbourne.

Ferrandon, J. (1948), Les lois de l'écoulement de filtration, Le Génie Civil, vol. 125.

Forchheimer, P. (1901), Wasserbewegung durch boden, Zeit. Ver.Deutsch.Ing.,45.

Forchheimer, P. (1924), Hydraulic, Leipzich-Berlin.

Goldstein, S. (1938), Modern development in fluid Dynamics, vol.2, Oxford Univ.Press.

Habid, J. & Thirriot, C. (1969), Ecarts à la loi de Darcy pour les écoulements d'eau à faible vitesse dans les argilles, Proc. XII Cong., IAHR, Kyoto, Japan, (Sept.)

Hannoura, A.A. (1978), Numerical and experimental modelling of unsteady flow in rockfill embankments, Ph.D.Thesis, Univ. of Windson, Windson, Ont.Canada.

Hannoura, A.A. & McCorquodale,J.A. (1978a) Virtual mass of coarse granular media, J.of WW,harbors and coastal Eng.Div., ASCE,vol.104, WWZ (May).

Hannoura, A.A. & McCorquodale, J.A. (1978b), Air-water flow in coarse granular media, J. Of Hyd.Div.,ASCE,Vol.104, HY7 (July).

Hannoura, A.A. & McCorquodale, J.A. (1979), Environmental wave forces on rubble-mound breakwaters, Proc. of Civil Engineers in the Ocean IV,ASCE, San Francisco, (Aug.).

Hannoura, A.A. (1980), Flow coefficients and the reliability of numerical models, Proc.of Hyd.Div.Sep.Conf.ASCE, Chicago, (July).

Harleman, D.R.F.; Melhorn, P.F. & Rumers, Jr., J.J. (1963), Dispersion permeability correlation, J. of Hyd.Div.,ASCE,vol.89, HY2 (March).

Irmay, S. (1958), On theoretical derivation of Darcy and Forchheimer formulas, A.G.U., Trans.,39,4.

Irmay, S. (1964), Theoretical models of flow through porous media, R.I.L.E.M. Symp. Transfer of water in Porous Media, Paris.

Izbash, S.V. (1931), O filtracii v kropnozernstom materiale, Izv. Nauchnosisted, Inst. Gidrotechniki (NIG), Leningrad, U.S.S.R.

Jaeger, C. (1956), Engineering fluid mechanics, Blackie, London.

Kalliontzis, C. (1981), A finite element and experimental non-Darcy groundwater flow study, Ph.D.Thesis, University of Leeds, England.

Karadi, G. & Nagy, I.V. (1961), Investigation into the validity of the linear seepage law, Proc.IX Cong. of the IAHR, Dubronik.

Khristianovicn, S.A. (1940), Motion of groundwater which does not conform to Darcy's law, U.S.S.R. Acad. of Sci., J. of Appl. Math. and Mech., T. IV, B.1.

Kirkham, C.E. (1967), Turbulent flow in porous media; an analytical and experimental model study, Water Research Foundation of Australia, Bull.No.11, (Febr.).

Kovacs, G. (1969), General characterization of different types of seepage, Proc.of the XIII Cong.,IAHR, Vol.4, Koyoto, Japan, (Sept.).

Kutilec, M. (1969), Non-Darcy flow of water in soils, IAHR, Symp.Fund.Transp.Phen. Porous Media, Haifa,vol.5.

Lane, K.S. (1964), Discussion of laminar and turbulent flow of water through sand, J.of Soil Mech.Div.,ASCE, vol.90, SM3 (May).

Le Mehaute, B. (1957), Etude des lois de l'écoulement périodique dans un massif perméable, la Houille Blanche, No.6,(Dec.).

Lindquist, E. (1933), On the flow of water through porous soil, 1st. Conf.of Large Dams, Stockholm.

Massarani, G. (1967), Critique de la validite de la loi Darcy en milieux poreux heterogen, CRAS,Vol.265, (Nov.).

McCorquodale, J.A. (1979), Finite Element Analysis of non-Darcy flow, Ph.D.Thesis, Univ. of Windsor, Windsor, Ont.,Canada.

McCorquodale, J.A.; Hannoura, A.A. & Nasser, M.S. (1978), Hydraulic conductivity of rockfill, J.of IAHR,Vol.2.

Missbach, A. (1937), Listy cukro var,55.

Morel-Seytoux, H.J. (1973), Two-phase flow in porous media, Advances in Hydrosc., 1973, Academic Press, New York, N.Y.

Mott, R.A. (1950), The laws of motion of particles in fluids and their application to the resistance of beds of solids to the passage of fluid, Conf.Proc.,(Leamington, Spa, England), Edward Arnold and Co., London, U.K.

Muskat, M. (1946), The flow of homogeneous fluids through porous media, J.W. Edwards, Inc.,Ann Arbor, Michigan.

Nasser, M.S. (1970), Radial non-Darcy flow through porous media, M.A.SC. Thesis, Univ. of Windsor, Windsor, Ont. Canada.

Nasser, M.S. (1974), Theoretical and experimental analysis of wave motion in rockfill structures, Ph.D.Tesis, Univ.of Windsor, Windsor, Ont. Canada.

Ng, H.C. (1969), An experimental study of steady non-Darcy flow in crushed rock, M.A.SC. Thesis, Univ.of Windsor, Windsor, Ont. Canada.

Olivier, H. (1967), Through and overflow rockfill dams-new design technique, Proc., Inst. of Civ.Eng., (Mar.).

O'Neill, B.K.; Nicklin, D.I.; Morgan, N.I. & Leung, L.S. (1972), The hydrodynamics of gas-liquid contacting in towers with fluidized packing, The Con. J. of Chem. Eng.,Vol.50, (Oct.).

Oseen, C.V. (1927), Neuere methoden und ergebnisse in der hydrodynamik, Leipzig.

Parking, A.K.; Lawson, J.D. & Trollope, D. H. (1966), Rockfill structures subject to waterflow, J. of Soil Mech.Div., ASCE, Vol.92, SM5, (Nov.).

Peters, D.C. (1974), Three dimensional flow through a porous media, Ph.D.Thesis, Univ. of Leeds.

Philip, J.R. (1969), Theory of infiltration, Advances in Hydrosc., Academic Press, New York, N.Y.

Polubarinova Kochina, P.Ya. (1962), Theory of groundwater movement, Princeton Univ. Press, Princeton, N.J.

Prickett, T.A. (1975), Modeling techniques for groundwater evaluation, Advances in Hydrosc.,Academic Press, New York, N.Y.

Prinze, E. (1930), Irrigation and hydraulic design, by Leliavsk,S.,Vol.1. 1st.Ed. Champman and Hall, London, U.K.

Reynolds, O. (1900), Papers on mechanical and physical subjects, Cambridge Univ. Press, London, U.K.

Rose, H.E. (1945), On the resistance coefficient-Reynolds number relationship for fluid flow through a bed of granular material, Inst.Mech.Engrs.,Proc.153.

Rose, H.E. (1950), Fluid flow through beds of granular material, Inst. of Physis, Conf.Proc., (Leamington, Spa, England), Edward Arnold and Company, London, U.K.

Rose, H.E. & Rizk, A.M.A. (1949), Further researches in fluid through beds of granular material, Proc., Inst.Mech.Eng., Vol.160.

Rouse, H. (1961), Fluid mechanics for hydraulic engineers, Dover Publications, New York, N.Y.

Rumer, R.R. & Drinker (1966), Resistance to laminar flow through porous media, J. of Hyd.Div., ASCE, Vol.92,HY5. (May).

Rumer, R.R. (1969), Resistance to flow in porous media, chap.3 in 'Flow through porous media' De Wiest (ed.), Acad.Press London.

Saunders,O.A. & Ford, U. (1940), Heat transfer in the flow of gas through a bed of solid particles, Iron and Steel Inst., J.141,291.

Scheidegger, A.E. (1956), Correlation tensors in statistical hydrodynamics in porous media, J. of Physics, Vol.34.

Scheidegger, A.E. (1960), The physics of flow through porous media, Univ. of Toronto Press, 2nd.Ed.

Schneebeli, G.P. (1955), Expériences sur la limité de validité de la loi de Darcy et l'apparition de la turbulence dans un écoulement de filtration, La Houille Blanche, Vol.10, No.2.

Slepicka, F. (1961), The laws of filtration and limits of their validity, IAHK 9th. Convention, Belgrade, Vol.2.

Sollitt, C.K. & Cross, R.H. (1972), Wave transmission through permeable breakwater, Proc. XIII Coastal Eng.Cong., ASCE, Vancouver, B.C., Canada.

Stark, K.P. (1968), Numerical solutions of Navier Stokes equations for flow through idealized porous materials, IASH, Vol.18, Tusson.

Stokes, G.G. (1851), Cambridge Transactions.

Sunada, D.K. (1965), Turbulent flow through porous media, Univ.of California, Berkeley Water Res.Centre, Contrib.No.103.

Swarzendruber, D. (1962), Non-Darcy flow behaviour in liquid-saturated porous media, J. of Geoph.R., Vol.67, No.13., (Dec.).

Teunissen, J.A.M. (1981), Mechanics of a fluid-gas mixtures in porous media, to appear in J.of Mech. of Materials.

Todd, D.K. (1959), Groundwater hydrology, John Wiley & Sons, New York, N.Y.

Trollope, D.H. (1971), Complex flow through porous media, Aust.Geomechanics J.,Vol.61.

Van der Kogel, H. (1977), Wave propagation in saturated porous media, Ph.D.Thesis, Cal.Inst.Techn., Pasadena, Cal.

Van der Kogel, H. (1979), Comparison of different field equations used by some authors, Delft Soil Mech.Lab., pers.communications.

Volker, R.E. (1969), Nonlinear flow in porous media by finite elements, J. of Hyd. Div.,ASCE, Vol.95, HY6, (Nov.).

Vreedenburgh, C.G.J. (1936), On the steady flow of water percolating through soils with homogeneous anisotropic permeability, Proc.Int.Conf.SMFE, Cambridge, Vol.1.

Verruyt, A. (1969), Elastic storage of aquifers, chap.8 in 'Flow through porous media', De Wiest (ed.), Acad.Press,London.

Ward, J.C. (1964), Turbulent flow in porous media, J, of Hyd.Div.,ASCE, Vol.90, HY5, (Sept.).

Watson, K.K. (1963), The permeability of an idealized two-dimensional porous medium using the Navier-Stokes equations, Proc. 4th. Aust.-N.Z.Conf. on Soil Mech. and found.Eng.

Whitaker, S. (1966), The equation of motion in porous media, J. Am.Inst.Chem.Engr., Vol.3.

Whittman, L, (1980), Filtrations- und Transportphänomene in porösen Medien. Veröf. Inst.für Bodenmech. und Felsmech., Univ. Karlsruhe, Heft86.

Wilkins, J.K. (1955), Flow of water through rockfill and its application to the design of dams, N.Z.Eng., (Nov.).

Wright, D.E. (1968), Nonlinear flow through granular media, J.of Hyd.Div., ASCE, Vol. 94, HY4, (July).

Wright, D.E. (1969), Nonlinear flow through granular media; Closure, J.of Hyd.Div., ASCE, Vol.95. HY5, (Sept.).

Yalin, S. & Franke, L. (1961), Experimental investigation of the laws of filter flow, IAHR, 9th. Conv., Dubrovnik.

Zampaglione, D. (1969), Turbulent seepage in filters,Proc.of the XIII Cong.IAHR, Kyoto, Japan., (Sept.).

Zeller, J. (1961), The significance of aquifer porosity in non-steady seepage flow with free surfaces IAHR 9th. Conv. Belgrade, Vol.2.

Proceedings of Euromech 143 / Delft / 2-4 September 1981

Finite element model for preventing high groundwater-level caused by the hydraulic power station at Iffezheim/Rhine

R.HOLLÄNDER & G.SCHULZ
Technische Universität Berlin, Germany

1 INTRODUCTION

As an example of the possibilities of finite element methods in groundwater hydrology, this contribution deals with a research project worked on in 1977 at the Institut für Wasserbau und Wasserwirtschaft of the Technical University of Berlin.

This project was chosen as a contribution here, because it includes different types of boundary conditions and because of its multifold geological and hydrological structure which had to be modeled.

In 1976 the Institute received an order from the Bundesanstalt für Wasserbau (BAW) at Karlsruhe to investigate the groundwater situation around the village of Greffern, which is located at Rhinekilometer 320 at the French-German border.

The power station at Iffezheim/Rhine, at km 334, raised the groundwater level in the area around Greffern and caused the danger of flooding basements of houses and plains used for agriculture.

The research was done by using a two-dimensional horizontal groundwater-model with the finite element method. It was completed by results of a former two-dimensional vertical investigation of the Technical University of Karlsruhe.

2 GEOLOGICAL SITUATION IN THE AREA OF GREFFERN

The Rhine-Plain is divided into the lowland and the upland which at the Greffern area raises only 2 m above the lowland. The upland consists of glacial deposits in which the river dug its bed during the holocene period. Therefore, the lowland consists of glacial sandy and rocky material covered with sedimental loams. Under the holocene deposits lie quarternary formations consisting of gravels and sand with silt-banks. The underlying pliocene formations at a depth of 50 m could be neglected because of their small k-value for further considerations as the vertical model at Karlsruhe had shown. The deposits in the entire area are of a very irregular kind, due to continuous changing of the meanders of the Rhine in the past.

The calibration of the model ended up with transmissivities from $T = 0{,}02\ m^2/s$ at the upland to $T = 0{,}07\ m^2/s$ at the river banks. A pumping test gave a medium of $T = 0{,}033\ m^2/s$ at the village of Greffern.

3 BOUNDARIES OF THE MODEL

The groundwater-flow does not follow the river-bend of the Rhine, but rather crosses the chosen northern and southern boundary of the model rectangularly. Therefore, both boundaries are suited for predefined piezometric heads as boundary conditions. The eastern boundary in the upland at a distance of 2.5 to 5 km off the Rhine River is assumed to be independent of phreatic fluctuations caused by the

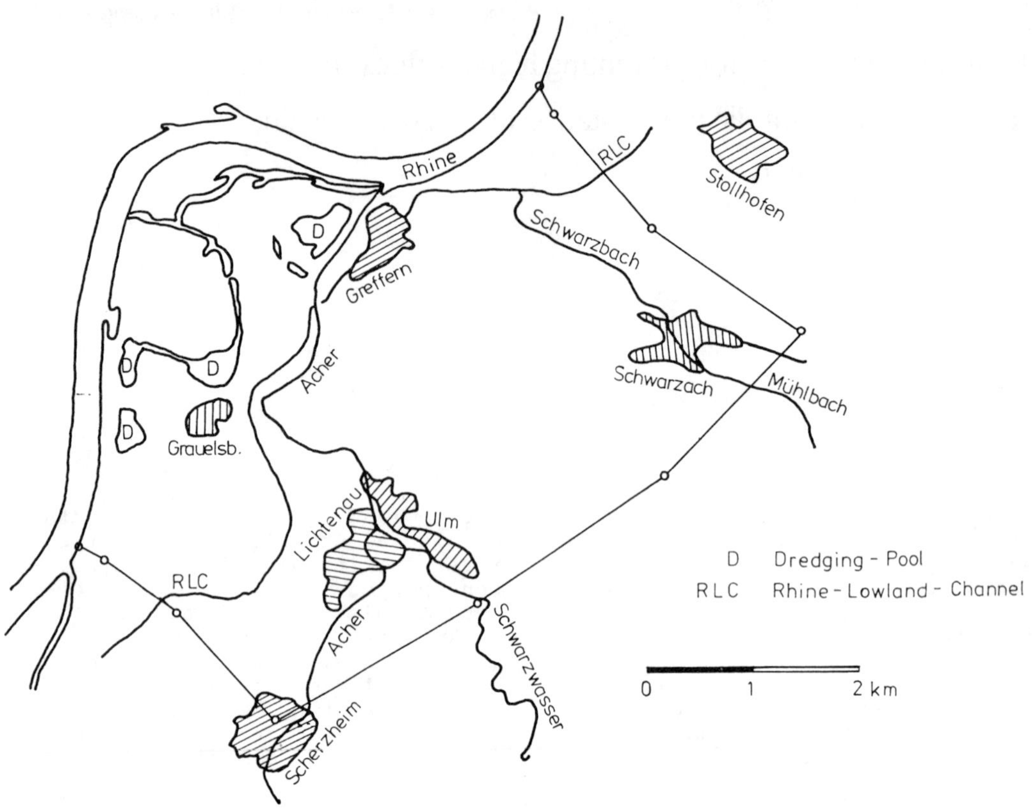

Horizontal Groundwater-Model of the Greffern-Area

River Rhine. This enabled us to take here also predefined heads as boundary conditions.

The River Rhine between kilometer 315 - 323 was taken as the western boundary of the model. Comparing the width of the river with the assumed depth of the aquifer underneath (50 m), a substantial underflow seems to be very unlikely so that the Rhine could be accepted as a hydraulic border and also as a boundary condition of the model.

With these boundaries, the model area amounted to 22 km^2.

4 HYDROLOGICAL SITUATION IN THE GREFFERN AREA BEFORE DAM CONSTRUCTION

As hydrologic parameters rivers, channels, dredging-pools, water works, bayous and rainfall have to be considered.

The most important hydrologic parameter is the Rhine with its floods. At high and medium water level it infiltrates, at low water level it exfiltrates from the groundwater. During calibration the Rhine was modeled by given head dependent on time.

The Rhine-lowland-channel (Rhein-Niederungs-Kanal) is always draining even at times of low groundwater level. It was simulated by different exfiltration rates for different parts of the channel.

The tributary Acher was simulated using the leaky-principle. Its effect in the upland is an infiltrating one. In the lowland it is influenced by the Rhine level. At high Rhine level it infiltrates, at low Rhine level it exfiltrates. The brook Schwarzbach flows into the Rhine-lowland-channel and was simulated by infiltration rates.

The effect of the widespread bayou-system which is connected at three points with the Rhine, cannot be defined easily, because of changing flow-directions and incomplete water level measurings. At high hydrologic conditions caused by rainfall and high Rhine level, the bayous near the Rhine infiltrate into the groundwater. When the Rhine level is medium or low, they show draining effects.

The bayous at a greater distance from the Rhine at the Rhine-bend prevailingly drain. The bayou near Grauelsbaum is influenced by the Rhine level and infiltrates to the near dredging-pools which drain.

Because of indefinite flow conditions, the entire bayou-system with connected dredging-pools was modeled using different infiltration and exfiltration rates for high, medium and low hydrologic conditions.

Only the large dredging-pool near Greffern was simulated by an enlargement of transmissivity.

Groundwater-recharge due to precipitation was considered using data obtained from a lysimeter near to Greffern. On triangular elements representing forest areas the lysimeter-data was diminished by 10 % considering higher interception.
All wells in the area (water works, chemical plants) were simulated by withdrawal at the corresponding nodes.

The values of all parameters were focused in the calibration phase by modeling a ten month-period containing the hydrologic conditions of high, medium and low Rhine level. The time-step was one day.

5 HYDROLOGIC SITUATION AFTER DAM CONSTRUCTION

With the construction of the dam and the power station at Iffezheim, measures were taken for the Acher, the dredging-pools, and the bayou-system. A drainage channel was built at the Rhine dams which I refer to as the Rhine-lateral-channel. In this channel the Acher was introduced and led into the Rhine behind Iffezheim so that the banked-up Rhine and the Acher were no more connected. Thus, the Acher can only drain the lowland.

The earlier connections from the Rhine to the bayou-system were shut and instead of that the bayous were opened to the Rhine-lateral-channel so that the bayou-system is now used for draining purposes simulated in the model by the leaky-principle.

The only exception is again the dredging-pool at Greffern which got an intake for the western bayou and an outlet to the Acher with a sill.

In case of high and medium Rhine level this sill is overflown and the dredging-pool with the connected bayou is draining. At low groundwater level the sill prevents a connection. In that case the dredging-pool is simulated by an enlargement of transmissivity.

The water level of the River Rhine after the dam construction raises between 6 and 2 m at low hydrologic conditions, between 4 1/2 and 1 m at medium, and between 3 1/2 and 1/2 m at high hydrologic conditions. In all these cases the Rhine now infiltrates into the groundwater. For the predicting computation the Rhine was simulated by given infiltration rates, obtained from the vertical model at Karlsruhe for the different hydrologic conditions.

6 PREVENTING MEASURES AND CONTROLLING

Different possibilities were used to reduce infiltration. A vertical screen was built in the Rhine dams at Greffern, which - together with the expected colmation - will reduce the amount of infiltration. The dredging-pool at Greffern, which is very deep, is also a good protection against oozing water. With the Acher, the Rhine-lateral-channel, and the Rhine-lowland-channel it reduces further the amount of infiltrating water. Some open-channel-drainage was constructed near the Rhine-bend, but it is not suitable as the main preventing system by two reasons: it would be necessary to cut the drains very deep into the ground and - the main point - it would be impossible to control the whole preventing system.

Therefore, a number of ten wells, including two emergency wells, were constructed forming a horseshoe around the deepest point in the village center of Greffern. All wells are equipped with two pumps. Therefore, each well has a capacity of 50 l/s in the first step and 100 l/s in the second step. For controlling the wells the installation of two gauges was suggested. One is located in the most affected area in the horseshoe, the other one is located outside the inner cone of water-table depression caused by wells. In this way a good controlling dependent on the hydrologic condition of the whole area can be achieved without too many switching operations.

7 PREDICTING COMPUTATIONS AND RESULTS

As the prediction should give information about long-term effects rather than about single events, the main computations were done for stationary conditions.

The vertical model at Karlsruhe supplied us with infiltration rates for

- a statistical high water level of the Rhine
- a statistical medium water level of the Rhine
- a statistical low water level of the Rhine.

These significant infiltration rates were gained considering certain periods of high, medium and low conditions.

Taking the connected groundwater levels, we considered these infiltration rates as "normal" for the hydrologic condition. Another group of computations were done with higher infiltration rates ("normal" multiplied by 2 and by 3) to estimate their effect and the effect of the wells on these extreme situations.

Computations with lower infiltration rates ("normal" multiplied by 0.5) were used to estimate the effect of colmation on the new Rhine level for the entire model area.

The results showed that in no case the groundwater level at low hydrologic conditions could be kept down at its former state, which had been very low. For medium and high hydrologic conditions the wells brought the satisfying effect that the groundwater-table could be kept downat nearly the same level as before dam construction. The model allowed the prediction that after colmation of the reservoir high groundwater-tables would be even lower than without the reservoir.

In addition to the stationary computations two unsteady cases were investigated. The effect of a flooding of the Greffern dredging-pool for agricultural purposes could be specified by instationary computations as well as the filling of the Greffern polder at extreme Rhine-floods. In the last case the pumping at the drainage wells could lower the groundwater-table at Greffern for about 1 m.

8 SUMMARY

A two-dimensional horizontal groundwater research of the Greffern area/Rhine was presented in which the finite element method had been used.

Emphasis was not laid on the numerical method but on the boundary conditions and hydrological parameters that had to be considered. in this model.

Dependent on the hydrologic situation, the boundary "Rhine" had to be simulated differently by a given head at one condition by prefixed infiltration rates at the other.

A dredging-pool was considered which depended only on groundwater level before dam construction, but became part of the drainage-system after dam construction.

Problems were caused by the bayou-system as definite data about changing flow direction was hard to get. With channels, waterworks and precipitation further hydrologic parameters had to be regarded.

Against the rise of the groundwater-table due to construction of a power station at the river the effect of preventing measures had been tested. The computations had shown that wells added to the open drainage-system would be suitable to prevent a substantial rise of the groundwater-table. The proposed measures were built in 1977 and have worked up till now very well.

Proceedings of Euromech 143 / Delft / 2-4 September 1981

From the pore scale to the macroscopic scale: Equations governing multiphase fluid flow through porous media

C.-M.MARLE
Université Pierre et Marie Curie, Paris, France

1 INTRODUCTION

Relative permeability equations, introduced originally by Wyckoff and Botset [10] and by Leverett [4], are widely used for the mathematical description of multiphase flow in porous media. These equations state that for given values of the saturations, the flow rate of each phase is proportional to the gradient of hydraulic charge in this phase. Moreover, the pressure differences between the phases (capillary pressures) are assumed to depend only on the values of the saturations. These equations have been introduced by analogy with the classical Darcy's equation, valid for single phase flows, which is in good agreement with experiments. But their theoretical foundations are uncertain, and their agreement with experiments, far from satisfactory. In fact, acceptable agreement is generally obtained only for steady-state flows. Some recent experiments, involving flows in which interfaces separating the phases are moving inside the pores [3] ,show clearly that relative permeability equations are unable to explain the obtained results. This is not surprising, as these equations state that for given values of the saturations and of the gradients of hydraulic charge in the various phases, the flow rates are the same as those in a corresponding steady-state flow. By using the relative permeability equations, one assumes that the motions of interfaces separating the phases, inside the pores, have no effect on the flow rates.

Several attempts to derive equations governing multiphase flows in porous media, valid at the macroscopic scale, from the corresponding equations valid at the pore scale, have been made in the past [2,8,9]. See also the bibliography of [5]. The present paper is another such attempt, in which phenomena occurring on interfaces are taken into account. New quantities are defined, and included in the final macroscopic equations. These quantities characterize, at the macroscopic scale, properties of the interfaces between phases and of their motion, inside the pores of the medium. The author believes that equations more realistic than the classical relative permeability equations, cannot be obtained without involving such quantities.

This paper gives only a brief account of the method used and of the results obtained. It is restricted to the case of the isothermal flow of two immiscible homogeneous fluids, in a perfectly rigid porous medium. The reader will find more details in [5], where in addition more general phenomena are included: arbitrary number of fluid phases, diffusion, matter exchanges between phases, adsorption in interfaces, heat transfer and chemical reactions.

2. STATEMENT OF THE PROBLEM, METHOD USED

We consider 2 pure fluids, indexed by i = a or b, flowing in the pores of a perfectly rigid porous medium. The solid part of the medium is indexed by s. Three types of interfaces are present: Σ_{is}, with i = a or b, between the fluid i and the solid; Σ_{ab} between the two fluids. Lines along which these interfaces intersect, and where the two fluids and the solid are contiguous, are designated by L. Mass excess per unit area of interface is neglected,but interfacial tension and energy are taken into account.

As we want to obtain a macroscopic entropy balance equation for the system, thermal phenomena are included, at least at the beginning. But at the end of the calculations it will be assumed that heat exchanges are sufficiently fast to make the temperature uniform and constant in the system.

Starting from fundamental balance equations, for matter, momentum and energy, valid at the pore scale, we derive the corresponding macroscopic balance equations, by taking mean values around each point. Then, a macroscopic entropy balance equation is derived from them. The entropy source appearing in it is, as usual in thermodynamics of irreversible processes [1], a sum of products of generalized forces and fluxes. Phenomenological equations, valid at the macroscopic scale,expressing linear relations betweenthe forces and the conjugate fluxes, are then written. The already obtained macroscopic balance equations and these phenomenological equations form the system proposed here for the mathematical description of the flow. However, at the present stage of this work, the number of unknown functions is larger than the number of equations obtained. It would be necessary to use other physical assumptions, in order to obtain a system involving as many equations as unknown functions.

3 EQUATIONS VALID AT THE PORE SCALE

All the functions, such as mass per unit volume, etc., appearing in the problem, may have jump discontinuities on interfaces Σ_{ab} or Σ_{is} (i = a or b). For this reason the classical matter, momentum and energy balance equations must be written with derivatives in the sense of distributions [7]. Then these equations may be split into several parts: parts valid inside each continuous fluid phase a, b, and inside the solid s; parts made of distributions concentrated on the interfaces Σ_{ab}, Σ_{as}, Σ_{bs}; and the last part made of distributions concentrated on the contact line L. Each of these parts must separately vanish. This leads to the classical balance equations valid for each fluid phase a and b, for the solid s, for interfaces, and to compatibility conditions on the contact lines:

Balance equations for the fluid phase i:

$$\{ \frac{\partial \rho_i}{\partial t} \} + \{ \operatorname{div} (\rho_i \vec{u}_i)\} = 0$$

$$\{ \frac{\partial}{\partial t} (\rho_i \vec{u}_i)\} + \{ \vec{\operatorname{Div}} (\rho_i \vec{u}_i \otimes \vec{u}_i - \bar{\bar{\tau}}_i)\} - \rho_i \vec{g} = 0$$

$$\{ \frac{\partial}{\partial t} [\rho_i (e_i + \frac{1}{2} |\vec{u}_i|^2)]\} - \rho_i \vec{u}_i . \vec{g}$$

$$+ \{\operatorname{div}[\rho_i (h_i + \frac{1}{2}|\vec{u}_i|^2)\vec{u}_i + \vec{q}_i - \vec{u}_i . \bar{\bar{\theta}}_i]\} = 0$$

Balance equation for the solid:

$$\rho_s \{ \frac{\partial e_s}{\partial t}\} + \{ \operatorname{div} \vec{q}_s \} = 0$$

In these equations the brackets { } around a derivative means that this derivative must be understood in the usual sense. Derivatives without brackets, occurring in the following equations, must be understood in the sense of distributions.

We write below balance equations for interfaces and contact lines together. By using formulae for derivatives of distributions concentrated on a surface, it is possible to deduce from them separate balance equations , for interfaces, and for contact lines (see [5]).We have assumed that there is no discontinuity of the tangential part of the velocity on interfaces ("no slip" assumption). This leads to:

$$\vec{w}_{ab} = \vec{u}_a = \vec{u}_b \quad \text{on } \Sigma_{ab}$$

$$\vec{u}_a = 0 \quad \text{on} \Sigma_{as} \; ; \quad \vec{u}_b = 0 \quad \text{on } \Sigma_{bs}$$

$$\left[- [\rho u]_a^b \, \vec{w}_{ab} . \vec{n}_a^b + [\rho \vec{u} \otimes \vec{u} - \bar{\bar{\tau}}]_a^b . \vec{n}_a^b \right] \delta_{\Sigma_{ab}}$$

$$- \vec{\operatorname{Div}} \left[\sigma_{ab} \, \bar{\bar{\pi}}_{\Sigma_{ab}} \, \delta_{\Sigma_{ab}} \right] = 0$$

$$\left[- [\rho(e + \frac{1}{2}|\vec{u}|^2)]_a^b \, \vec{w}_{ab} . \vec{n}_a^b + [\vec{q}]_a^s . \vec{n}_a^s \, \delta_{\Sigma_{as}}\right.$$

$$+ [\vec{q}]_b^s . \vec{n}_b^s \, \delta_{\Sigma_{bs}} + \frac{\partial}{\partial t}(\varepsilon_{ab} \delta_{\Sigma_{ab}} + \varepsilon_{as} \delta_{\Sigma_{as}} + \varepsilon_{bs} \delta_{\Sigma_{bs}})$$

$$+ \left[[\rho(h + \frac{1}{2}|\vec{u}|^2)\vec{u} + \vec{q} - \vec{u} . \bar{\bar{\theta}}]_a^b . \vec{n}_a^b\right] \delta_{\Sigma_{ab}}$$

$$+ \operatorname{div} \left[(\varepsilon_{ab} \vec{w}_{ab} - \sigma_{ab} \, \bar{\bar{\pi}}_{\Sigma_{ab}} \, \vec{w}_{ab}) \delta_{\Sigma_{ab}} \right] = 0$$

4 MACROSCOPIC EQUATIONS

They are obtained by taking the convolution product of equations valid at the pore level with a weight function m (this idea is due to Matheron [6]). The function m , of space variables only, is assumed positive, continuous and with continuous derivatives of all orders, with compact support, and such that its integral, over the whole space, is equal to 1. By definition the convolution product of m with a function f(x, t) of space and time variables is:

$$(f * m) (x, t) = \int_{\mathbb{R}^3} f(y, t) \, m(x - y) \, dy$$

This leads to the following macroscopic equations:

$$\frac{\partial}{\partial t} (\Phi \, S_i \, R_i) + \operatorname{div} (\Phi \, S_i \, R_i \, \vec{V}_i) = 0$$

$$\frac{\partial}{\partial t} A_{ab} + \operatorname{div} \vec{V}_{ab} + \mathfrak{R}_{ab} \, Z_{ab} - Z_{ab,s} = 0$$

$$\frac{\partial}{\partial t} A_{is} = Z_{is,j}$$

$$\Phi S_i R_i (\frac{\partial \vec{V}_i}{\partial t} + \vec{V}_i . \overline{\overline{\text{Grad}}}\, \vec{V}_i) - \text{Div}\, \overline{\overline{\theta}}_i$$

$$+ \Phi S_i (\vec{\text{grad}}\, P_i - R_i \vec{g}) + \vec{\mathcal{F}}_{ij} + \vec{\mathcal{F}}_{is}$$

$$+ P_i \vec{\text{grad}} (\Phi S_i) = 0$$

$$- \vec{\text{Div}}\, \overline{\overline{\theta}}_{ab} - A_{ab} \vec{\text{grad}}\, \mathcal{J}_{ab}$$

$$- (\vec{\mathcal{F}}_{ab} + \vec{\mathcal{F}}_{ba} - \vec{\mathcal{F}}_{ab,s} + \mathcal{J}_{ab} \vec{\text{grad}}\, A_{ab}) = 0$$

All the macroscopic quantities appearing in these equations are related by well dened formulae to the corresponding quantities valid at the pore scale (see [5]).

Similarly, macroscopic internal energy balance equations for the fluid phases, the solid and the interfaces, are obtained:

$$\frac{\partial}{\partial t} (\Phi S_i R_i E_i) + P_i \frac{\partial (\Phi S_i)}{\partial t} - \overline{\overline{\mathcal{C}}}_i : \overline{\overline{\text{Grad}}}\, \vec{V}_i$$

$$+ \text{div} (\Phi S_i R_i E_i \vec{V}_i + \vec{Q}_i) + X_{ij} - \vec{\mathcal{F}}_{ij} . \vec{V}_i$$

$$+ X_{is} - \vec{\mathcal{F}}_{is} . \vec{V}_i = 0$$

$$\frac{\partial}{\partial t} \left[(1 - \Phi) R_s E_s \right] + \text{div}\, \vec{Q}_s + \vec{X}_{sa} + \vec{X}_{sb} = 0$$

$$\frac{\partial}{\partial t} (A_{ab} E_{ab}) + \text{div} (A_{ab} E_{ab} \vec{V}_{ab} + \vec{Q}_{ab})$$

$$- \overline{\overline{\text{Grad}}}\, \vec{V}_{ab} : \overline{\overline{\mathcal{C}}}_{ab} + X_{ab,s} - \mathcal{J}_{ab} Z_{ab,s}$$

$$+ \vec{V}_{ab} . (\vec{\mathcal{F}}_{ab} + \vec{\mathcal{F}}_{ba} - \vec{\mathcal{F}}_{ab,s})$$

$$- \left[X_{ab} + X_{ba} + (P_a - P_b) Z_{ab} \right] = 0$$

$$\frac{\partial}{\partial t} (A_{is} E_{is}) + \text{div} (\vec{Q}_{is}) + X_{is,j} - \mathcal{J}_{is} Z_{is,j}$$

$$- (X_{is} + X_{si}) = 0$$

The energy balance equation on the contact line L leads to the macroscopic equation:

$$(\mathcal{J}_{ab} Z_{ab,s} - X_{ab,s}) + (\mathcal{J}_{as} Z_{as,b} - X_{as,b})$$

$$+ (\mathcal{J}_{bs} Z_{bs,a} - X_{bs,a}) = 0$$

5 THE ENTROPY SOURCE

By using the macroscopic internal energy balance equations, and assuming that the usual state equations between the various thermodynamic functions remain valid at the macroscopic scale, balance equations for entropy of the two fluid phases, the solid and the interfaces, are obtained. As usual in thermodynamics of irreversible systems, the total entropy balance equation may be written under the following form: the time derivative of the entropy contained in a unit volume of the system, plus the divergence of an entropy flux, is equal to the entropy source per unit volume of the system. This equation is too long to be written in full here (see 5). We will indicate only the expression of the entropy source, after simplifications obtained by assuming that the temperature in the whole system is practically uniform. After multiplication by the temperature T, the entropy source per unit volume of the system may be written as:

$$\overline{\overline{\text{Grad}}}\, \vec{V}_a : \overline{\overline{\theta}}_a + \overline{\overline{\text{Grad}}}\, \vec{V}_b : \overline{\overline{\theta}}_b + \overline{\overline{\text{Grad}}}\, \vec{V}_{ab} . \overline{\overline{\theta}}_{ab}$$

$$+ \vec{V}_a . \left[\vec{\mathcal{F}}_{as} + \vec{\mathcal{F}}_{ab} + P_a \vec{\text{grad}} (\Phi S_a) \right]$$

$$+ \vec{V}_b . \left[\vec{\mathcal{F}}_{bs} + \vec{\mathcal{F}}_{ba} + P_b \vec{\text{grad}} (\Phi S_b) \right]$$

$$- \vec{V}_{ab} . \left[\vec{\mathcal{F}}_{ab} + \vec{\mathcal{F}}_{ba} - \vec{\mathcal{F}}_{ab,s} + \mathcal{J}_{ab} \vec{\text{grad}}\, A_{ab} \right]$$

$$- \vec{V}_{ab} . \vec{\text{grad}}\, \mathcal{J}_{ab}$$

$$+ Z_{ab} (\mathcal{R}_{ab} \mathcal{J}_{ab} + P_a - P_b)$$

$$- (\mathcal{J}_{ab} Z_{ab,s} + \mathcal{J}_{as} Z_{as,b} + \mathcal{J}_{bs} Z_{bs,a})$$

Most of the terms appearing in this expression have a clear physical meaning: the first three terms correspond to the entropy created by the non uniformity of the macroscopic velocities of the fluid phases and of the interface. It is a kind of viscosity effect, at the macroscopic scale. The three terms after correspond to the entropy created by the frictions at the pore scale, which act against the velosities of the two phases and of the interface. The term involving the mean curvature of the interface $\mathcal{R}_{ab}$, is interesting, as it involves the desequilibrium between the pressure difference between the two phases and the capillary forces. The two other terms, one involving the gradient of the interfacial tension, by its scalar product with the macroscopic interface velocity, and the other the macroscopic quantities $Z_{ab,s}$, $Z_{bs,a}$ and $Z_{as,b}$, are not so clearly understood at the present stage of this work. In the former, the interfacial velocity occurs only because we have made here a **simplifying assumption**: we have neglected the mass excess of the interface. When this mass excess is not neglected, the interfacial velocity $\vec{V}_{ab}$ is replaced in this term by another quantity $\vec{Y}_{ab}$, related but not equal to $\vec{V}_{ab}$ (see [5]). The quantities $Z_{ab,s}$, $Z_{as,b}$ and $Z_{bs,a}$ appearing in the last term are related to the velocity of the contact line L, and the physical meaning of this term has probably something to do with the displacement of this line. All these points are to be clarified.

6 THE PHENOMENOLOGICAL EQUATIONS

We assume, for simplicity, that the macroscopic viscous stress tensors of the two fluid phases and of the interface, $\bar{\bar{\Theta}}_a$, $\bar{\bar{\Theta}}_b$, and $\bar{\bar{\Theta}}_{ab}$, are negligible. We assume also that the macroscopic interfacial tension $\mathcal{S}_{ab}$ is constant and uniform in space. Then the fifth macroscopic equation established in paragraph 4 becomes:

$$\vec{\mathcal{F}}_{ab} + \vec{\mathcal{F}}_{ba} - \vec{\mathcal{F}}_{ab,s} + \mathcal{S}_{ab}\, \vec{\text{grad}}\, A_{ab} = 0$$

We assume, as usual in thermodynamics of irreversible processes [1], that there are linear relations between the generalized forces and fluxes appearing in the entropy source. We assume also that cross effects, (effect of a generalized force on generalized fluxes other than its conjugate), are negligible. Then these relations may be written as:

$$\vec{\mathcal{F}}_{is} + \vec{\mathcal{F}}_{ij} + P_i\, \vec{\text{grad}}\, (\Phi\, S_i) = B_i\, \vec{V}_i$$

$$\mathcal{R}_{ab}\mathcal{S}_{ab} + P_a - P_b = C\, Z_{ab}$$

Moreover, it is easy to see that:

$$Z_{ab} = \frac{\partial(\Phi\, S_a)}{\partial t}$$

The macroscopic balance equations obtained in paragraph 4, and the present phenomenological equations, lead to the system:

$$\frac{\partial}{\partial t}(\Phi\, S_i\, R_i) + \text{div}\,(\Phi\, S_i\, R_i\, \vec{V}_i) = 0$$

$$S_a + S_b = 1$$

$$\Phi\, S_i\, R_i\left(\frac{\partial \vec{V}_i}{\partial t} + \vec{V}_i . \bar{\bar{\text{Grad}}}\, \vec{V}_i\right) + \Phi\, S_i(\vec{\text{grad}}\, P_i - R_i\, \vec{g}) + B_i\, \vec{V}_i = 0$$

$$P_a - P_b = -\mathcal{R}_{ab}\mathcal{S}_{ab} + C\, \frac{\partial(\Phi\, S_a)}{\partial t}$$

It may be seen that when the forces of inertia are neglected, the third equation of this system becomes:

$$\vec{V}_i = -\frac{\Phi S_i}{B_i}\,(\vec{\text{grad}}\, P_i - R_i\, \vec{g})$$

This is exactly the relative permeability equation for phase i (i = a or b).

The above system differs from the usual one, first by the terms coming from the forces of inertia modifying the relative permeability equations, second by its last equation: the pressure difference between the two fluids has no more the same value as in a steady-state flow; there is a term proportional to the time derivative of the saturation S_a. This is the easiest way to take into account the motion of the interfaces between the two fluids, inside the pores.

The above system has 10 scalar equations (remember that the index i takes two values a and b, and that a vector equation is equivalent to 3 scalar equations). If we assume that the mass per unit volume R_a and R_b of phases a and b, are known functions of the corresponding pressures P_a and P_b, and that the phenomenological coefficients B_i and C are known functions of the other parameters, there are 11 independent scalar unknown functions in this system: the two saturations S_a and S_b, the two pressures P_a and P_b, the three components of the fluid velocities $\vec{V}_a$ and $\vec{V}_b$, and the macroscopic mean curvature of the interface $\mathcal{R}_{ab}$. Thus one equation is missing. If we assume that $\mathcal{R}_{ab}$ has, for given values of the saturations, the same value as at equilibrium, then the system has as many equations as unknown functions, and it may be used for a mathematical description of the flow.

7 CONCLUSION

All the macroscopic equations obtained in paragraph 4, have not been used in paragraph 6, and some macroscopic quantities such as the interfacial areas A_{ab}, A_{as} and A_{bs} per unit volume of porous medium, the macroscopic interfacial velocity $\vec{V}_{ab}$, do not appear in the final system. This shows that improvements remain to be done. But there is a reasonable hope that a mathematical formulation of multiphase flows in porous media, better than the one presently used, may be obtained by the method proposed here.

8 LIST OF SYMBOLS

P.S. means pore scale, and M.S. macroscopic

scale.

- e P.S. internal energy per unit weight
- h P.S. enthalpy per unit weight
- p P.S. pressure
- $\vec{q}$ P.S. heat flux
- $\vec{n}$ unit vector normal to interface .The superscript indicates the phase towards which $\vec{n}$ is directed
- $\vec{u}$ P.S. velocity
- $\vec{w}_{ab}$ P.S. velocity of Σ_{ab}

- A M.S. interfacial area per unit volume of porous medium
- B phenomenological coefficient
- C phenomenological coefficient
- E M.S. internal energy per unit weight
- L contact line between the two fluid phases and the solid
- P M.S. pressure
- $\vec{Q}$ M.S. heat flux
- R M.S. mass per unit volume (of phase)
- S M.S. saturation (fraction of pore space occupied by the phase indicated by the subscript)
- $\vec{V}$ M.S. velocity
- X_{ij}, X_{is}, X_{si} M.S. heat leaving phase i for the first two, solid s for the third, respectively through Σ_{ij} for the first, Σ_{is} for the two others, per unit time and unit volume of porous medium
- $X_{ab,s}$, $X_{is,j}$ M.S. heat leaving Σ_{ab} (resp. Σ_{is}) through L, per unit time and unit volume of porous medium
- Z_{ab} M.S. scalar quantity related to the velocity of Σ_{ab}, and equal to the time derivative of (ΦS_a)
- $Z_{ab,s}$, $Z_{is,j}$ M.S. scalar quantities related to the velocity of line L

- $\vec{\mathcal{F}}_{ij}$, $\vec{\mathcal{F}}_{is}$ M.S. force exerted by phase i on interface Σ_{ij} (resp. Σ_{is}) per unit volume of porous medium
- $\vec{\mathcal{F}}_{ab,s}$ M.S. force exerted by interface Σ_{ab} on contact line L, per unit volume of porous medium
- $\bar{\bar{\mathfrak{I}}}$ unit rank 2 tensor
- $\mathcal{R}_{ab}$ M.S. mean curvature of Σ_{ab}
- $\mathcal{S}$ M.S. interfacial tension
- $\bar{\bar{\mathcal{C}}}$ M.S. stress tensor

- δ_Σ Dirac distribution of interface Σ
- ε^Σ P.S. interfacial internal energy per unit area
- $\bar{\bar{\theta}} = \bar{\bar{\tau}} + p\,\bar{\bar{\mathfrak{I}}}$ P.S. viscous stress tensor
- π_Σ orthogonal projection operator on the plane tangent to interface Σ
- ρ P.S. mass per unit volume
- σ P.S. interfacial tension
- $\bar{\bar{\tau}}$ P.S. stress tensor

- $\bar{\bar{\Theta}}_i = \bar{\bar{\mathcal{C}}}_i + \Phi S_i P_i \bar{\bar{\mathfrak{I}}}$ M.S. viscous stress tensor for phase i
- $\bar{\bar{\Theta}}_{ab} = \bar{\bar{\mathcal{C}}}_{ab} - A_{ab}\mathcal{S}_{ab}\bar{\bar{\mathfrak{I}}}$ M.S. viscous stress tensor for interface Σ_{ab}
- Σ interface
- Φ porosity

Subscripts and superscripts:

- a,b, for fluid phases a,b
- i,j, for fluid phases a or b, b or a (when i = a, then j = b)
- ab for interface Σ_{ab}
- as, bs, for interfaces Σ_{as}, Σ_{bs}

Symbols for operators:

- $\vec{\mathrm{grad}}$ is the gradient operator (for scalar functions).
- div is the divergence operator (for vector fields).
- $\bar{\bar{\mathrm{Grad}}}$ is the gradient operator for vector fields.
- $\vec{\mathrm{Div}}$ is the divergence operator for rank 2 tensor fields.

Special signs:

An arrow $\rightarrow$ is put above letters indicating vector fields.

Two bars $=$ are put above letters indicating rank two tensor fields.

A quantity between brackets, with a subscript and a superscript, such as $[f]_a^b$, means the value of this quantity on side of the superscript, minus the value of the same quantity on side of the subscript.

9 REFERENCES

1 De Groot, S.R. & Mazur, P. 1962, Non equilibrium thermodynamics.Amsterdam,NorthHoll.

2 Ene, H.I. 1981, On the thermodynamics of mixtures, Int. J. Engng. Sci., to appear.

3 Jacquin, Ch. 1981, Conditions aux limites et lois d'écoulement en réservoir fissuré, Rapport interne IFP 28831.

4 Leverett, M.C. 1938, Flow of oil-water mixtures through unconsolidated sands, Trans. AIME 132: 149.

5 Marle, C.-M. 1981, On macroscopic equations governing multiphase flow with diffusion and chemical reactions in porous media, Int. J. Engng. Sci., to appear.

6 Matheron, G.1967, Les variables régionalisées et leur estimation. Paris, thèse.

7 Schwartz, L. 1965, Méthodes mathématiques pour les sciences physiques.Paris,Hermann.

8 Slattery, J.C. 1970, Two-phase flow through porous media, A.I.Ch.E. Journal 16, 3: 345-352.

9 Whitaker, S.1973, The transport equations for multiphase systems,Chem.Engng.Sci28:139.

10 Wyckoff,R.D.&Botset,H.G.1936,The flow of gas-liquid mixtures in sands,Physics 7:325.

Proceedings of Euromech 143 / Delft / 2-4 September 1981

Unsteady groundwater flow in a stream-fractured aquifer system

H.ÖNDER
Middle East Technical University, Ankara, Turkey

1 INTRODUCTION

Variations in piezometric head in a stream-aquifer system due to a rise or decline of the water level in the stream has been investigated by many researchers. The problems of bank storage due to flood stages in streams, determination of aquifer parameters, and groundwater contributions to stream flow had been motivations in the studies of streamaquifer systems. In those investigations the aquifer is taken as a granular medium. However, a widely encountered type of aquifer, which is of a world-wide occurrence as water-bearing formation, is fractured (or fissured) rock. This type of formation is commonly qualified as a heterogeneous medium in which the low permeable blocks of primary porosity are separated by highly permeable fissures of low fissure volume. It is usually assumed that the conducting properties of the aquifer is associated mainly with the fracture permeability while the storage properties are related to the primary porosity of the blocks. (Strelsova-Adams 1978)

One of the conceptual models to study the flow in fractured formations is the double porosity medium which was first considered by Barenblatt et. al (1960). This approach has been followed closely by other investigators (Warren & Root 1968, Odeh 1965, Kazemi 1959, Kazemi et al. 1969 Streltsova 1976, Duguid & Lee 1977).

In this study, a linear stream-fractured aquifer system is considered and a numerical solution is presented.

2 GOVERNING EQUATIONS

In the application of the double porosity conceptual model, introduced by Barenblatt et al. (1960), naturally fractured medium is seperated into two-overlapping continuum, each filling the entire space. The flow takes place through both fractures and pores of the blocks. The fluid transfer between blocks and fractures is taken intc acount. At each point of the space, following the continuum concept, two sets of medium and flow properties are introduced, the first set being for fissure flow and the second for the flow in blocks. The representative elementary volume at the point considered, consists of a large number of porous blocks, as well as large number of fissures having random distribution, orientation and size.

Following the assumptions of Barenblatt et.al (1960) the general differential equation describing the one-dimensional fissure flow in a confined aquifer is obtained by combining Darcy's law and the continuity equation as follows:

$$T_1 \frac{\partial^2 s_1}{\partial x^2} = S_1 \frac{\partial s_1}{\partial t} + v_d \qquad (1)$$

where

v_d = is rate of water drained from blocks

$T_1 = K_1 b$ is coefficient of transmissibility of fissures

K_1 = is hydraulic conductivity of fissures

b = is the thickness of the aquifer

s_1 = is mean drawdown in the fissures

S_2 = is coefficient of storage of fissures

Combining again Darcy's law and continuity equation, and using Barenblatt et al. (1960) assumptions the differental equation of the flow in the pores of blocks is

$$v_d = S_2 \frac{\partial s_2}{\partial t} \qquad (2)$$

The quasi-steady transfer of water between pores of blocks and fissures is given by

$$v_d = \frac{K_2}{\ell}(s_1-s_2)=\varepsilon(s_1-s_2) \quad (3)$$

where

K_2 is hydraulic conductivity of blocks

ℓ is a linear parameter

s_2 in mean drawdown in blocks

$$\varepsilon = \frac{K_2}{S_2\ell}$$

Introducing the following dimensionless variables

$$y_1 = s_1/s_o \quad (4a)$$

$$y_2 = s_2/s_o \quad (4b)$$

$$x_D = x/L \quad (4c)$$

$$t_D = \frac{T_1 t}{S_1 L^2}$$

Eq.(1) & (2) can be reduced to the form

$$\frac{\partial^2 y_1}{x_D^2} = \frac{\partial y_1}{\partial t_D} + \eta \frac{\partial y_2}{\partial t_D} \quad (5)$$

$$\text{where} \quad \eta = \frac{S_2}{S_1} \quad (6)$$

Similarly, Eq(2) & (3) reduces to

$$\frac{\partial y^2}{\partial t_D} = \delta(y_1-y_2) \quad (7)$$

$$\text{where} \quad \delta = \frac{\varepsilon S_1 L^2}{T_1} \quad (8)$$

Solving Eq.(7) by laplace transform technique as in Streltsova (1976) with condition that $y_1=y_2= 0$ when $t_o=0$ and making in Eq.(5) leads to

$$\frac{\partial^2 y_1}{\partial x_D^2} = \frac{\partial y_1}{t_D} + \eta\delta \int_0^{t_D} e^{-\delta(t_D-\tau)} \left.\frac{\partial y_1}{\partial t_D}\right|_{t_o=\tau} d\tau \quad (9)$$

or integrating the last term by parts

$$\frac{\partial^2 y_1}{\partial x_D^2} \quad \frac{\partial y_1}{\partial t_D} + \eta\delta y_1 - \eta\delta \int_o^{t_D} e^{-\delta(t_D-\tau)} y_1(\tau)\,d\tau \quad (10)$$

is obtained.

Initial & boundary conditions that should be satisfied by s_1 for the idealized problem deputed in Fig. 1 are:

$$s_1(x,0) = 0 \quad (11\ a)$$

$$s_1(0,t) = s_o \quad (11\ b)$$

$$s_1(L,t) = 0 \quad (11\ c)$$

or corresponding conditions for y_1

$$y_1(X_D,0) = 0 \quad (12\ a)$$

$$y_1(0, t_D)= 1 \quad (12\ b)$$

$$y_1(1, t_D)= 0 \quad (12\ c)$$

The solution of integro-differential equation, (10), subjected to initial and boundary conditions expressed in Eq.(12) will be searched for.

3 METHOD OF SOLUTION

To integrate the integro-differential equation given above an attempt is made to apply the Method of Lines (Liskovets 1965) This method has been used successfully to investigate various problems in engineering science (Koussiz & Watson 1980, Varoğlu & Finn 1979). In this method, the partial differential equation is reduced to a set of ordinary differential equations in space variable by discretizing in time. The set of ordinary differential equations thus obtained can be solved by techniques more efficient and powerful than those available for solving partial differential equations.

To apply the Method of lines in the solution of Eq.(10), the integral in last term on the right hand side is first replaced by a convenient integration formula. (The subscripts for y,t, and x will be omitted in the following to prevent confusion. Therefore it should be understood that y is dimensionless fissure drawndown, t is dimensionless time, and x is dimensionless space variable). Let the value of this integral at any time t_n is I_n

$$I_n = \int_o^{t_n} e^{-\delta(t_n-\tau)} y(\tau)d\tau \quad (13)$$

Using trapezoidal rule of integration, (Carnahan et.al 1969) which is the simpliest integration formula, Eq.(13) can be written as :

$$I_n = \frac{\Delta t}{2} \sum_{k=1}^{n} [y_{k-1} e^{-\delta(t_n-\tau_{k-1})} - y_k e^{-\delta(t_n-\tau_k)}] \quad (14)$$

where Δt is time increment and $y_k=y(x,\tau_k)$, with $0\leq\tau_k\leq t_n$. Secondly the time derivative in Eq.(10) is replaced by forward difference representation:

$$\frac{\partial y(x,t)}{\partial t^2} = \frac{y_{n+1} - y_n}{\Delta t} \tag{15}$$

Furthermore the time level for the evaluation of the second derivative in Eq.(10) has to be specified. By selecting time level t_n, the lefthand side member of Eq.(10) may be written as

$$\frac{\partial^2 y}{\partial x^2} = \frac{\partial^2 y_n}{\partial x^2} \tag{16}$$

Substituting Eqs.(14), (15) and (16) in Eq.(10), the following set of second order ordinary differential equations in x for y is obtained.

$$\frac{d^2 y_n}{dx^2} = \frac{y_{n+1} - y_n}{\Delta t} + \eta\delta y_n - \eta\delta \frac{\Delta t}{2}\sum_{k=1}^{n}\left[y_{k-1} e^{-\delta(t_n-\tau_{k-1})} - y_k e^{-\delta(t_n-\tau_k)}\right] \tag{17}$$

In the solution of Eq.(17), a simple difference scheme, which consists of approximating the second derivative given in Eq.(16) by central difference expression has been used. Thus Eq.(16) is written as

$$\frac{d^2 y_n}{dx^2} = \frac{y_n^{i-1} - 2y_n^i + y_n^{i+1}}{(\Delta x)^2} \tag{18}$$

in which supercripts denotes position, Substitution of Eq.(18) into Eq.(17) reduces to

$$y_{n+1}^i = \alpha_1(y_n^{i-1} + y_n^{i+1}) + \alpha_4 y_n^i + \alpha_3 \sum_{k=1}^{n}\left[y_{k-1}^i e^{-\delta(t_n-\tau_{k-1})} + y_k^i e^{-\delta(t_n-\tau_k)}\right] \tag{19}$$

in which

$$\left.\begin{aligned} \alpha_1 &= \Delta t/(\Delta x)^2 \\ \alpha_2 &= \eta\delta\Delta t \\ \alpha_3 &= \eta\delta \frac{(\Delta t)^2}{2} = \alpha_2 \frac{\Delta t}{2} \\ \alpha_4 &= 2\alpha_1 + 1 - \alpha_2 \end{aligned}\right\} \tag{20}$$

In Eq.(19) the value of y at time t_{n+1} is given in terms of the values of y at times $0 \leqslant t \leqslant t_n$. Therefore it is a simple explicit scheme. It should be noted that the time increment, Δt, used in the approximation of the integral given by Eq.(13) and of the time derivative given by Eq.(15) are same.

4 NUMERICAL EXAMPLE

A fractured aquifer is assumed to be intersected by two parallel surface stream (Fig.1). Initially the aquifer is in equilibrium with the streams. At time t=0 the level in the stream on the left hand side, is assumed to increase instantaneously by s_o. The following hydraulic and geometric properties are considered.

$T_1 = 5 \times 10^{-4}\ m^2/s$

$S_1 = 1 \times 10^{-4}$

$\varepsilon = 5 \times 10^{-6}\ m$

$L = 1 \times 10^3\ m$

$S_2 = 1 \times 10^{-2}$

$S_o = 1m.$

For dimensionless time and space increments $\Delta t = 2\text{x}10^{-2}$, $\Delta x = 0.1$ has been used the piezometric head variations in the stream are calculated. Computations are carried out on an IBM 370/145 computer. Numerical results are presented in Fig.(2) and (4)

In the initial phase, during which the drawdown has not reach the aquifer end, the aquifer appears effectively infinite. In this case, if the blocks were impervious the aquifer would behave as if it were homogeneous and $\eta \to 0$. An analytical solution is available for this flow (Hantush, 1964) and the dimensionless drawdown is given by

$$y = \text{erfc}\,\frac{x}{2\sqrt{t}} \tag{21}$$

The numerical solution is checked with this analytical solution and a close agreement is found for the times during which drawdown has not reached the aquifer end. (Fig. 4).

5 DISCUSSION AND CONCLUSION

It should be noted that the results presented in this paper are the outcome of some limited number of numerical experimentions. Therefore the discussion and conclusion should be considered with a certain reservation.

About the behaviour of fractured aquifer, Fig. 3 indicates that, the presence of fracture modifies substantially the flow mechanism. The two curves for finite aquifer reveals that the delayed reaction of porous blocks reduces the fissure drawdown.

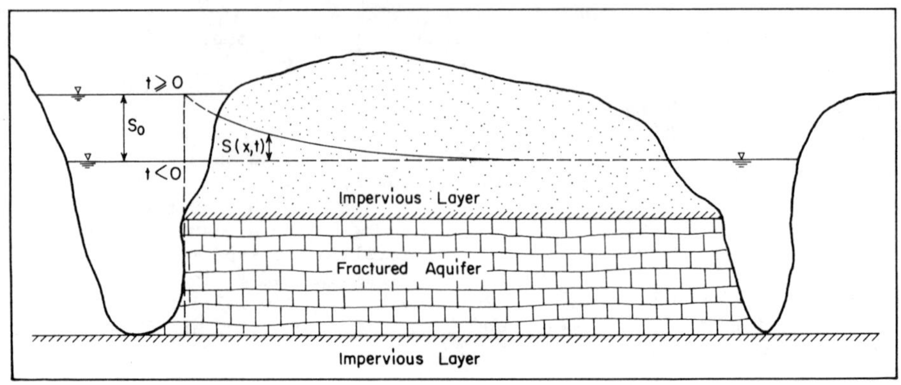

Fig.1 Schematic representation of a finite aquifer between two parallel stream

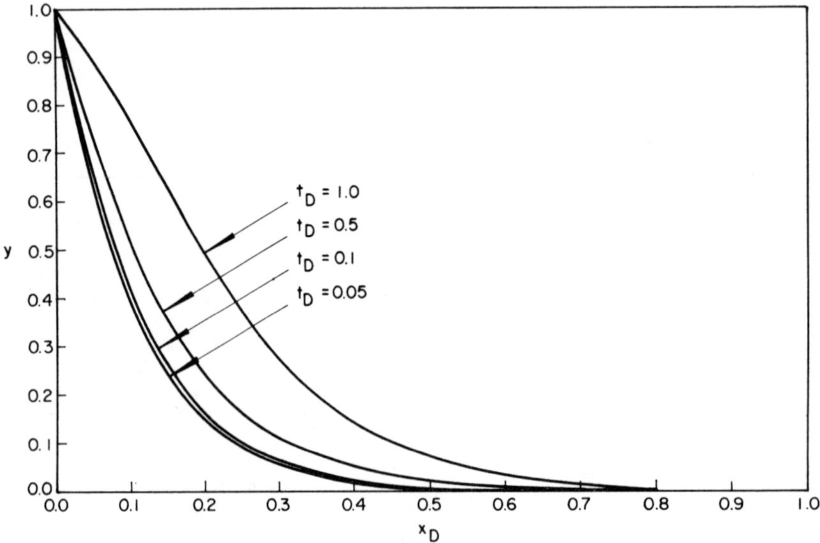

Fig.2 Variation of dimensionless drawdown with dimensionless distance

At $t_D = 1.0$, although the flow in the homogeneous aquifer reaches the steady state condition, the flow in the fractured aquifer with $\eta = 100$ is transient. In Fig.4 it can be seen that at early times, the response of fracture to pressure change is very rapid. However at $t_D \approx 0.0025$ the effect of transfer of water between blocks and fissures is felt. When the pressure in the fractures and in the blocks are readjusted, a rapid change in the slope of drawdown-time curve is observed again. In late times it seems that the behavior of fractured aquifer approaches that of a homogeneous aquifer. As for the numerical thechnique used, it is simple and easy to program. It can certainly be improved. In the approximation of the integral given by Eq.(13) a more sophisticated integration formula could be used. If backward difference representation is used in Eq.(15) and since the time level for the evaluation of the second

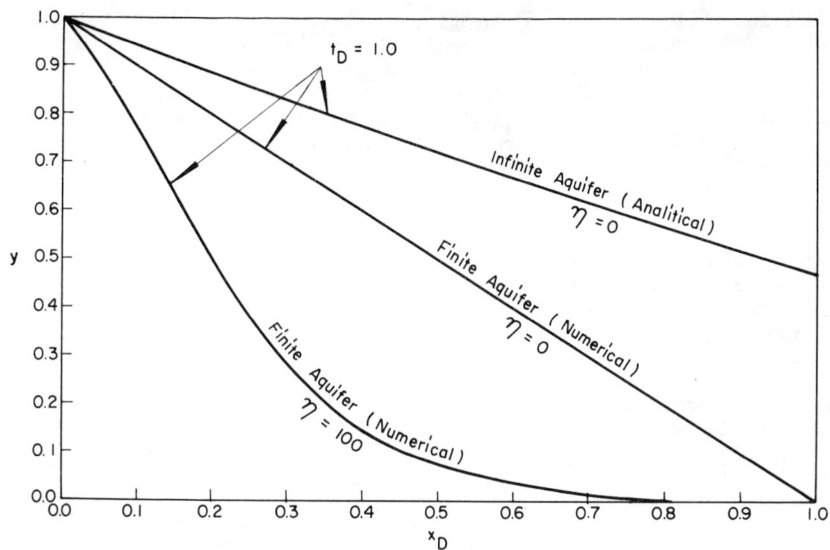

Fig.3 Variation of dimensionless drawdown with dimensionless distance (t_D=1.0)

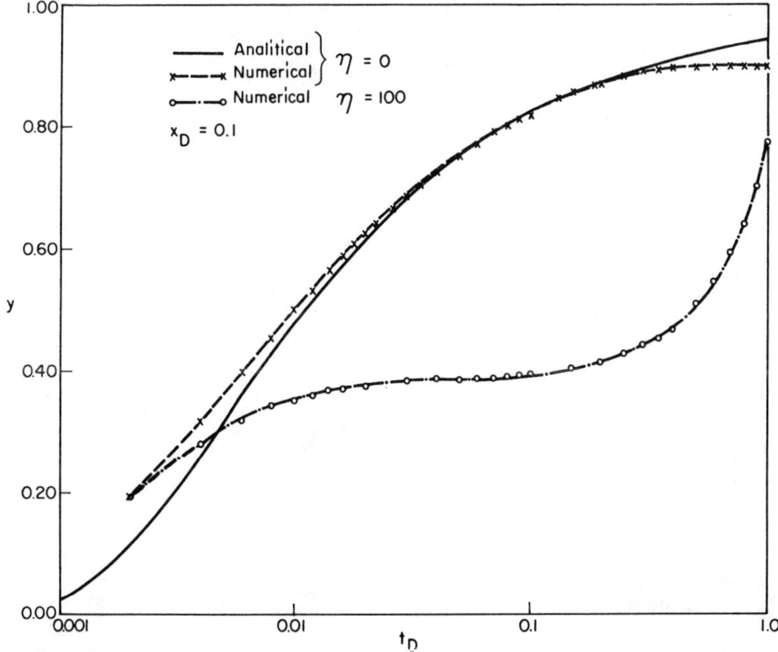

Fig.4 Variation of dimensionless drawdown with dimensionless time at point $x_D = \frac{x}{L} = 0.1$

derivative given by Eq.(16) would still be t_n, then, the resulting set of second order ordinary differential equations could not be solved by the program which is used in this work, and an appropriate numerical technique should be used. The deviation between numerical and solution for homogeneous aquifer at early times is reduced when smaller dimensionless and time increaments.

The boundary conditions considered in this work are rather idealized. Therefore it would worth to solve the problem for other boundary conditions that would be encountered more commanly in the nature.

The validity of any solution will be measured by the degree of correspondance of the predicted flow behavior with the actual aquifer response. Therefore a field observation would be helpful to check both the assumptions and solution techniques.

6 REFERENCES

Barenblatt, G.I., I.P.Zheltov, & I.N. Kochina 1960, Basic concepts in the theory of seepage of homogeneous liquids in fissured rocks (Strata), J.Appl. Math and Mech. 24: 5, 852-864 (1286-1303).

Carnahan, B., H.A.Luther & J.O. Wilkes 1969, Applied numerical methods, NewYork, John Wiley & Sons, Inc.

Duguid, J.O. & P.C.Y. Lee 1977, Flow in fractured porous media, Wat.Res.Research, 13 : 3, 558-566.

Hanstush, M.S. 1964, Hydraulics of wells, In V.T. Chow (ed.) Advanced in hydrosciences Vol 1. London, Academic Press.

Kazami, H., M.S. Seth, G.W. Thomas 1969, The interpretation of interference tests in naturally fractured reservoirs with uniform fracture distribution, Soc.of Pet.Eng. J., 563-472.

Kazemi, H. 1969, Pressure transient analysis of naturally fractured reservoirs with uniform fracture distribution, Soc. of Pet.Eng. J. Dec., 451-462

Koussis, A.D. & R.W. Watson 1980, Groundwater flow computation by method of lines, ASCE J. of Irr. and Drainage Division. IR1, 1-8

Liskovets, O.A. 1965, The method of lines (Review), Dif. Eqns. 1: 12, 1308-1323 (1662-1678)

Odeh, A.S. 1965, Unsteady-state behavior of naturally fractured reservoirs, Soc. of Pet. Eng. J. March, 60-64

Streltsova-Adams, T.D. 1978, well hydraulic in heterogeneous aquifer formation. In V.T. Chow (ed), Advances in hydrosciences. Vol. 11. London, Academic Press.

Varoğlu, E., & W.D.L. Finn, 1979, Variable domain finite elements for some stefan type problems of Mechanics

Warren, J.E. & P.J. Root 1963, The behaviour of naturally fractured reservoirs, Soc. of Pet. Eng. J.Sept., 245-255.

Proceedings of Euromech 143 / Delft / 2-4 September 1981

Piping due to flow towards ditches and holes

J.B.SELLMEIJER
Delft Soil Mechanics Laboratory, Netherlands

INTRODUCTION

The problem of piping is recently becoming more and more a topic of common interest. Especially in the Netherlands, where large parts of the country are protected from flooding by dikes the phenomenon of piping is important. A fundamental model to predict its appearance is not yet available. For design purposes some rules are in use, based on statistical methods. The ones of Bligh (1910) and Lane (1935), are most commonly applied.

It is becoming emphasized nowadays, that these rules are not sufficient for every piping problem met in structures of increasing complexity. A better understanding of the piping mechanism or at least the behaviour of the mechanism is necessary. Therefore, the "Delft Soil Mechanics Laboratory" under the direction of the "Netherlands Water Defenses Research Centre" is carrying out an extensive investigation on the phenomenon of piping.

In the laboratory model tests are run for some typical geometries. Different types of sand are used at various porosities. During a test the discharge and a large number of pore pressures are measured. For the present only the results on homogeneous sand are discussed. Sand mixtures show much more scatter in the results than more homogeneous material. The degree of mixing seems to play an important role, to be studied in a later stage.

The models in the experiments are obviously on a much smaller scale than the prototype situation. However, the zone, where piping is initiated, is relatively small and is present in the model on full scale. Therefore, the interpretation of the results only affects the determination of scale rules for the measured pressure drop when piping occurs. Provided of course, that the zone in the model, where piping is initiated, has identical properties as the one in the prototype situation.

GEOMETRY OF THE MODEL

The model tests are carried out in a tank, filled with sand and covered by clay. In figure 1 a cross section of the geometry is shown. Two tanks of different size are used. The length l is equal to 5.75 m and 1.90 m; the height h 1.50 m and 0.50 m respectively. Both tanks have the same width of 0.50 m. At the upstream side a filter is placed, where the water flows into the model. In order to create a gradient over the whole length of the tank a filtercloth is applied along the downstream face. Its presence does not effect the nature of piping, but extremely interferes with the interpretation of the results. Therefore, it is better being omitted. The width f of the filtercloth is 0.01 m.

The sand in the tank is overlain by a clay layer. In this layer two typical kinds of gap are made in order to create an area, sensitive to piping. One is a ditch over the whole width of the tank; the other a hole in the middle of the width. The distance from the filter to the center of the ditch or hole is denoted by "a". The radius of the hole or half the width of the ditch is indicated by r. The height d of the gap in the clay is only important in case of a hole.

FLOW TOWARDS A DITCH

It is already stated, that the zone, where piping is initiated, is small and present

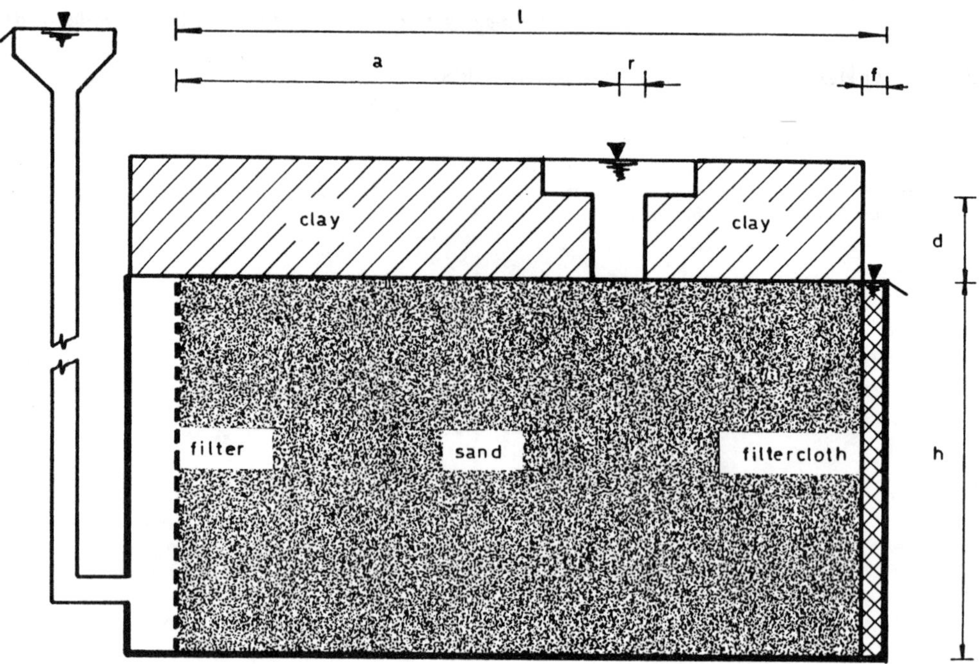

Fig.1: Cross section of the model

on full scale in the model. Therefore in case of flow towards a ditch one should not scale the width of the ditch too severely. Otherwise the parameter r might influence the nature of the piping phenomenon. Thus it is best to take a+r=l, r being sufficiently large. A great number of tests has been carried out in this way. The scaling effect has been checked by comparing the results in the small tank with those in the big one. The scaling procedure is as follows. The steady state flow solutions are determined for the two geometries to be compared. The condition, that the areas, sensitive to piping, are identical, yields a relation between the pressure drops in the two geometries. That means, that the pressure drop during piping in a certain situation may be predicted from model test measurements. The results of the tests and the scaling procedure is reported by Sellmeijer (1980) and De Wit et al. (1981). The agreement between theory and measurements is satisfactory.

At present some tests are run to study the influence of small values of the parameter r in order to quantify the minimum size of a model test. Values of r of 0.05 m and 0.10 m are considered. At the first glance the influence looks not yet very severe

In a later stage more detailed results will be reported.

FLOW TOWARDS A HOLE

The piping phenomenon in case of flow towards a small hole is more complicated than in case of flow towards a ditch. The pressure drop depends not only on the stability of the sand at the hole, but also on the resistance of the fluidized sand in the hole. Therefore, the scaling procedure used by Sellmeijer (1980) is insufficient. This procedure was the same as used in the case of ditches, but with extra scaling factor the radius r. The agreement is more or less casual.

This type of model test is started, while the hole in the clay only contains water. At a certain pressure drop the critical phase at the sand surface is reached. This is noticed by the presence of sand boils and a sudden increase of the discharge through the hole. By De Wit et al. (1981) it is noticed, that the scaling procedure in the case of ditches also may be used for this stage. The radius of holes usually is small so that sand transported by internal erosion will form an extra resistance due to fluidized sand in the hole. That means, that the pressure drop may increase further until the whole hole is filled with fluidized sand.

To measure the behaviour of fluidized sand some additional tests have been run in a perspex tube with a diameter of 0.04 m. Figure 2 shows a sketch of the test setup. The discharge and head at several places along the tube have been measured. As to be expected, the hydraulic gradient is uniform all over the fluidized sand sample. The relation between the specific discharge and the hydraulic gradient is given in figure 3. Q is the discharge, r the radius of the tube; ϕ is the hydraulic head over a certain height d of the tube.

The roughness of the inside wall of the tube has been varied. A smooth wall has been applied, a wall plastered with clay and one plastered with sand. It turned out that the material of the inside wall did not effect the results. The relation between the specific discharge and the hydraulic gradient may be represented by the simple equation,

$$\frac{Q}{\pi r^2 k_0} = \ln \{\alpha \frac{d}{\phi}\} \qquad (1)$$

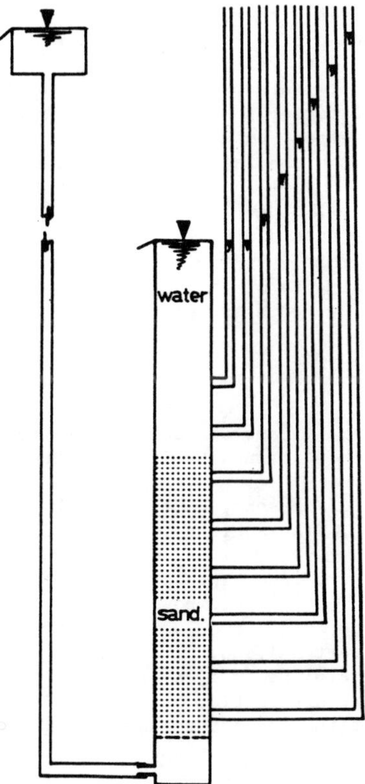

Fig. 2: Setup for the fluidization tests

where k_0 : scale parameter of the specific discharge,

α : scale parameter of the hydraulic gradient.

For the specific sand tested (beach type), k_0 = 9 mm/s and α = 0.9 .

In order to set up a prediction procedure for piping, it will be assumed, that during the initial stage the hole is filled with sand of the same conditions as elsewhere. A moderate pressure drop then will cause flow to and through the hole, to be determined by the steady state flow equation. Increasing values of the pressure drop will lead to fluidization of the sand in the hole. This situation is to be determined as follows. In the field the steady state flow equation is used. In the hole the relation between the specific discharge and the hydraulic gradient (1) is applied. At the interface between field and hole both solutions are linked by the continuity conditions for the head and discharge. Further increase of the pressure drop finally leads to piping along the clay layer.

The appearance of piping now may be predicted in the same way as described in case of flow towards ditches. In the procedure the pressure drop until the hole has to be considered, of course. Therefore it is convenient to measure directly this value by filling the hole only by water in the initial stage. In this way four laboratory tests have been run in the big tank. The diameter of the hole r was varied (0.04 m and 0.10 m) for different values of the distance from the filter to the hole "a", being 2.40 m and 4.50 m.
In De Wit et al. (1981) it is already mentionned, that the prediction method for the internal erosion phase (in the test called boiling phase) is satisfactory.

It is interesting to predict test results in case the hole is filled with sand during the initial stage. Calculations in accordance with the above described method yield the result presented in tabel I. The fluidized sand resistance apparently increases the hydraulic head by about 50%. Unfortunately for this piping phase the measurements during the tests are less suitable. In tests 2 and 4 the diameter of the hole was 0.10 m and the height of the hole 0.12 m. Their ratio was too large, resulting in an accumulation of sand at the downstream side of the hole, while at the upstream side pure water ran off. The extra resistance of fluidized sand therefore could not become very effective.

test	1	2	3	4
measured head in boiling phase	300	375	595	740
predicted head in piping phase	454	548	910	1093

Tabel 1: Predicted hydraulic head in mm.

In tests 1 and 3 the diameter of the hole was 0.04 m. Here the ratio of diameter and height was small enough to assure fluidization all over the hole. Tabel 2 shows, that the measured head during piping agrees well with the predicted one in tabel 1.

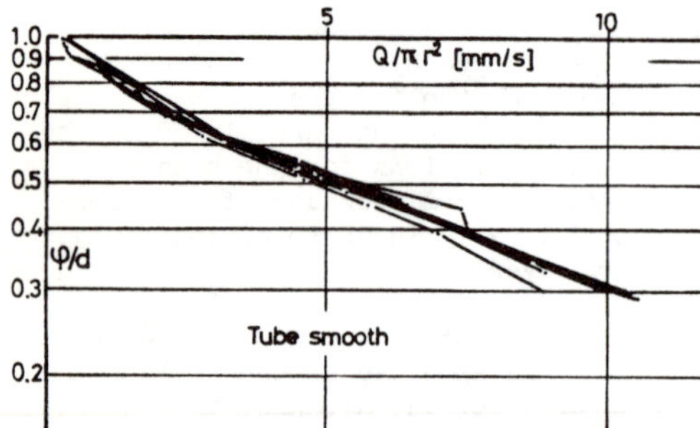

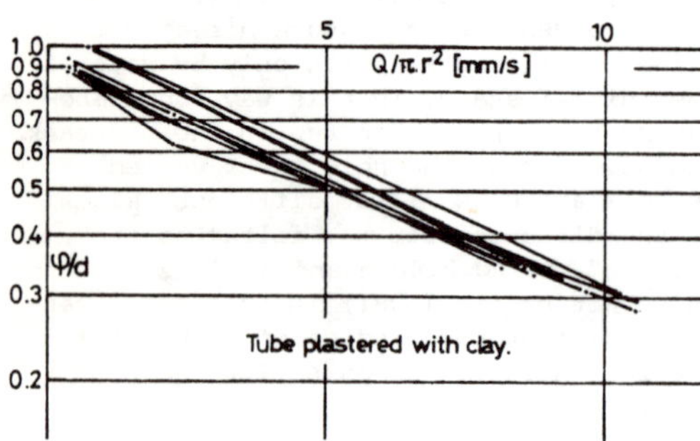

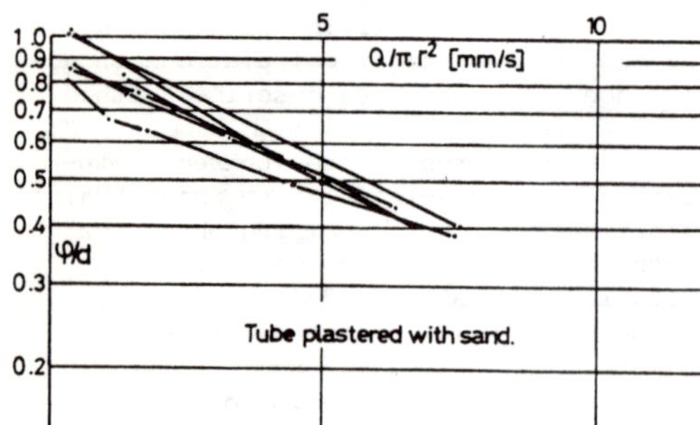

Fig. 3: Relation between specific discharge and pressure gradient during fluidization.

test	1	3
measured head in piping phase	470	860

Tabel 2: Measured hydraulic head in mm.

However, the measured specific discharge is considerably higher than the predicted one. This is quite natural. For, the hole is filled with sand, transported by the internal erosion mechanism. The increase of discharge by the internal erosion is not included in the model. If the specific discharge plays a minor role - it is certainly not controversial to suppose this - then the prediction method may be applied to cases of flow towards holes, independant of the initial stage at the hole. More detailed tests are planned.

CONCLUSIONS

- The phenomenon of piping towards holes may be devided into two mechanisms,
  - internal erosion from the hole in upstream direction;
  - fluidization in the hole, which causes an extra resistance.

  The phenomenon of piping towards ditches contains only the first mechanism.
- A model is presented to predict the appearance of piping towards holes, making use of laboratory tests. The part of it, which concerns the internal erosion, is equal to the model in case of piping towards ditches.
- The measurements during laboratory tests were not entirely sufficient in order to check all aspects of the derived model. More specific tests are planned.

REFERENCES

W.G. Bligh, 1910, Dams Barrages and Weirs on porous Foundations, Engineering News, p. 708;

E.W. Lane, 1935, Masonry Dams on Earth Foundations, Transactions of the ASCE, no. 100;

J.B. Sellmeijer, 1980, Interpretation of Laboratory Tests on Piping, Tribute to Professor de Josselin de Jong, LGM-Mededelingen, Part XXI, No. 2, June;

J.M. de Wit, J.B. Sellmeijer, A. Penning, 1981, Laboratory Testing on Piping, Proceedings of the tenth international Conference on Soil and Foundation Engineering, Part I, p. 517.

2. Multiphase flow

Proceedings of Euromech 143 / Delft / 2-4 September 1981

The simultaneous flow of fresh and salt water in aquifers of large horizontal extension determined by shear flow and vortex theory

G.DE JOSSELIN DE JONG
Retired from University of Delft, Netherlands

SUMMARY

An interface motion equation is derived, taking into account the complete Edelman shear flow conditions and the Dietz-Dupuit approximations. A solution is verified with a result from exact vortex theory.

1 INTRODUCTION

Shear flow

When two fluids (1,2) of different specific weight γ_1, γ_2 respectively, occupying an aquifer, are separated by a sharp interface there exists a *shearflow* at every point of the interface, where it has a tilt angle α with the horizontal. The first description of this phenomenon is by Edelman (1940), who showed that a difference in q_{s1}, q_{s2}, the specific discharge components parallel to the interface, must occur in order to guarantee that along the interface the pressures in the two fluids at either side is equal. A small error, unfortunately, crept into the end of the derivation, leading to the *incorrect* expression

$$[q_{s1}-(\gamma_2/\gamma_1)q_{s2}]=(\kappa/\mu)(\gamma_2-\gamma_1)\sin\alpha \qquad (1.1)$$

where: κ is intrinsic permeability,
μ is dynamic viscosity.

In a later publication, Edelman (1947, pg 59), removes the error and gives the *correct* equation:

$$(q_{s1}-q_{s2})=(\kappa/\mu)(\gamma_2-\gamma_1)\sin\alpha \qquad (1.2)$$

Also Edelman (1940, 1949) mentions, that at every point of the interface, the normal specific discharge components q_{n1}, q_{n2} in the two fluids at either side must be equal in order to satisfy continuity. In formula

$$(q_{n1}-q_{n2}) = 0 \qquad (1.3)$$

Edelman's work has escaped international attention, because it was written in Dutch. Better known is the work of Hubbert (1940), who also describes shear flow at a sharp interface. His treatise starts with a clarifying description of the mechanism of groundwater flow and its relation to pressures, expressed in terms of equilibrium of forces. From page 842 on, however, the analysis is confused by the presupposition, that groundwater flow is always subject to potentials and the treatment continues with a form of Darcy's law, that is uncapable to describe variable density flow, correctly. Applying an apparently rigourous reasoning an equation (192) is derived for the shearflow, similar to (1.1) here. It would be interesting to know, when this was corrected.

At the time it was not recognised, that density differences create rotation in the flow and that therefore a description with potentials is impossible. Lusczynski (1961) attempted to formulate heads, that could serve as potentials. Here, heads are not recommendable for practical use, because they are pseudo-potentials whose gradients are related to the flow only in particular directions. This is shown by Bear (1975) in his equation (9.5.19).

Using the formulation of Darcy's law in terms of pressures (9.5.6), Bear (1975) derives a correct expression for the shearflow (9.5.7) in the case of combined density and viscosity difference. The same flow relations in terms of pressures served as a basis to the vortex theory for variable density flow, de Josselin de Jong (1960). In this and later papers (1977, 1979) it is shown, that rotation exists proportional to the horizontal component of the density gradient and how the specific discharge can be computed in the entire aquifer by locating vortices in the regions of rotation. By this theory it is possible to treat any distribution of densities, also of the gradual transition zones, that exist between miscible fluids because of diffusion and dispersion. The theory is verified in the papers by reducing a gradual transition zone to a sharp interface and establishing the magnitude of the shearflow. The solutions then become singular and produce discontinuities of magnitudes as given by the equations (1.2) and (1.3), here.

Horizontally extended aquifers

Dietz (1953) treats the combined flow of fluids with different properties in elongated, confined aquifers. The analysis starts with the concept of shearflow created by the equality of pressure on either side of a sharp interface. Then the correct values of the shearflow components, parallel to the aquifer boundaries, are attributed with a Dupuit assumption to the specific discharge components parallel to the aquifer in every plane normal to the aquifer. Finally the partial differential equation (21.a) is obtained for the motion of the interface in course of time.

Bear (1975) pg 535 improves the Dietz-Dupuit analysis by removing a small inaccuracy mentioned by Dietz (1953) pg 88. This leads to Bear's eq.(9.5.64) which is similar to Dietz's (21.a) except for a commutation of a number of terms with the second $\partial/\partial x$ of $\partial^2\eta/\partial x^2$. Comparison of the two papers is somewhat impeded, because in Bear's (9.5.64) the numerator between braces is misprinted. The reader can readily perform the correction by reworking the analysis. Such a correction can be verified with Bear's (9.7.20), which is obtained with a similar analysis.

The work of Dietz and also eq.(9.5.64) of Bear describes an aquifer, that is tilted with respect to the horizon, and the coordinate system is parallel and normal to the aquifer.Other, horizontal coordinates are used in Bear's treatment leading to (9.7.20) and that equation is therefore better suited for comparison with relations developed in this paper.

In the analysis of Dietz (1953) and Bear's (1975) version of it, the discontinuous character of the flow at the interface is only accounted for in a direction parallel to the aquifer. It is an objective of this paper to show, how the analysis is changed by considering relations of the kind (1.2) (1.3), such that the character of the discontinuities in the flow, both parallel and normal to the interface, is taken into account. It appears, that f.i. introduction of aquifer anisotropy changes the interface motion equation only slightly. As example the solution for the interface motion is considered, when the interface starts as a straight line.

A second objective of this paper is to compare this approximate solution, with an exact solution obtained by vortex theory. For a confined aquifer this theory requires the introduction of an infinite series of image vortices, which is sometimes believed to impede practical application, when the aquifer is elongated and the interface inclination angle is small, because of the integrations required in the analysis. It will be shown here that for the case of parallel, impervious boundaries, the solution is simplified, because the integrals have closed solutions.

The use of vortex theory is required, when the interface inclination angle is not small, as f.i. in the viscinity of wells, and the Dietz-Dupuit approximation is insufficient. In that case a computer is indispensable for establishing the integrations numerically. A description of the application of vortex theory to computer analysis is given by Haitjema (1977). Other programs are in course of development.

The general objective of this paper is to reconcile the Dietz-Dupuit approximation with the complete Edelman shearflow conditions and vortex theory.

2 FLOW CONDITIONS AT AN INTERFACE

The case is considered here of a sharp interface between fresh and salt water, fluid 1,2 respectively. The specific weights are then $\gamma_1<\gamma_2$ and the viscosities are $\mu_1<\mu_2$. The difference in both properties is of the order of several procents. The intrinsic permeabilities of the anisotropic aquifer are κ_h, κ_v in horizontal and vertical directions respectively, and these are considered to be the principal directions. The fluids are miscible and replacement in the pore space is complete such that κ_h and κ_v are the same for both fluids.

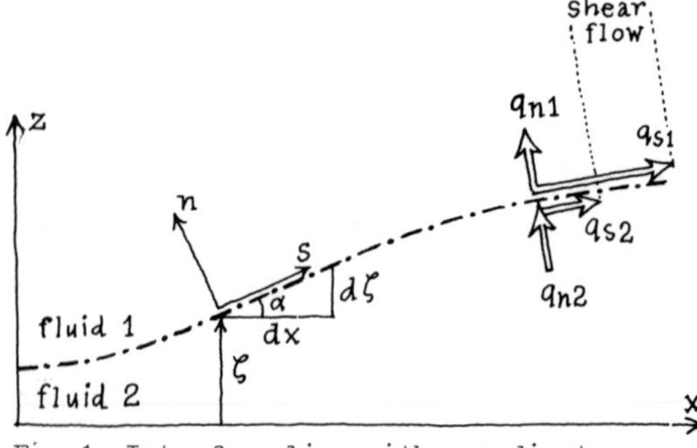

Fig.1. Interface line with coordinates n,s and specific discharges $q_{n1}=q_{n2}$; $q_{s1}>q_{s2}$.

The z-coordinate is vertical, the x,y coordinates are horizontal. The height of the interface is called ζ. So the equation for the interface is $z=\zeta(x,y,t)$, where $\zeta(x,y,t)$ is in general a function of x,y and the time t.

First the two dimensional case is considered that the situations in all planes, y=constant, are equal. The interface has then a cylindrical surface, whose intersection line with the arbitrary z,x plane is called the interface line here. This line is represented by $z=\zeta(x,t)$.

The angle of the interface line with the x-coordinate is α (see fig. 1), and so relations exist of the form

$$\tan\alpha=(\partial\zeta/\partial x) \tag{2.1}$$
$$\sin\alpha=(\partial\zeta/\partial x)/[1+(\partial\zeta/\partial x)^2] \tag{2.2}$$
$$\cos\alpha=1/[1+(\partial\zeta/\partial x)^2] \tag{2.3}$$

At the interface local, orthogonal coordinates n,s respectively normal and along the interface line are considered (fig.1). The coordinates n,s are taken positive, when pointing in positive z,x directions respectively. The components q_n, q_s of the specific discharge in the n,s directions are related to the components q_x, q_z in horizontal and vertical directions by

$$q_n=-q_x\sin\alpha+q_z\cos\alpha \tag{2.4}$$
$$q_s=+q_x\cos\alpha+q_z\sin\alpha \tag{2.5}$$

and the inversion

$$q_x=q_s\cos\alpha-q_n\sin\alpha \tag{2.6}$$
$$q_z=q_s\sin\alpha+q_n\cos\alpha \tag{2.7}$$

Continuity

In order to satisfy *continuity*, the normal specific discharge components q_{n1}, q_{n2} on either side of the interface are equal. This is expressed by (1.3), a relation that is not changed by differences in fluid properties or anisotropy of the pore space. So the normal component can be called q_n and with (2.4) can be written as

$$q_n=q_{n1}=-q_{x1}\sin\alpha+q_{z1}\cos\alpha= \\ =q_{n2}=-q_{x2}\sin\alpha+q_{z2}\cos\alpha \tag{2.8}$$

Multiplication with μ_1, μ_2 respectively gives

$$(\mu_1-\mu_2)q_n=-(\mu_1q_{x1}-\mu_2q_{x2})\sin\alpha+ \\ +(\mu_1q_{z1}-\mu_2q_{z2})\cos\alpha \tag{2.9}$$

Equilibrium

In order to satisfy *equilibrium* the fluid pressures p_1, p_2 in the fluids at either side of the interface are equal, if they are miscible in the manner of fresh and salt water. So the relation exists

$$p_1-p_2=0 \tag{2.10}$$

Since this is true for all points along the interface line, also the derivative of (2.10) with respect to s has to be zero or

$$(\partial p_1/\partial s)-(\partial p_2/\partial s)=0 \tag{2.11}$$

Transforming (2.11) in x,z directions gives

$$[(\partial p_1/\partial x)-(\partial p_2/\partial x)]\cos\alpha+ \\ [(\partial p_1/\partial z)-(\partial p_2/\partial z)]\sin\alpha=0 \tag{2.12}$$

In order to express this relation (2.12) in terms of specific discharge components, the relations between flow and pressure developed from equilibrium of forces are used. These are for the ith fluid:

$$(\mu_i/\kappa_h)q_{xi}=-(\partial p_i/\partial x) \tag{2.13}$$
$$(\mu_i/\kappa_v)q_{zi}=-(\partial p_i/\partial z)-\gamma_i \tag{2.14}$$

Eliminating p_1, p_2 from (2.12),(2.13) and (2.14) gives

$$(\mu_1q_{x1}-\mu_2q_{x2})(\cos\alpha/\kappa_h)+ \\ (\mu_1q_{z1}-\mu_2q_{z2})(\sin\alpha/\kappa_v)=(\gamma_2-\gamma_1)\sin\alpha \tag{2.15}$$

Discontinuities in the flow components

The relations (2.9),(2.15) can be solved to give

$$(\mu_1q_{x1}-\mu_2q_{x2})[(\cos^2\alpha/\kappa_h)+(\sin^2\alpha/\kappa_v)]= \\ =(\gamma_2-\gamma_1)\sin\alpha\cos\alpha-(\mu_1-\mu_2)q_n(\sin\alpha/\kappa_v) \tag{2.16}$$

$$(\mu_1q_{z1}-\mu_2q_{z2})[(\cos^2\alpha/\kappa_h)+(\sin^2\alpha/\kappa_v)]= \\ =(\gamma_2-\gamma_1)\sin^2\alpha+(\mu_1-\mu_2)q_n(\cos\alpha/\kappa_h) \tag{2.17}$$

These relations give an impression of the discontinuities in the specific discharge components at the interface.

The normal specific discharge component q_n is somewhat alien in these relations. In section 3 it turns out, that it is convenient to have q_n converted into a term which contains $\partial\zeta/\partial t$, the vertical velocity of the interface, in the following manner.

When the interface moves upwards with a velocity $\partial\zeta/\partial t$, a surface of area A of the interface plane moves during a time interval dt, through a volume of the aquifer of magnitude $\cos\alpha(\partial\zeta/\partial t)dt.A$. (see fig. 2). The amount of fluid in this aquifervolume is ε times the volume, when ε is the porosity of the aquifer. So the amount of fluid displaced is $\varepsilon\cos\alpha(\partial\zeta/\partial t)dt.A$.

The specific discharge q_n is defined as the volume of fluid passing normal to the interface, through a unit area during unit time. So the amount of fluid passing through the area A during a time dt is $q_n A dt$. Comparing the two volumes gives

$$q_n=\varepsilon\cos\alpha(\partial\zeta/\partial t) \tag{2.18}$$

Using this value for q_n in (2.16),(2.17) and applying (2.2),(2.3) results in

$$(\mu_1 q_{x1}-\mu_2 q_{x2})= [(\gamma_2-\gamma_1)-\varepsilon(\mu_1-\mu_2)\frac{1}{\kappa_v}\frac{\partial\zeta}{\partial t}]\frac{\kappa_h\frac{\partial\zeta}{\partial x}}{1+\frac{\kappa_h}{\kappa_v}(\frac{\partial\zeta}{\partial x})^2} \tag{2.19}$$

$$(\mu_1 q_{z1}-\mu_2 q_{z2})= [(\gamma_2-\gamma_1)+\varepsilon(\mu_1-\mu_2)\frac{\frac{\partial\zeta}{\partial t}}{\kappa_h(\frac{\partial\zeta}{\partial x})^2}]\frac{\kappa_h(\frac{\partial\zeta}{\partial x})^2}{1+\frac{\kappa_h}{\kappa_v}(\frac{\partial\zeta}{\partial x})^2} \tag{2.20}$$

At this point it may be remarked, that the relations (2.19),(2.20) are valid for any value of α in the interval $-\pi/2\leqslant\alpha\leqslant\pi/2$. Beyond that interval the interface is unstable, because heavier fluid is above lighter.

When the viscosities are equal, such that $\mu_1=\mu_2=\mu$, equations (2.16),(2.17) reduce to

$$(q_{x1}-q_{x2})=\frac{(\gamma_2-\gamma_1)\sin\alpha\cos\alpha}{\mu[(\cos^2\alpha/\kappa_h)+(\sin^2\alpha/\kappa_v)]} \tag{2.21}$$

$$(q_{z1}-q_{z2})=\frac{(\gamma_2-\gamma_1)\sin^2\alpha}{\mu[(\cos^2\alpha/\kappa_h)+(\sin^2\alpha/\kappa_v)]} \tag{2.22}$$

Using the tensor character of permeability, it is possible to show that the term $[(\cos^2\alpha/\kappa_h)+(\sin^2\alpha/\kappa_v)]$ represents the reciproval of the intrinsic permeability, κ_s, in s-direction. Using further (2.5) it is found, that

$$(q_{s1}-q_{s2})=(\kappa_s/\mu)(\gamma_2-\gamma_1)\sin\alpha, \tag{2.23}$$

a relation for the shearflow reminding of (1.2).

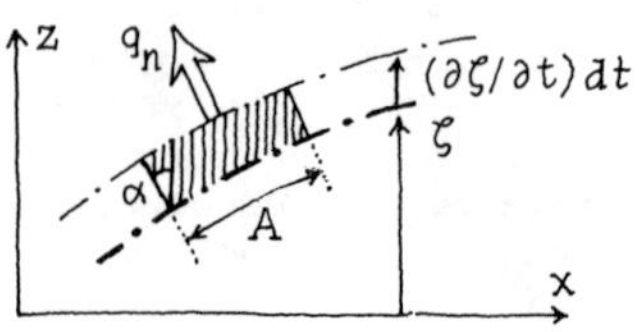

Fig. 2. Interface moving upwards

3. INTERFACE MOTION EQUATION

Continuity in x-direction

Let the aquifer have its upperboundary at $z=Z_1$ and its lower boundary at $z=Z_2$. The height occupied by fluid 1 is $(Z_1-\zeta)$ and for fluid 2 it is $(\zeta-Z_2)$, see fig. 3.

The horizontal specific discharge components q_{x1}, q_{x2} are as an approximation assumed to be constant in a vertical plane. Per unit width the discharges Q_{x1}, Q_{x2} of fluids 1,2 resp. are then

$$Q_{x1}=q_{x1}(Z_1-\zeta) \tag{3.1}$$

$$Q_{x2}=q_{x2}(\zeta-Z_2). \tag{3.2}$$

The total discharge of both fluids together over unit width is called, Q_x. Then we have

$$Q_x=Q_{x1}+Q_{x2}= =q_{x1}(Z_1-\zeta)+q_{x2}(\zeta-Z_2). \tag{3.3}$$

Let there be infiltration of fluid 1 through the upperboundary adding a volume S_1, per unit horizontal area and unit time to the discharge Q_{x1} and simirlarly S_2 to Q_{x2} from below. Continuity requires then, that the total discharge Q_x satisfies the relation

$$(\partial Q_x/\partial x)=S_1+S_2 \tag{3.4}$$

When the interface moves upwards with a velocity $(\partial\zeta/\partial t)$, the volume of fluid 1 added to the discharge Q_{x1} over a unit horizontal area and unit time is $\varepsilon(\partial\zeta/\partial t)$, with ε the aquifer porosity. The same volume is subtracted from Q_{x2}. Continuity of each fluid separately requires

$$(\partial Q_{x1}/\partial x)=S_1+\varepsilon(\partial\zeta/\partial t) \tag{3.5}$$

$$(\partial Q_{x2}/\partial x)=S_2-\varepsilon(\partial\zeta/\partial t) \tag{3.6}$$

Equation in ζ alone

When Q_x, S_1, S_2 are known from boundary conditions, a relation in the unknown variable ζ alone is obtained by eliminating q_{xi}, Q_{xi} from equations (2.19) and (3.1) through (3.6). This was the reason to replace q_n by expression (2.18) involving ζ.

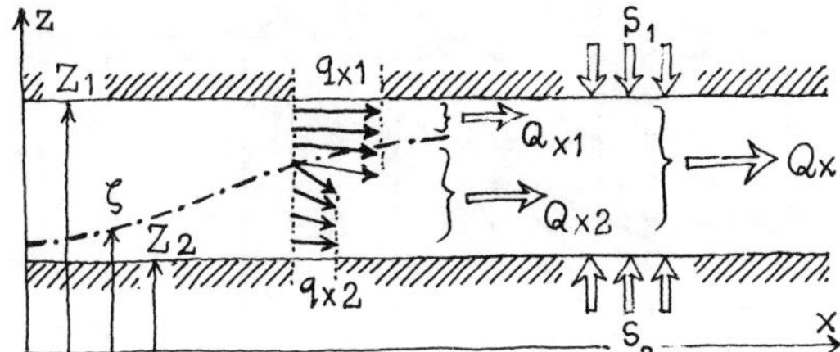

Fig. 3. Aquifer with discharges.

By using the abbreviations

$$\Gamma=(\gamma_2-\gamma_1)\kappa_h/\mu_1 \qquad (3.6)$$

$$\Lambda=\varepsilon[1-(\mu_2/\mu_1)](\kappa_h/\kappa_v)(\partial\zeta/\partial t) \qquad (3.7)$$

$$\Phi=(\kappa_h/\kappa_v)(\partial\zeta/\partial x)^2 \qquad (3.8)$$

equation (2.19) becomes

$$(\mu_1 q_{x1}-\mu_2 q_{x2})=\mu_1(\Gamma-\Lambda)\frac{\partial\zeta}{\partial x}/(1+\Phi) \qquad (3.9)$$

and solving with (3.3) for q_{x1}, q_{x2} gives

$$q_{x1}[\mu_1(\zeta-Z_2)+\mu_2(Z_1-\zeta)]= =\mu_2 Q_x+\mu_1(\Gamma-\Lambda)(\zeta-Z_2)\frac{\partial\zeta}{\partial x}/(1+\Phi) \qquad (3.10)$$

$$q_{x2}[\mu_1(\zeta-Z_2)+\mu_2(Z_1-\zeta)]= =\mu_1 Q_x-\mu_1(\Gamma-\Lambda)(Z_1-\zeta)\frac{\partial\zeta}{\partial x}/(1+\Phi) \qquad (3.11)$$

It is now possible to obtain a relation in ζ alone, either by eliminating q_{x1} and Q_{x1} by (3.10) in (3.1) and using (3.5), or by eliminating q_{x2} and Q_{x2} by (3.11) in (3.2) and using (3.6). The two expressions are similar but not identical. Averaging them gives

$$\varepsilon(\partial\zeta/\partial t)+\tfrac{1}{2}(S_1-S_2)=$$

$$\frac{\partial}{\partial x}\left\{\tfrac{1}{2}Q_x\frac{\mu_2(Z_1-\zeta)-\mu_1(\zeta-Z_2)}{\mu_2(Z_1-\zeta)+\mu_1(\zeta-Z_2)}+\mu_1(\Gamma-\Lambda)\frac{(Z_1-\zeta)(\zeta-Z_2)(\partial\zeta/\partial x)}{[\mu_2(Z_1-\zeta)+\mu_1(\zeta-Z_2)](1+\Phi)}\right\} \qquad (3.12)$$

This is the *interface motion equation*, comparable with Bear's (9.7.20). When the location of the interface is known at a particular moment, ζ is known as a function of x and the upwards motion ∂ζ/∂t of the interface can be determined with (3.12) for every point of it.

In the case of fresh, salt groundwater the difference between μ_1 and μ_2 is small and introducing $\mu_1=\mu_2=\mu$ gives

$$\varepsilon(\partial\zeta/\partial t)+\tfrac{1}{2}(S_1-S_2)=$$

$$\frac{\partial}{\partial x}\left\{Q_x\frac{(Z_1+Z_2-2\zeta)}{2(Z_1-Z_2)}+(\Gamma-\Lambda)\frac{(Z_1-\zeta)(\zeta-Z_2)}{(Z_1-Z_2)(1+\Phi)}\frac{\partial\zeta}{\partial x}\right\} \qquad (3.13)$$

In this last expression Λ is not reduced to zero, since from (3.7) it appears that the difference $[1-(\mu_2/\mu_1)]$ may be overruled by the anisotropy proportion (κ_h/κ_v), which can be quite large in horizontally deposited aquifers. However, since Λ contains (∂ζ/∂t), it is inappropriate to incorporate it in a first determination of (∂ζ/∂t). The procedure is to consider Λ as a correction term, set equal to zero at first and adjusted iteratively afterwards.

The factor Φ in (3.13) also depends on anisotropy, according to (3.8). Here the proportion (κ_h/κ_v) may be large enough to balance the interface inclination factor $(\partial\zeta/\partial x)^2$ and to require that Φ is taken into account from the beginning.

Three dimensional and axial symmetric cases

In the general threedimensional case, the two horizontal coordinates x,y are involved. The total discharge is then a vector $\vec{Q}$ with components Q_x, Q_y. Continuity of the total discharge requires instead of (3.4)

$$\nabla\vec{Q}=S_1+S_2 ,$$

where ∇ stands for (∂/∂x)+(∂/∂y).
The shear flow equations are

$$q_{n1}-q_{n2}=0$$

$$q_{s1}-q_{s2}=\Gamma|\nabla\zeta|(1+|\nabla\zeta|^2)^{\frac{1}{2}}/(1+\frac{\kappa_h}{\kappa_v}|\nabla\zeta|^2)$$

$$q_{t1}-q_{t2}=0 ,$$

where t is the horizontal direction on the interface. The same analysis as above produces the following interface motion equation for $\mu_1=\mu_2=\mu$:

$$\varepsilon(\partial\zeta/\partial t)+\tfrac{1}{2}(S_1-S_2)=$$

$$\nabla\left[\vec{Q}(Z_1+Z_2-2\zeta)/2(Z_1-Z_2)\right]+$$

$$+\Gamma\nabla\left\{\frac{(Z_1-\zeta)(\zeta-Z_2)\nabla\zeta}{(Z_1-Z_2)[1+(\kappa_h/\kappa_v)|\nabla\zeta|^2]}\right\} \qquad (3.14)$$

In the axial symmetric case, with r the horizontal radial coordinate, the interface motion equation, with $S_1=S_2=0$, is

$$\varepsilon(\partial\zeta/\partial t)=(Q_r/2\pi r)\frac{\partial}{\partial r}[(Z_1+Z_2-2\zeta)/2(Z_1-Z_2)]+$$

$$+\Gamma\frac{1}{r}\frac{\partial}{\partial r}\left\{r\frac{(Z_1-\zeta)(\zeta-Z_2)(\partial\zeta/\partial r)}{(Z_1-Z_2)[1+(\kappa_h/\kappa_v)(\partial\zeta/\partial r)^2]}\right\} \qquad (3.15)$$

A difference between the equations here and those developed in literature is in the term containing anisotropy: (κ_h/κ_v).

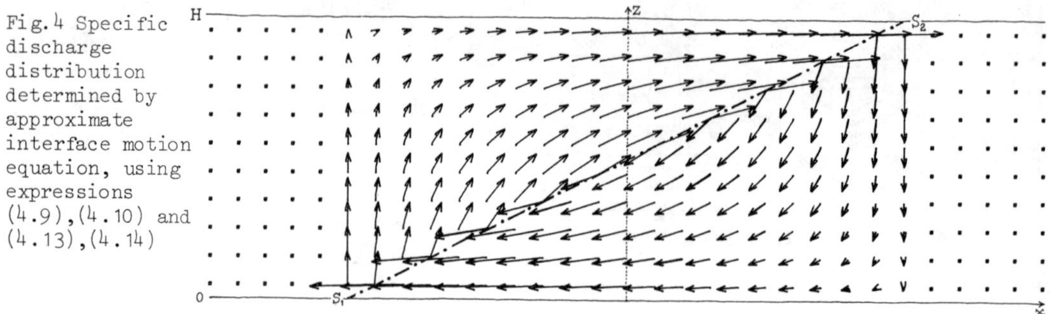

Fig.4 Specific discharge distribution determined by approximate interface motion equation, using expressions (4.9),(4.10) and (4.13),(4.14)

Character of the interface motion equation

The equations (3.12), (3.13), (3.14), (3.15) describing the motion of the interface in different cases, have the character of non-linear diffusion equations with a convection term. The character of solutions satisfying such equations have been studied by Peletier and van Duyn (1977), van Duyn (1979). For a detailed account of the mathematical aspects concerning solutions, the reader is referred to the contribution of van Duyn in these proceedings.

4 EXAMPLE OF A SOLUTION

In order to show a solution of the interface motion equation (3.12), this equation is reduced by considering a simplified case. The aquifer boundaries are at constant heights: $Z_1=H$ and $Z_2=0$. There is neither infiltration, nor convection, so

$$S_1=S_2=Q_x=(\partial Z_1/\partial x)=(\partial Z_2/\partial x)=0 \qquad (4.1)$$

The viscosity differences are small enough to disregard Λ throughout. Then (3.12) is

$$\varepsilon H\frac{\partial\zeta}{\partial t}=\Gamma\frac{\partial}{\partial x}\left\{(H-\zeta)\zeta\frac{\partial\zeta}{\partial x}/\left[1+\frac{\kappa_h}{\kappa_v}\left(\frac{\partial\zeta}{\partial x}\right)^2\right]\right\} \qquad (4.2)$$

The motion is considered of an interface, that at time t is a straight line with inclination angle $\alpha=\text{arc tan}\, f$, such that it is represented by the relation

$$\zeta(t)=fx+\tfrac{1}{2}H \qquad (4.3)$$

Its intersection points S^1, S^2 with the boundaries have coordinates

$$x_{S1}=-(H/2f)\,; \qquad x_{S2}=+(H/2f) \qquad (4.4)$$

For all $x<x_{S1}$ the aquifer is filled with fresh water, with salt water for $x>x_{S2}$.

Since from (4.3), $\partial\zeta/\partial x=f=\text{constant}$ with respect to x, relation (4.2) gives

$$\varepsilon H(\partial\zeta/\partial t)=-2\Gamma f^3x/\left(1+\frac{\kappa_h}{\kappa_v}f^2\right) \qquad (4.5)$$

This result shows, that the increase of the interface height ζ is linear in x and therefore the interface apparently turns as a rigid line. This means, that in equation (4.3) for the interface, the time dependence is only in f, being f(t) a function of t alone. So differentiating (4.3) with respect to t, gives

$$(\partial\zeta/\partial t)=(df/dt)x \qquad (4.6)$$

Combining with (4.5) gives a differential equation in f and t

$$\varepsilon H(df/dt)=-2\Gamma f^3/[1+(\kappa_h/\kappa_v)f^2]. \qquad (4.7)$$

Solving by separation of variables gives

$$\frac{\kappa_h}{2\kappa_v}\ln\frac{f}{f_o}-\frac{1}{4f^2}+\frac{1}{4f_o{}^2}=-\frac{\Gamma}{\varepsilon H}(t-t_o)\,, \qquad (4.8)$$

where integration constants are added, such that at time t_o, f equals f_o. The result (4.8) shows, that the inclination angle α decreases in course of time.

Using (3.10),(3.11) gives, for $\mu_1=\mu_2=\mu$ and $Q_x=\Lambda=0$, the values of q_{x1}, q_{x2} at the interface

$$q_{x1}H=\Gamma(fx+\tfrac{1}{2}H)f/[1+(\kappa_h/\kappa_v)f^2] \qquad (4.9)$$

$$q_{x2}H=\Gamma(fx-\tfrac{1}{2}H)f/[1+(\kappa_h/\kappa_v)f^2] \qquad (4.10)$$

These values are assumed to be constants over every vertical.

In order to find the components q_{z1}, q_{z2} at the interface, (2.18) is combined with (2.8) to give $\varepsilon(\partial\zeta/\partial t)=-fq_{xi}+q_{zi}$ and use of (4.5) gives with (4.9),(4.10), for $z=\zeta$:

$$q_{\zeta 1}H=\Gamma(-fx+\tfrac{1}{2}H)f^2/[1+(\kappa_h/\kappa_v)f^2]. \qquad (4.11)$$

$$q_{\zeta 2}H=\Gamma(-fx-\tfrac{1}{2}H)f^2/[1+(\kappa_h/\kappa_v)f^2] \qquad (4.12)$$

In order to satisfy continuity in the fresh and salt water regions, the components q_{z1}, q_{z2} are distributed linearly over the heights: $\zeta<z<H$ and $0<z<\zeta$ respectively. Using (4.3) it is found that

$$q_{z1}=\frac{H-z}{H-\zeta}q_{\zeta 1}=\Gamma\frac{H-z}{H}f^2/[1+(\kappa_h/\kappa_v)f^2] \qquad (4.13)$$

$$q_{z2}=\frac{z}{\zeta}q_{\zeta 2}=\Gamma\frac{-z}{H}f^2/[1+(\kappa_h/\kappa_v)f^2] \qquad (4.14)$$

The relations (4.9),(4.10) and (4.13)(4.14) were used to construct fig. 4.

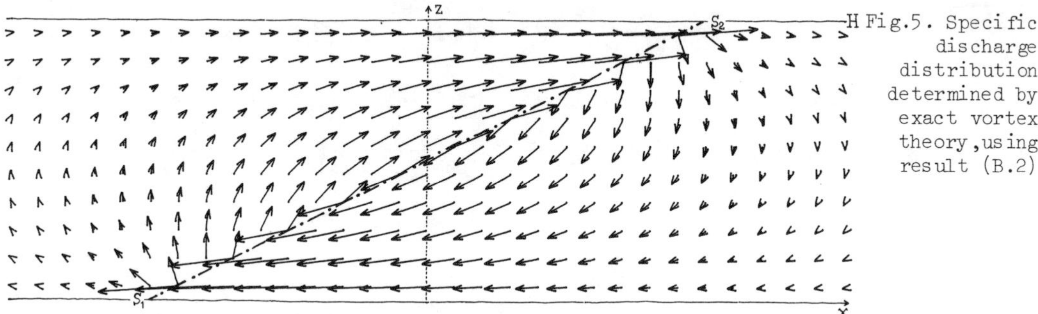

Fig.5. Specific discharge distribution determined by exact vortex theory, using result (B.2)

EXACT SOLUTION BY VORTEX THEORY

Vortices in confined aquifer

According to vortex theory the specific discharge distribution, associated with a section ds at a point S of an inclined interface, can be determined by locating a vortex of strength Γdz_s in S, with dz_s the vertical height difference spanned by ds. The vortex has a horizontal axis parallel to the interface and creates a rotation, that turns the interface towards the horizontal.

Each vortex contributes to a stream function Ψ, whose derivatives are the specific discharge components

$q_x=-(\partial\Psi/\partial z)$; $q_z=+(\partial\Psi/\partial x)$ (5.1)

The contribution of the vortex in ds to Ψ is in two dimensions:

$(\Gamma/2\pi)dz_s \ln(r_{ps})$ (5.2)

In this expression r_{ps} is the distance, between P and S, defined by $r_{ps}=[(x_p-x_s)^2+(z_p-z_s)^2]^{\frac{1}{2}}$, with x_s, z_s the coordinates of the point S, where the vortex is located, and x_p, z_p the coordinates of a point P, where the discharge is considered. The contribution (5.2) to Ψ is reminding of the contribution to a potential Φ by a two-dimensional well of magnitude Γdz_s.

In a confined aquifer, with impermeable boundaries at z=0 and z=H, an infinite row of image vortices has to be introduced in the manner of image wells. Using a result of Muskat (1937), sect 9.8 concerning an infinite row of wells, each vortex contributes to Ψ by an amount $d\Psi$ given by

$$d\Psi=-\frac{\Gamma}{2\pi}\cdot\tfrac{1}{2}\ln\frac{\cosh(X_p-X_s)-\cos(Z_p-Z_s)}{\cosh(X_p-X_s)-\cos(Z_p+Z_s)}\,dz_s \quad (5.3)$$

where $X_j=\pi x_j/H$; $Z_j=\pi z_j/H$ (5.4)

An interface element between S^1 and S^2 requires integration of (5.3) over z_s from $z_s{}^1$ to $z_s{}^2$. Using Euler's relation to modify (5.3), gives for Ψ the value (A.1) in table A. Since x_s is a function of z_s integration of (A.1) is difficult. In order to simplify integration, the discharges in P are combined as follows

$q_{xp}-iq_{zp}=-(\partial\Psi/\partial z_p)-i(\partial\Psi/\partial x_p)$ (5.5)

Using (A.1) this gives (A.2). Subtracting 1 from the first term between braces and adding 1 to the second term, (A.3) is produced.

Let the interface interval between S^1 and S^2 have the inclination angle α=arctanf. Then the infinitesimal distance dx_s along the interface equals dz_s/f and using (5.4) it is possible to write dz_s in (A.3) as

$dz_s=(H/\pi)(dX_s\pm idZ_s)(f/1\pm if)$ (5.6)

Using this in (A.3) permits integration producing the result (A.4).

The right side of (A.4) is separated in a real and an imaginary part by applying Euler's equation twice. This gives (B.1) in Table B, using abbreviations R_{ps}, θ_{ps}, ξ_{ps}, η_{ps} defined there. The result (B.1) contains the values for q_x and q_z in a point P, separately and in terms of real quantities. For a straight section of the interface, the value of f is constant and q_x, q_z are found by introducing the integration limit points S^1, S^2, with coordinates

TABLE A

$$\Psi=-\frac{\Gamma}{4\pi}\int_{S^1}^{S^2}\ln\frac{[e^{(X_p+iZ_p)}-e^{(X_s+iZ_s)}][e^{(X_p-iZ_p)}-e^{(X_s-iZ_s)}]}{[e^{(X_p+iZ_p)}-e^{(X_s-iZ_s)}][e^{(X_p-iZ_p)}-e^{(X_s+iZ_s)}]}dz_s \quad \text{(A.1)}$$

$$q_{xp}-iq_{zp}=\frac{i\Gamma}{2H}\int_{S^1}^{S^2}\left\{[e^{(X_p+iZ_p)}/(e^{(X_p+iZ_p)}-e^{(X_s+iZ_s)})]-[e^{(X_p+iZ_p)}/(e^{(X_p+iZ_p)}-e^{(X_s-iZ_s)})]\right\}dz_s \quad \text{(A.2)}$$

$$q_{xp}-iq_{zp}=\frac{i\Gamma}{2H}\int_{S^1}^{S^2}\left\{[e^{(X_s+iZ_s)}/(e^{(X_p+iZ_p)}-e^{(X_s+iZ_s)})]-[e^{(X_s-iZ_s)}/(e^{(X_p+iZ_p)}-e^{(X_s-iZ_s)})]\right\}dz_s \quad \text{(A.3)}$$

$$q_{xp}-iq_{zp}=\frac{\Gamma}{2\pi}\left\{-\frac{if}{1+if}\ln[e^{(X_p+iZ_p)}-e^{(X_s+iZ_s)}]+\frac{if}{1-if}\ln[e^{(X_p+iZ_p)}-e^{(X_s-iZ_s)}]\right\}_{S^1}^{S^2} \quad \text{(A.4)}$$

TABLE B $R_{ps}=[\xi_{ps}^2+\eta_{ps}^2]^{\frac{1}{2}}$; $R^*_{ps}=[\xi_{ps}^2+\eta^{*2}_{ps}]^{\frac{1}{2}}$; $\theta_{ps}=\arctan(\eta_{ps}/\xi_{ps})$; $\theta^*_{ps}=\arctan(\eta^*_{ps}/\xi_{ps})$

with

$\xi_{ps}=(e^{X_p}\cos Z_p-e^{X_s}\cos Z_s)$; $\eta_{ps}=(e^{X_p}\sin Z_p-e^{X_s}\sin Z_s)$; $\eta^*_{ps}=(e^{X_p}\sin Z_p+e^{X_s}\sin Z_s)$

$$q_{xp}-iq_{zp}=\frac{\Gamma}{2\pi}\frac{f}{(1+f^2)}\left\{-(i+f)\ln R_{ps}+(i-f)\ln R^*_{ps}+(1-if)\theta_{ps}-(1+if)\theta^*_{ps}\right\}_{S1}^{S2} \qquad \text{(B.1)}$$

$$q_{xp}-iq_{zp}=(\Gamma/\pi)(f/1+f^2)\left\{-f\ln(R_{ps}2/R_{ps}1)-if\theta_{ps}2+if\theta_{ps}1+\langle(1-if)\pi\rangle\right\} \qquad \text{(B.2)}$$

$$q_{xp}-iq_{zp}=(\Gamma/H)(f/1+f^2)\left\{-\tfrac{1}{2}H+fx_p+ifz_p+\langle(1-if)H\rangle\right\} \qquad \text{(B.3)}$$

---

x_s1, z_s1 and x_s2, z_s2 the result (B.1) is comparable to Haitjema's (1977) equation (9) derived for a straight interface element AB in an infinite aquifer. The procedure for combining such elements to analyse a curved interface, described by Haitjema, can be applied similarly to (B.1).

In (B.1) angles θ are involved, defined by arcs of tangents. These angles have to be chosen in the interval $0<\theta<\pi$, when θ^*_{ps} is concerned and θ_{ps}, when P is in fluid 2. When P is in fluid 1, and y_p is between y_s1 and y_s2, the value of $(\theta_{ps}2-\theta_{ps}1)$ has to be increased with 2π. For a justification see De Josselin de Jong (1960), p3750-3753.

Comparison with example of section 4

In the example the interface is a straight line with limit points S^1S^2 on the boundaries, such that $z_{s1}=0$, $z_s2=H$ and x_{s1}, x_s2 given by (4.4) The boundary values for ξ_{ps} and η_{ps} are then:

$$\xi_{ps}1=e^{X_p}\cos Z_p-e^{-\frac{1}{2}\pi/f}$$
$$\xi_{ps}2=e^{X_p}\cos Z_p+e^{\frac{1}{2}\pi/f} \qquad (5.7)$$
$$\eta_{ps}1=\eta^*_{ps}1=\eta_{ps}2=\eta^*_{ps}2=e^{X_p}\sin Z_p$$

Because of the identity of the η's, (B.1) reduces to (B.2), where the term between $\langle\rangle$ has to be added, if fluid 1 is concerned. The values of (B.2) were used to construct the specific discharge distribution of fig. 5.

When the point of consideration P is located in the region $x_{s1}<x_p<x_s2$ such that

$$e^{-\frac{1}{2}\pi/f}\ll e^{X_p}\ll e^{\frac{1}{2}\pi/f}, \qquad (5.8)$$

the variables can be approximated by

$R_{ps}2\approx e^{\frac{1}{2}\pi/f}$; $R_{ps}1\approx e^{X_p}$; $\theta_{ps}2\approx 0$; $\theta_{ps}1\approx Z_p$

and using (5.4), (B.2) reduces to (B.3).

Separation of real and imaginary parts shows, that the result (B.3) gives the same values as (4.9), (4.10) and (4.13),(4.14), when $\kappa_h=\kappa_v$ and the meaning of the brackets $\langle\rangle$ is taken into account. In the case of anisotropy, such that $\kappa_h\neq\kappa_v$, the vortex theory is similar, only the z-coordinate has to be multiplied by a factor $(\kappa_h/\kappa_v)^{\frac{1}{2}}$ throughout. The only consequence is, that the factor $(1+f^2)$ in (B.3) is replaced by $[1+(\kappa_h/\kappa_v)f^2]$.

REFERENCES

Bear,J. 1975, Dynamics of fluids in porous media, Book 2nd ed.,Am.Elsevier Publ.

Dietz,D.N. 1953, A theoretical approach to the problem of encroaching and bypasing edgewater. Proc.Kon.Ned.Academie van Wetenschappen, series B,56, no.1,83-92.

Van Duyn,C.J. & L.A. Peletier, 1977, A class of similarity solutions of the nonlinear diffusion equation. Nonlinear Analysis, Theory, Methods and Applications, 1, 223-233.

Van Duyn,C.J. 1979, Nonlinear diffusion problems, Dr. Thesis, Leiden.

Edelman,J.H. 1940, Strooming van zoet en zout grondwater, Rapport 1940 inzake de watervoorziening van Amsterdam. Bijlage 2 8-14.

Edelman,J.H. 1947, Over de berekening van grondwaterstromingen, Dr. Thesis, Delft.

Haitjema,H.M., 1977. Numerical application of vortices to multiple fluid flow in porous media. Delft Progr.Rep.2,237-248.

Hubbert,M.K. 1940, The theory of groundwater flow, vol XLVIII, 8, 785-944.

De Josselin de Jong, G. 1960, Singularity distributions for the analysis of multiple fluid flow through porous media, J.Gephys.Res., vol.65, 11, 3739-3758.

De Josselin de Jong, G. 1977, Review of vortex theory for multiple fluid flow, Delft Progr.Rep. 2, 225-236.

de Josselin de Jong, G. 1979, Vortex theory for multiple fluid flow in three dimensions, Delft Progr.Rep. 4, 87-102.

Muskat, M. 1937, The flow of homogeneous fluids through porous media, Book, Mc Graw Hill.

Proceedings of Euromech 143 / Delft / 2-4 September 1981

Some fundamental properties of the simultaneous flow of fresh and salt groundwater in horizontally extended aquifers

C.J.VAN DUYN
Delft Soil Mechanics Laboratory, Netherlands

1. INTRODUCTION

In this paper we shall discuss some fundamental properties of solutions of a nonlinear partial differential equation which arises in the study of the simultaneous flow of fresh and salt groundwater in horizontally extended aquifers. This differential equation is an approximate version of the interface motion equation derived by de Josselin de Jong in the previous paper (equation 3.12). There he gives a general treatment of the movement of fresh and salt groundwater and he obtains an equation describing the interface between the two fluids as a function of position and time for the case that the horizontal flow component is constant over the height.

Here we shall make a number of additional assumptions in order to simplify the mathematical description. Using de Josselin de Jong's notation, we assume through out this paper that

$\mu_1 = \mu_2$;

$S_1 = S_2 = 0$;

$Z_1 = H$ (constant), $Z_2 = 0$;

$(\kappa_h/\kappa_v)(\partial\zeta/\partial x)^2 << 1.$

Then the interface motion equation in the one dimensional case reduces to

$$\varepsilon\frac{\partial\zeta}{\partial t} + \frac{Q_x}{H}\frac{\partial\zeta}{\partial x} = \frac{\Gamma}{H}\frac{\partial}{\partial x}\left(\zeta(H-\zeta)\frac{\partial\zeta}{\partial x}\right). \qquad (1.1)$$

This equation is similar to equations derived by Bear(1972) and, at a much earlier date, by Dietz(1953) who considered the case of a water-oil interface.

Given the first three assumptions, de Josselin de Jong's equation (3.12) reduces to his equation (4.2). This is a nonlinear convection-diffusion equation with a diffusion coefficient of the form: $\zeta(H-\zeta)/\{1+(\kappa_h/\kappa_v)(\partial\zeta/\partial x)^2\}$.

This equation is new and, to the author's best knowledge, the theory for it has not yet been developed. However if we neglect $(\kappa_h/\kappa_v)(\partial\zeta/\partial x)^2$ in the denominator, then (4.2) reduces to equation (1.1) of this paper, which is of a familiar type and therefore its solutions can be studied here. In a forthcoming paper we shall drop the fourth restriction and study the behaviour of solutions of the complete equation (4.2).

The diffusion coefficient of equation (1.1) has the form $\zeta(H-\zeta)$. Thus at points in the aquifer where $\zeta = 0$ (only fresh water) or $\zeta = H$ (only salt water) the diffusion coefficient vanishes and (1.1) reduces to a first order differential equation. We say that at those points the equation degenerates. It can be shown, that because of this degeneracy, singularities in the form of jumps in derivatives of solutions of (1.1) occur: see van Duyn & Peletier (1977), van Duyn(1979). In section 2 an explicit solution of equation (1.1) is constructed which shows precisely the above mentioned behaviour.

We shall consider the initial value problem of equation (1.1). Thus given a fresh-salt water distribution at some time $t = t_o$, we are interested to know how the interface evolves for $t > t_o$. To deal with this question, the following plan is carried out. Section 2: An exact solution of the one dimensional interface motion equation (1.1) is given in the form of a similarity solution. Section 3: Some of the general properties of solutions of equation (1.1) are discussed and compared with the exact solution from Section 2. In particular the asymptotic behaviour for large time ($t \to \infty$) is investigated. Section 4: The two dimensional axial symmetric case is considered here. Using a shooting method, the numerical solution of a two dimensional initial value problem is constructed.

2. AN EXACT SOLUTION

In order to get some notion of the behaviour of solutions of equation (1.1), we give in this section an example of an explicit solution. To derive this solution we tranform equation (1.1) three times. First we define new variables in equation (1.1) that absorb the physical constants. Let

$$s = (x/2)\sqrt{(\varepsilon/(\Gamma H))}, \quad u = \zeta/H,$$

and

$$\lambda = Q_x/\{2H\sqrt{(\varepsilon\Gamma H)}\}.$$

Then (1.1) becomes:

$$\frac{\partial u}{\partial t} + \lambda\frac{\partial u}{\partial s} = \frac{\partial}{\partial s}\{\frac{1}{4}u(1-u)\frac{\partial u}{\partial s}\}, \tag{2.1}$$

in the domain $-\infty<s<+\infty$ and $t>0$.
Since in general Q_x is still a function of t, we have in (2.1) $\lambda = \lambda(t)$.
Next we choose new independent variables such that the convection term in (2.1) disappears. Set

$$z = s - \int_0^t \lambda(p)dp,$$

and

$$v(z,t) = v(s-\int_0^t \lambda(p)dp, t) = u(s,t).$$

Then v satisfies the nonlinear diffusion equation

$$\frac{\partial v}{\partial t} = \frac{\partial}{\partial z}\{\frac{1}{4}v(1-v)\frac{\partial v}{\partial z}\}, \tag{2.2}$$

in the domain $-\infty<z<+\infty$ and $t>0$.
Finally we reduce the number of independent variables in equation (2.2) by introducing a similarity variable. Set

$$\eta = z/\sqrt{(t+a)},$$

where a is some positive constant, and look for a similarity solution f which only depends on the single variable η: $v(z,t) = f(\eta)$. Then f satisfies the ordinary differential equation:

$$\frac{d}{d\eta}\{\frac{1}{4}f(1-f)\frac{df}{d\eta}\} + \frac{1}{2}\eta\frac{df}{d\eta} = 0, \tag{2.3}$$

for $-\infty<\eta<\infty$.

To solve this equation we set at the boundaries:

$$f(-\infty) = 0 \text{ and } f(+\infty) = 1.$$

The solution of this boundary value problem is given by Philip(1959) and has the form:

$$f(\eta) = \begin{cases} 0 & \text{for } -\infty<\eta\leq-\frac{1}{2}, \\ \frac{1}{2}+\eta & \text{for } -\frac{1}{2}<\eta<\frac{1}{2}, \\ 1 & \text{for } \frac{1}{2}\leq\eta<\infty. \end{cases} \tag{2.4}$$

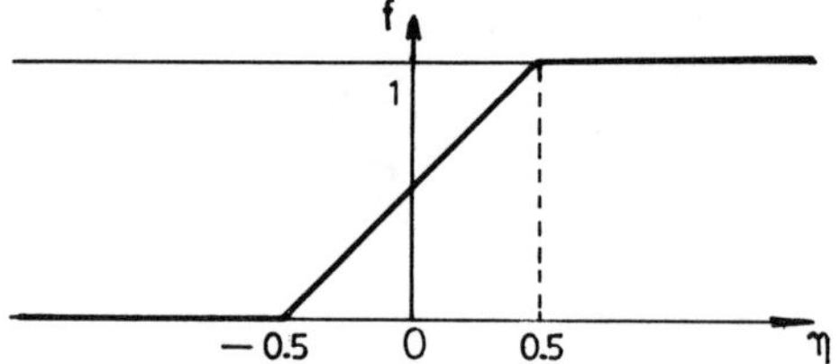

Fig. 2.1. Similarity solution (2.4).

This solution defines two intersection points: $\eta = -\frac{1}{2}$ with the bottom of the aquifer and $\eta = +\frac{1}{2}$ with the top of the aquifer and f varies linear with η in between. Observe that f is a smooth function when $0<f<1$, but its derivative has a jump when $f\downarrow 0$ and $f\uparrow 1$ (Fig. 2.1).
In terms of the original variables x and t, we find that the intersection points become curves in the (x,t)-plane. The point $\eta = -\frac{1}{2}$ corresponds to

$$x = -\sqrt{\{\Gamma H(t+a)/\varepsilon\}} + (\int_0^t Q_x(p)dp)/(\varepsilon H) \underline{\underline{\text{def}}}\ S_f(t), \tag{2.5}$$

and the point $\eta = +\frac{1}{2}$ corresponds to:

$$x = +\sqrt{\{\Gamma H(t+a)/\varepsilon\}} + (\int_0^t Q_x(p)dp)/(\varepsilon H) \underline{\underline{\text{def}}}\ S_s(t). \tag{2.6}$$

The curves $S_f(t)$ and $S_s(t)$ form the boundary of a region in the (x,t)-plane where both fresh and salt water are present. For $x<S_f(t)$ only fresh water is present and for $x>S_s(t)$ only salt water is present (Fig. 2.2 and 2.3).
Since f is linear in η between the intersection points, we find that the corresponding interface, denoted by ζ_p, is linear in x between the curves $S_f(t)$ and $S_s(t)$ such that

$$\zeta_p(x,t) = 0 \text{ for } x \leq S_f(t),$$

and

$$\zeta_p(x,t) = H \text{ for } x \geq S_s(t).$$

Fig. 2.2 and 2.3 show that the position of the intersection points of an interface with the top and the bottom of the aquifer vary in time. The velocity at which they travel can be calculated directly from (2.5) and (2.6).

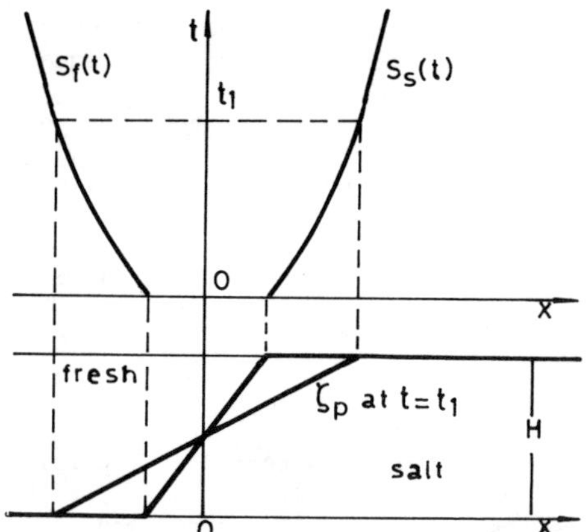

Fig. 2.2. Salt-fresh distribution: $Q_x=0$.

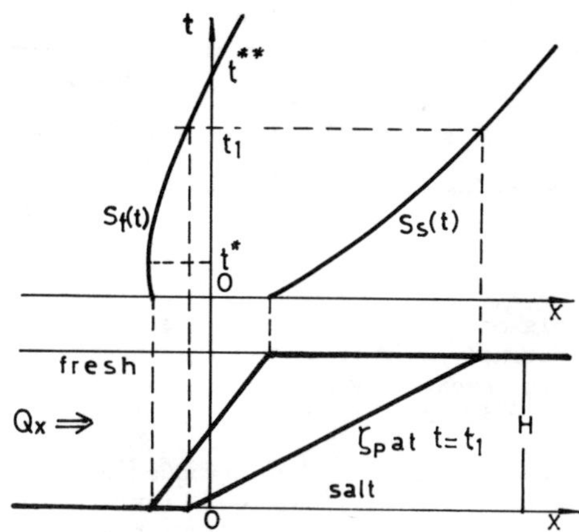

Fig. 2.3. Salt-fresh distribution: $Q_x > 0$.

We find:

$$\frac{dS_f(t)}{dt} = -\tfrac{1}{2}\sqrt{\{\Gamma H/((t+a)\varepsilon)\}} + Q_x(t)/(\varepsilon H), \qquad (2.7)$$

for the bottom point, and

$$\frac{dS_s(t)}{dt} = +\tfrac{1}{2}\sqrt{\{\Gamma H/((t+a)\varepsilon)\}} + Q_x(t)/(\varepsilon H), \qquad (2.8)$$

for the top point.

When $Q_x = 0$ (Fig. 2.2), we observe that $S_f(t)$ is a decreasing function of t and $S_s(t)$ increases with t. In the case that Q_x is positive and constant (Fig. 2.3), $S_f(t)$ reaches a minimum at

$$t^* = H^3\Gamma/(4Q_x^2\varepsilon) - a,$$

and it becomes positive for

$$t > t^{**} = \Gamma H^3\varepsilon/(2Q_x^2) + \tfrac{1}{2}\sqrt{\{\Gamma^2H^6\varepsilon^2/Q_x^4 + 4\Gamma H^3\varepsilon a/Q_x^2\}}.$$

In both cases the solutions correspond to the initial distribution:

$$\zeta_0(x)=\begin{cases} 0 & \text{for } -\infty<x\leq S_f(0), \\ H(x-S_f(0))/(S_s(0)-S_f(0)) & \text{for } S_f(0)<x<S_s(0), \\ H & \text{for } S_s(0)\leq x<S_f(0). \end{cases} \qquad (2.9)$$

Strictly speaking, the above discussed solution $\zeta = \zeta_p$ does not satisfy equation (1.1) at the points $S_f(t)$ and $S_s(t)$, since there the derivative with respect to x has a jump. For instance at the point $x = S_s(t)$,

$$\lim_{x\uparrow S_s(t)} \frac{\partial\zeta_p(x,t)}{\partial x} = \tfrac{1}{2}\sqrt{\{H\varepsilon/((t+a)\Gamma)\}}, \qquad (2.10)$$

and

$$\lim_{x\downarrow S_s(t)} \frac{\partial\xi_p(x,t)}{\partial x} = 0.$$

Therefore we call ζ_p a weak solution, which satisfies equation (1.1) almost everywhere in the domain $-\infty<x<\infty$, $t>0$.
Finally we observe that $\zeta_p(x,t)$ becomes flatter as time increases. Thus if initially $(\kappa_h/\kappa_v)(\partial\zeta_0/\partial x)^2<<1$, then this inequality will hold for all $t>0$.

3. SOME GENERAL PROPERTIES

In this section we shall discuss some properties and mathematical peculiarities of solutions of the initial value problem of equation (1.1). Since (1.1) can always be transformed into the nonlinear diffusion equation (2.2) by eliminating the convection term, we restrict ourselves to the initial value problem

$$\frac{\partial v}{\partial t} = \frac{\partial}{\partial z}\{\frac{1}{4}v(1-v)\frac{\partial v}{\partial z}\}, \qquad (3.1)$$

in the domain $-\infty<z<\infty$, $t>0$ and

$$v(z,0) = v_0(z), \qquad (3.2)$$

for $-\infty<z<\infty$.

With respect to the initial value $v_0(z)$, we suppose that it vanishes for large negative values of z and that it is 1 for large positive values of z, Fig. 3.1.

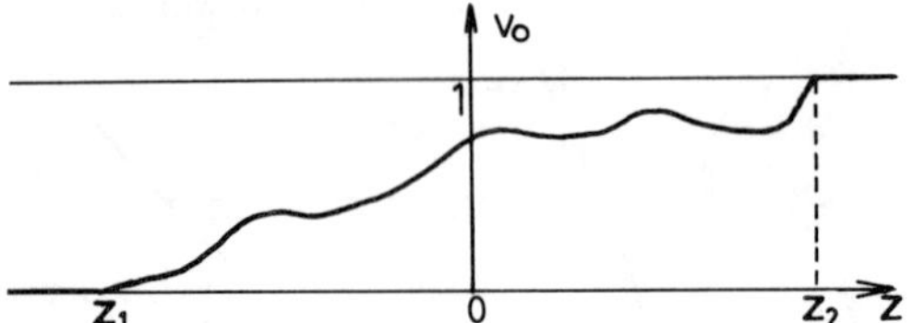

Fig. 3.1. Example of an initial value (3.2)

This means that we consider situations where initially only fresh water is present for large negative values of x and only salt water is present for large positive values of x.

As mentioned in the introduction, the mathematical difficulties which arise when studying problem (3.1), (3.2) stem from the fact that equation (3.1) is degenerate parabolic: at points where $0<v(z,t)<1$ the diffusion coefficient is positive and (3.1) is parabolic, but at points where $v = 0$ or $v = 1$, the diffusion coefficient vanishes and (3.1) reduces to a first order differential equation. As we can see from the example in Section 2, this leads to singularities in the form of jumps in the derivatives of the solutions.

In the mathematical literature, a certain class of degenerate parabolic equations has extensively been studied. These equations are nonlinear diffusion equations in which the diffusion coefficient vanishes only at $v = 0$. A well-known example of an equation from this class is the so called 'Porous Media Equation'

$$\frac{\partial v}{\partial t} = \frac{\partial^2}{\partial z^2}(v^m), \quad m > 1. \tag{3.3}$$

Here the diffusion coefficient is $m v^{m-1}$ and thus (3.3) degenerates at points where $v = 0$. We know now how to define a class of solutions of equations like (3.3), including solutions which have singularities in the form of discontinuous derivatives. Then using the mathematical theory developed for these equations, one can prove that the solutions have properties which one expects on physical grounds: Aronson (1969, 1970), Peletier (1971), Knerr (1977), Caffarelli & Friedman (1979).

With respect to equation (3.1) we have the following situation. Van Duyn & Peletier (1977) studied a special class of solutions (similarity solutions) and van Duyn (1979) considered the large time behaviour of solutions of the initial value problem (3.1), (3.2). Moreover, since for values of v close to zero

$\frac{1}{4}v(1-v) \approx \frac{1}{4}v$, it can be shown that solutions of (3.1) behave the same as solutions of (3.3) for $m = 2$ and $v \to 0$. This also applies to the case where $v \to 1$, which can be seen by making the transformation $\bar{v} = 1-v$ in (3.1) and by comparing the equation for $\bar{v}$ with (3.3).

Therefore many of the methods and results obtained for solutions of equation (3.3), with $m = 2$, are applicable to solutions of the initial value problem (3.1), (3.2). In what follows we shall discuss some of these results and the reader is referred to the above mentioned authors for the detailed mathematical proofs.

3.1 Speed of propagation

In Section 2 we constructed an explicit solution of problem (3.1), (3.2) for the case that $v_0(z)$ is the transformed of (2.9). Given this solution, we could define two functions, $S_f(t)$ and $S_s(t)$, which form the boundaries of a region in the (x,t)-plane where both the fluids (salt and fresh) are present (Fig. 2.2 and 2.3). As mentioned there, these curves correspond to the intersection points of the interface with the bottom and the top of the aquifer. Their speed of propagation could be calculated and is given by (2.7) and (2.8).

Here we are dealing with an arbitrary initial value $v_0(z)$ as shown in Fig. 3.1. Therefore we cannot give an explicit expression for the position of the intersection points as a function of time. However, one can prove a number of interesting facts about these curves.

Let $\bar{\theta}_f(t)$ and $\bar{\theta}_s(t)$ denote the position of the intersection points of v with 0 and 1 at time t. Then $\bar{\theta}_f$ and $\bar{\theta}_s$ form continuous curves with the property that

$$v(z,t) = 0 \text{ for } -\infty<z\leq\bar{\theta}_f(t),\ t>0,$$

$$0<v(z,t)<1 \text{ for } \bar{\theta}_f(t)<z<\bar{\theta}_s(t),\ t>0,$$

$$v(z,t) = 1 \text{ for } \bar{\theta}_s(t)\leq z<\infty.$$

Now dependent on the form of the initial value near the points z_1 and z_2, different situations can occur.

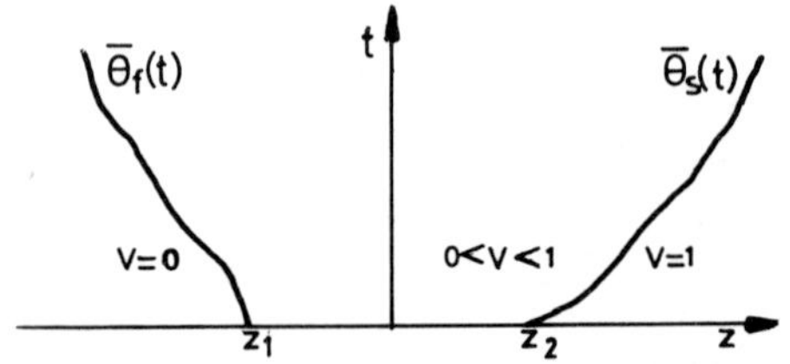

Fig. 3.2. Solution with moving boundaries.

If $dv_0(z_1)/dz > 0$ and $dv_0(z_2)/dz > 0$, as is the case in Fig. 3.1, then a subdivision of the (z,t)-plane as in Fig. 3.2 is found. The functions $\bar\theta_f(t)$ and $\bar\theta_s(t)$ are continuously differentiable for $t > 0$ and $d\bar\theta_f(t)/dt < 0$ and $d\bar\theta_s(t)dt > 0$. These derivatives, which represent the propagation speed at which the intersection points move along the z - axis, can be shown to satisfy the equations

$$-\frac{d\bar\theta_f(t)}{dt} = \frac{1}{4}\lim_{z\downarrow\bar\theta_f(t)} \frac{\partial v(z,t)}{\partial z}, \qquad (3.4)$$

and

$$+\frac{d\bar\theta_s(t)}{dt} = \frac{1}{4}\lim_{z\uparrow\bar\theta_s(t)} \frac{\partial v(z,t)}{\partial z}. \qquad (3.5)$$

Note that the above mentioned situation also occurs in the example form Section 2.

If on the other hand $v_0 \to 0$ and $v_0 \to 1$ sufficiently fast (more precisely: if

$v_0(z) \leq C|z-z_1|^2$ and $v_0(z) \geq 1-C|z-z_2|^2$,

with C some positive constant), then a subdivision of the (z,t)-plane as in Fig. 3.3 will arise.

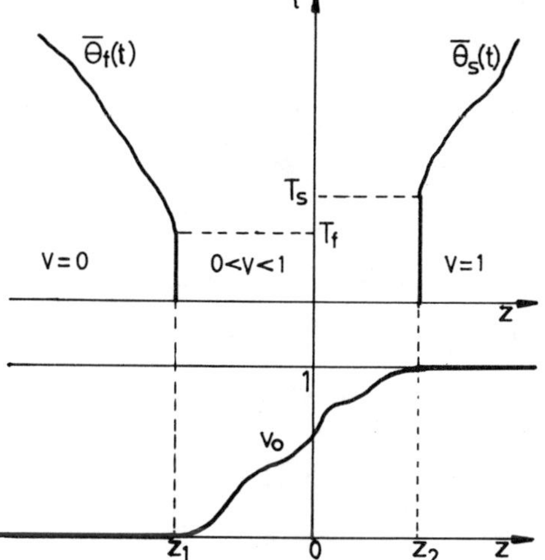

Fig. 3.3. Initial value and distribution of v in the (z,t)-plane with partial constant boundaries.

The functions $\bar\theta_f$ and $\bar\theta_s$ remain constant for a while and then start to move. For $t > T_f$ we have $d\bar\theta_f/dt < 0$ and for $t > T_s$ we have $d\bar\theta_s/dt < 0$. Further for $t \neq T_f$, $\bar\theta_f$ satisfies (3.4) and for $t \neq T_s$, $\bar\theta_s$ satisfies (3.5).

To interpret equations (3.4) and (3.5) we return to the original variables x,t and ζ. If θ_f and θ_s denote the curves $\bar\theta_f$ and $\bar\theta_s$ in the (x,t)-plane, then (3.4) and (3.5) transform into

$$\frac{d\theta_f(t)}{dt} = -\frac{\Gamma}{\varepsilon}\lim_{x\downarrow\theta_f(t)} \frac{\partial\zeta(x,t)}{\partial x} + \frac{Q_x}{\varepsilon H}, \qquad (3.6)$$

and

$$\frac{d\theta_s(t)}{dt} = +\frac{\Gamma}{\varepsilon}\lim_{x\uparrow\theta_s(t)} \frac{\partial\zeta(x,t)}{\partial x} + \frac{Q_x}{\varepsilon H}. \qquad (3.7)$$

The right hand side of (3.6) is the x-component of the specific discharge of the salt water at $x = \theta_f(t)$ divided by ε and the right hand side of (3.7) is the x-component of the specific discharge of the fresh water at $x = \theta_s(t)$ divided by ε. These equations can also be derived from equations (2.21), (3.1)-(3.3) in de Josselin de Jong's paper. However here they appear as the result of a pure mathematical theory.

Finally, it will be clear that if we substitute $\zeta = \zeta_p$ into (3.6) and (3.7), we find equations (2.7) and (2.8).

3.2 Large time behaviour

Next we consider the asymptotic behaviour as $t \to \infty$ of solutions of the initial value problem (3.1), (3.2). We demonstrate that an arbitrary solution will converge towards the similarity solution (2.4) as $t \to \infty$ and we give an expression for the rate of convergence.

In the proof of this property we use the so-called maximum principle. By the maximum principle for solutions of (3.1), (3.2) we mean the following: let $v_{01}(z)$ and $v_{02}(z)$ be two initial values and let $v_1(z,t)$ and $v_2(z,t)$ denote the corresponding solutions. If $v_{01}(z) \geq v_{02}(z)$ for $-\infty<z<\infty$, then $v_1(z,t) \geq v_2(z,t)$ for $-\infty<z<\infty$, $t > 0$. Thus the maximum principle says that if initially one interface is above the other, then this is true for all $t > 0$.

To prove the convergence result we observe that $f(\eta) = f(z/\sqrt{(t+a)})$, with f given by (2.4), satisfies (3.1) for any $a > 0$. For $a = 1$, this solution corresponds to the initial value:

$$f(z) = \begin{cases} 0 & \text{for } -\infty<z\leq-\tfrac{1}{2}, \\ +\tfrac{1}{2}+z & \text{for } -\tfrac{1}{2}<z<\tfrac{1}{2}, \\ 1 & \text{for } \tfrac{1}{2}<z<\infty. \end{cases}$$

Now suppose we have an initial value as in Fig. 3.4.

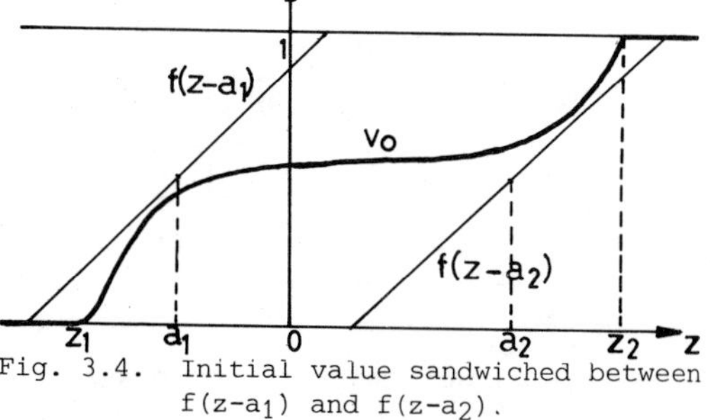

Fig. 3.4. Initial value sandwiched between $f(z-a_1)$ and $f(z-a_2)$.

Then it is clear that we can always find two numbers a_1 and a_2 such that

$$f(z-a_1) \geq v_0(z) \quad \text{for} \quad -\infty<z<\infty, \tag{3.8}$$

$$f(z-a_2) \leq v_0(z) \quad \text{for} \quad -\infty<z<\infty. \tag{3.9}$$

Next we use the maximum principle to obtain from (3.8) and (3.9)

$$f((z-a_1)/\sqrt{(t+1)}) \geq v(z,t) \geq f((z-a_2)/\sqrt{(t+1)}),$$

for $-\infty<z<\infty$, $t>0$.

From these inequalities it follows that

$$|v(z,t) - f((z-a_3)/\sqrt{(t+1)})| \leq f((z-a_1)/\sqrt{(t+1)}) - f((z-a_2)/\sqrt{(t+1)}), \tag{3.10}$$

for any $a_3 \in (a_1, a_2)$.
The right hand side of (3.10) can be estimated by

$$f((z-a_1)/\sqrt{(t+1)}) - f((z-a_2)/\sqrt{(t+1)}) \leq (a_2-a_1)/\sqrt{(t+1)}.$$

So finally we obtain the estimate:

$$|v(z,t) - f((z-a_3)/\sqrt{(t+1)})| \leq (a_2-a_1)/\sqrt{(t+1)}.$$

This inequality shows that $v(z,t)$ will tend towards a similarity solution as $t\to\infty$ with a convergence rate of $O(t^{-\frac{1}{2}})$.
In terms of the original variables x,t and ζ, we have

$$|\zeta(x,t) - \zeta_p(x-2a_3\sqrt{(\Gamma H/\varepsilon)},t)| \leq H(a_2-a_1)/\sqrt{(t+1)}.$$

Thus an arbitrary interface between the fresh and salt water will converge towards the explicit solution from Section 2 as $t\to\infty$. This implies that interfaces will become flatter as time increases (Fig. 2.2 and 2.3) and that eventually the straight line will be approximated.

4. THE AXIAL SYMMETRIC CASE

Sofar, we studied the interface motion equation only in one space dimension: ζ was a function of x alone. This was done just for mathematical convenience.
Here we shall consider the two dimensional case in it simplest form, the axial symmetric case. Then $\zeta = \zeta(r,t)$ where $r = \sqrt{(x^2+y^2)}$.
We base our analysis on equation (3.15) of de Josselin de Jong. Again we make the additional four assumptions as mentioned in the introduction. Then the interface motion equation reduces to

$$\varepsilon\frac{\partial\zeta}{\partial t} + \frac{Q_r}{2\pi rH}\frac{\partial\zeta}{\partial r} = \frac{\Gamma}{H}\frac{1}{r}\frac{\partial}{\partial r}\{\zeta(H-\zeta)r\frac{\partial\zeta}{\partial r}\}. \tag{4.1}$$

To illustrate some properties of its solutions, we construct a similarity solution in this section and demonstrate a number of interesting facts about it. This similarity solution is found numerically by means of a shooting method.
In what follows we shall assume that the total discharge Q_r is constant.
As in Section 2, we define new variables that absorb the physical constants.
Let

$$\rho = r\sqrt{(\varepsilon/(\Gamma H))}, \quad u = \zeta/H,$$

and

$$\lambda = Q_r/(2\pi H^2\varepsilon\Gamma).$$

Then (4.1) becomes

$$\frac{\partial u}{\partial t} = \frac{1}{\rho}\frac{\partial}{\partial\rho}\{u(1-u)\,\rho\frac{\partial u}{\partial\rho}\} - \lambda\frac{1}{\rho}\frac{\partial u}{\partial\rho}. \tag{4.2}$$

Next we reduce the number of independent variables in (4.2) by making a similarity transformation. Set

$$\xi = \rho/\sqrt{(t+a)}.$$

Then, if we look for solutions of the form

$$u(\rho,t) = g(\xi) = g(\rho/\sqrt{(t+a)}),$$

we find that g satisfies the ordinary differential equation

$$\frac{d}{d\xi}\{g(1-g)\,\xi\frac{dg}{d\xi}\} + \tfrac{1}{2}\xi^2\frac{dg}{d\xi} - \lambda\frac{dg}{d\xi} = 0, \tag{4.3}$$

for $0<\xi<\infty$.

At the boundaries we set

$$g(0) = 0 \text{ and } g(\xi) = 1. \tag{4.4}$$

In this case, we are not able to solve the boundary value problem explicitly. However, using a similar technique as Graven & Peletier (1970) and Atkinson & Peletier (1974), it is possible to obtain the following existence result.
Given any $\lambda>0$, there exist numbers ξ_1 and ξ_2, with $0<\xi_1<\sqrt{(2\lambda)}<\xi_2<\infty$, and there exists a solution $g(\xi)$ such that

$$\begin{aligned} g(\xi) &= 0 && \text{for } 0<\xi\leq\xi_1, \\ dg(\xi)/d\xi &> 0 && \text{for } \xi_1<\xi<\xi_2, \\ g(\xi) &= 1 && \text{for } \xi_2\leq\xi<\infty. \end{aligned} \tag{4.5}$$

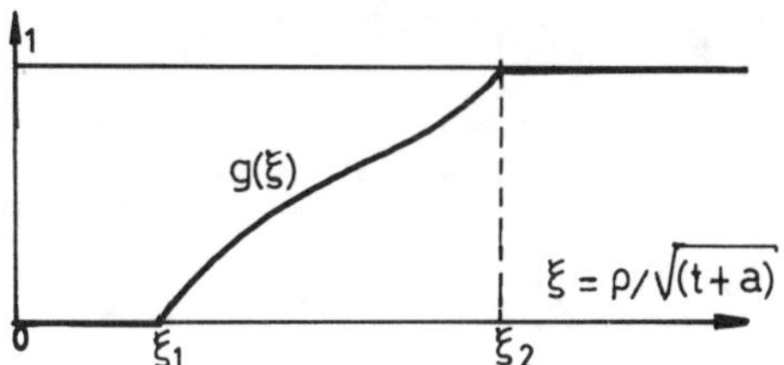

Fig. 4.1. Schematic representation of similarity solution (4.5).

The points ξ_1 and ξ_2 correspond to circles in the (x,y)-plane.
At $\xi = \xi_1$:

$$\sqrt{(x^2+y^2)} = \xi_1 \sqrt{(\Gamma H/\varepsilon)(t+a)} \underset{=}{\mathrm{def}}\, R_f(t). \qquad (4.6)$$

At $\xi = \xi_2$:

$$\sqrt{(x^2+y^2)} = \xi_2 \sqrt{(\Gamma H/\varepsilon)(t+a)} \underset{=}{\mathrm{def}}\, R_s(t). \qquad (4.7)$$

The physical interpretation of this solution is the following, see Fig. 4.2. Fresh water is being injected at the origin, over the full height of the aquifer at a constant rate Q_r. Then, for each $t \geq 0$, there are two circles with radius $R_f(t)$ and $R_s(t)$ such that only fresh water is present inside the first circle and only salt water is present outside the second circle,

Thus although we do not know the exact form of the solutions, this existence result gives us some idea of the kind of solutions one can expect, Fig. 4.1. Note that again we find discontinuities in the derivative as $g \downarrow 0$ and $g \uparrow 1$.

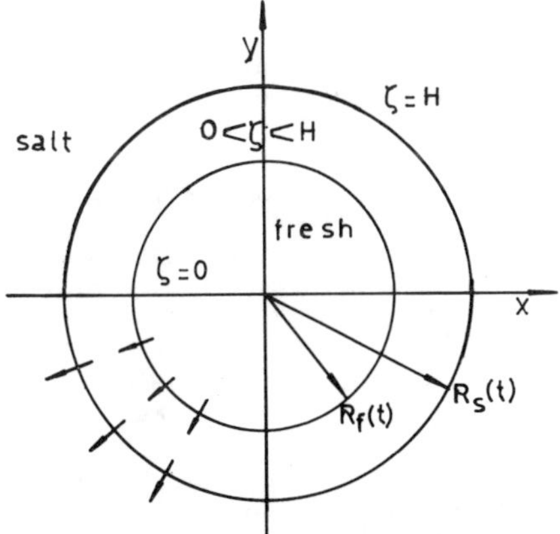

Fig. 4.2. Distribution of fresh and salt water at time $t \geq 0$.

The two circles become larger as time increases. Their speed of expansion is found by differentiating (4.6) and (4.7) with respect to t. This gives

$$\frac{dR_f(t)}{dt} = \tfrac{1}{2}\xi_1\sqrt{\{\Gamma H/(\varepsilon(t+a))\}},$$

and

$$\frac{dR_s(t)}{dt} = \tfrac{1}{2}\xi_2\sqrt{\{\Gamma H/(\varepsilon(t+a))\}}.$$

Notice that the outer circle moves faster then the inner circle and that both the expressions tend to zero as $t \to \infty$.

To solve equation (4.3) numerically is not an easy task because the interval where $dg/d\xi > 0$, which is bounded by ξ_1 and ξ_2, is apriori unknown. Therefore we transform this boundary value problem to one which is defined on a fixed and known interval. We execute this transformation step by step.

First we remark that it follows from the existence proof that

$$\lim_{\xi \downarrow \xi_1} g(\xi)\{1-g(\xi)\}\frac{dg(\xi)}{d\xi} = 0, \qquad (4.8)$$

and

$$\lim_{\xi \uparrow \xi_2} g(\xi)\{1-g(\xi)\}\frac{dg(\xi)}{d\xi} = 0. \qquad (4.9)$$

Next we integrate (4.3) with respect to ξ. Using (4.8) this gives

$$g(1-g)\xi\frac{dg}{d\xi} + \int_{\xi_1}^{\xi}\{\tfrac{1}{2}s^2-\lambda\}\frac{dg(s)}{ds}\,ds = 0 \qquad (4.10)$$

Since $dg(\xi)/d\xi > 0$ on the interval (ξ_1, ξ_2) we can define the inverse of g: i.e. we consider ξ as a function g.
Let

$$\xi = \sigma(g) \quad \text{for } 0 \leq g \leq 1.$$

Then

$$\xi_1 = \sigma(0), \quad \xi_2 = \sigma(1) \text{ and}$$
$$dg/d\xi = +(d\sigma/dg)^{-1}.$$

We use these expressions in (4.10) and obtain the equation

$$g(1-g)\sigma = \frac{d\sigma}{dg}\int_0^g\{\lambda-\tfrac{1}{2}\sigma^2(s)\}ds, \qquad (4.11)$$

on the domain $0 \leq g \leq 1$.

Next we multiply both sides of (4.11) by σ and find a differential equation for $\tfrac{1}{2}\sigma^2$. In this equation we set

$$z = \int_0^g\{\lambda-\tfrac{1}{2}\sigma^2(s)\}ds. \qquad (4.12)$$

Then it is easy to check that z satisfies

$$z\frac{d^2z}{dg^2} = 2\,g(1-g)\,(\frac{dz}{dg}-\lambda), \qquad (4.13)$$

on $0 \leq g \leq 1$.
It follows directly from definition (4.12) that

$$z(0) = 0, \tag{4.14}$$

and

$$\frac{dz(0)}{dg} = \lambda - \tfrac{1}{2}\sigma^2(0) = \lambda - \tfrac{1}{2}\xi_1^2. \tag{4.15}$$

We consider (4.14) and (4.15) as initial values for equation (4.13) and we solve this problem numerically.

In deriving the equation for z, we used only property (4.8). From equation (4.10) one can see that the other property (4.9) corresponds to $z(1) = 0$. In order to obtain a numerical solution of the boundary value problem (4.3), (4.4), we first solve the initial value problem (4.14)-(4.15) in such a way that $z(1) = 0$. This can be done by means of a shooting method which gives a unique value for ξ_1. Any other choice of ξ_1 will give such an initial derivative that the corresponding solution of the initial value problem will not satisfy $z(1) = 0$. Having found the correct solution z, we can work our way back and determine the solution $g(\xi)$ of (4.3) and (4.4) which has now the desired properties.

The constant ξ_2 can be found from the derivative of z at $g = 1$:

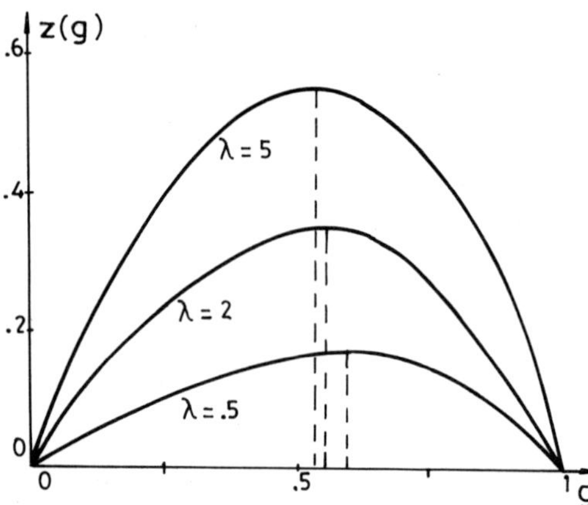

Fig. 4.3. Solution z(g) of the initial value problem for $\lambda = \frac{1}{2}$, 2 and 5.

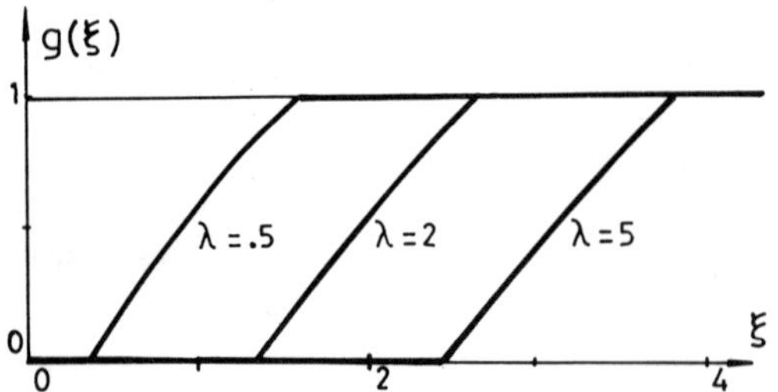

Fig. 4.4. Solution $g(\xi)$ of the boundary value problem for $\lambda = \frac{1}{2}$, 2 and 5.

$$\frac{dz(1)}{dg} = \lambda - \tfrac{1}{2}\sigma^2(1) = \lambda - \tfrac{1}{2}\xi_2^2.$$

Fig. 4.3 gives the numerical solution z of the initial value problem for three values of λ. Fig. 4.4 gives the corresponding solutions of the boundary value problem.

ACKNOWLEDGEMENT

The author whishes to acknowledge a number of stimulating discussions with Professor G. de Josselin de Jong.

5. REFERENCES

Aronson, D.G. 1969, Regularity properties of flows through porous media, SIAM J. Appl.Math. 17:461-467.

Aronson, D.G. 1970, Regularity properties of flows through porous media: A counter-example, SIAM J. Appl.Math. 19:299-307.

Aronson, D.G. 1970, Regularity properties of flows through porous media: The interface, Arch.Rational Mech.Anal. 54:373-392.

Atkinson, F.V. & L.A. Peletier 1974, Similarity solutions of the nonlinear diffusion equation, Arch.Rational Mech.Anal. 54:373-392.

Bear, J. 1972, Dynamics of Fluids in Porous Media, New York: American Elzevier Publishing Company Inc.

Caffarelli, L.A. & A. Friedman 1979, Regularity of the free boundary for the one-dimensional flow of gas in a porous media, preprint.

Dietz, D.N. 1953, A theoretical approach to the problem of encroaching and by-passing edge water, Proc.Ned.Akad.Wetensch.Series B, 56:83-92.

Duyn, C.J. van & L.A. Peletier 1977, A class of similarity solutions of the nonlinear diffusion equation, Nonlinear Analysis, Theory, Methods and Applications, 1:223-233.

Duyn, C.J. van 1979, On the diffusion of immiscible fluids in porous media, SIAM J. Math.Anal. 10:486-497.

Graven, A.H. & L.A. Peletier 1972, Similarity solutions for degenerate quasilinear parabolic equations, J.Math.Anal.Appl. 38:73-81.

Knerr, B.F. 1977, The porous media equation in one dimension, Trans.Amer.Math.Soc. 234:381-415.

Peletier, L.A. 1971, Asymptotic behaviour of solutions of the porous media equation, SIAM J. Appl.Math. 21:542-551.

Philip, J.R. 1960, General method of exact solution of the concentration dependent diffusion equation, Australian J.Phys. 13:1-12.

Proceedings of Euromech 143 / Delft / 2-4 September 1981

Simulation of two-dimensional saturated/unsaturated flows with an exact water balance

U.HORNUNG & W.MESSING
Universität Münster, Germany

1 SUMMARY

An efficient method to simulate water flows in a two-dimensional rectangular region is described. It has the following properties:
- there is no error in the water balance,
- inequality boundary conditions, occuring e.g. at seepage faces, are handled economically.

The method is tested for an analytical solution of an artificial flow problem, representing infiltration from a point source with discharge at a constant rate.

2 FLOW PROBLEM

2.1 Flow equation

Unsteady horizontal water flows in an isotropic homogeneous porous medium may be described by Richards' equation

$$\frac{\partial\Theta}{\partial t} = \text{div}\ (K(\psi) \cdot \text{grad}\ \psi), \qquad (1)$$

where Θ is the water content, K the hydraulic conductivity and ψ the pressure head. After Kirchhoff's transformation (Remson et al., 1971) with

$$v = \int_0^{\psi} K(s)ds \qquad (2)$$

equation (1) can be written in the equivalent form

$$\frac{\partial\Theta}{\partial t} = \Delta v\,, \qquad (3)$$

in which Θ is now considered as a function of v. Using the specific moisture capacity $\frac{d\Theta}{d\psi}$ and $C(v) = \frac{d\Theta}{d\psi} \cdot \frac{1}{K}$, equation (3) has the quasilinear form

$$C(v) \cdot \frac{\partial v}{\partial t} = \Delta v\,. \qquad (4)$$

2.2 Boundary conditions

To define a flow problem completely initial and boundary conditions have to be prescribed. Therefore, let the boundary consist of the parts Γ_D, Γ_N and Γ_I, on which the conditions are:

$$\begin{array}{lll} \text{a)}\ v = v_D & \text{on } \Gamma_D & \\ \text{b)}\ q = q_N & \text{on } \Gamma_N & \\ \text{c)}\ \left.\begin{array}{l} v \geq v_{min} \\ q \leq q_{max} \\ (v-v_{min})\cdot(q-q_{max}) = 0 \end{array}\right\} & \text{on } \Gamma_I\ . & (5) \end{array}$$

Here v_D, q_N, v_{min}, q_{max} are given functions of space and time. q means the boundary flux, that is $q = -n_x \cdot \partial_x v - n_y \cdot \partial_y v$, where (n_x, n_y) is the outer normal vector. Inequality boundary condition like 5c) appear as atmospheric boundary conditons (Neuman et al., 1975) or at seepage faces with $v \leq 0$, $q \geq 0$ and $v \cdot q = 0$.

3 NUMERICAL METHOD

3.1 Spatial discretisation

For the numerical solution the flow problem discribed above is discretized with respect to space and time. We use finite differences for spatial discretisation. A net of grid points is spread over the whole region as shown in figure 1. Functions on the rectangle are represented by their values at the grid points, and partial derivatives like ∂_{xx} are approximated in the usual way by differences δ_x^2 etc.

3.2 Time discretisation

In order to simulate saturated/unsaturated flow we use a backward time discretisation in the form of the implicit scheme

$$\frac{\Theta(v^{k+1})-\Theta(v^k)}{\Delta t^k} = \delta_x^2\, v^{k+1} + \delta_y^2\, v^{k+1} . \qquad (6)$$

This discretisation has an exact water balance, i.e. the variation of the water content in the whole region is the amount of water, which crosses the boundary during the time interval considered.

Because of the nonlinear Θ-v-relation equations (6) are nonlinear and a special numerical algorithm has to be applied to solve them. Therefore, the problem is usually linearized by starting from equation (4) and considering a discretisation like

$$C(v^*) \cdot \frac{v^{k+1}-v^k}{\Delta t^k} = \delta_x^2\, v^{k+1} + \delta_y^2\, v^{k+1} . \qquad (7)$$

To assure the algebraic problem to be linear the specific moisture content is calculated from $v^* = v^k$, or by extrapolation or by a predictor step which estimates $v^* = v^{k+1/2}$. The disadvantage of these linearisations is an error in the water balance, which can be expressed by the difference

$$\sum_{i,j}\left[\Theta_{ij}^{k+1} - \Theta_{ij}^{k} - C(v_{ij}^*)\cdot(v_{ij}^{k+1} - v_{ij}^{k})\right]\Delta x \Delta y \qquad (8)$$

This quantity is small, if the differences $v_{ij}^{k+1} - v_{ij}^{k}$ are small, which is usually achieved by using small time steps.

In order to measure the efficiency of the FID (Fully Implicit Discretisation) equation (6), it is compared with the most simple LID (Linearized Implicit Discretisation) equation (7) with $v^* = v^k$.

3.3 ADI-algorithm

For every time step a set of equations has to be solved, which is the most time consuming part of a simulation method. For FID these equations become nonlinear, in the case of LID they are linear. Since the nonlinear equations (6) have to be solved by iteration, an iterative method has been chosen for (7), namely an ADI (Alternating Direction Implicit) method. This choice was stimulated by the ADIPIT method, used by Vauclin et al. (1975) and by ADI methods in Hornung and Messing (1980). A general description of ADI methods is given in Remson et al. (1971: 204-219).

The algorithm to solve the linear equations (7) can shortly be written as follows:

Sweep in x-direction

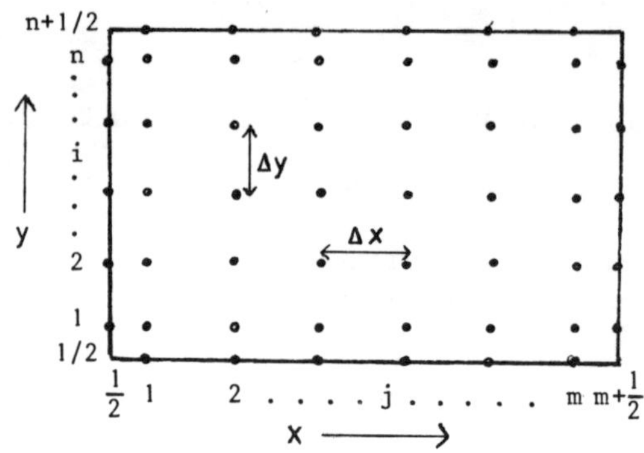

Fig. 1 Spatial discretisation

$$\omega_x^\nu\cdot(v^{k+1,2\nu+1} - v^{k+1,2\nu}) = \delta_x^2\, v^{k+1,2\nu+1} + \delta_y^2\, v^{k+1,2\nu} - C^k\cdot(v^{k+1,2\nu+1} - v^k)/\Delta t^k$$

Sweep in y-direction

$$\omega_y^\nu\cdot(v^{k+1,2\nu+2} - v^{k+1,2\nu+1}) = \delta_x^2\, v^{k+1,2\nu+1} + \delta_y^2\, v^{k+1,2\nu+2} - C^k\cdot(v^{k+1,2\nu+2} - v^k)/\Delta t^k$$

with the iteration-index $\nu = 0,1,2,\ldots$. This algorithm is refered to as LADI (Linear ADI).

In order to solve FID (6) the iteration was extended to the nonlinearity in such a way, that convergence was practically assured. The algorithm NADI (Nonlinear ADI) is:

Sweep in x-direction

$$\omega_x^\nu\cdot(v^{k+1,2\nu+1} - v^{k+1,2\nu}) = \delta_x^2\, v^{k+1,2\nu+1} + \delta_y^2\, v^{k+1,2\nu} - \left[\Theta^{k+1,2\nu} + C^{k+1,2\nu}\cdot(v^{k+1,2\nu+1} - v^{k+1,2\nu}) - \Theta^k\right]/\Delta t^k$$

Sweep in y-direction

$$\omega_y^\nu\cdot(v^{k+1,2\nu+2} - v^{k+1,2\nu+1}) = \delta_x^2\, v^{k+1,2\nu+1} + \delta_y^2\, v^{k+1,2\nu+2} - \left[\Theta^{k+1,2\nu+1} + C^{k+1,2\nu+1}\cdot(v^{k+1,2\nu+2} - v^{k+1,2\nu+1}) - \Theta^k\right]/\Delta t^k$$

In both cases of LADI and NADI the iteration was started by an extrapolation

$$v^{k+1,0} = v^k + (v^k - v^{k-1})\cdot \Delta t^k/\Delta t^{k-1}$$

to get a first estimate for v^{k+1}. The iteration was stopped, if the difference $|v^{k+1,2\nu+2} - v^{k+1,2\nu}|$ was small enough ($\leq 10^{-5}$ in our case). Then $v^{k+1} = v^{k+1,2\nu+2}$ is seen to solve (6) and (7) respectively. The problem in (6) and (7) reduces to solve a set of tridiagonal systems. The efficiency of this procedure depends mainly on the choice of the ADI-parameters ω_x^ν and ω_y^ν. They are chosen in cycles of length s (s=4 in our case). In the case of LADI, for each type of boundary conditions and every s there exists an unique set of iteration parameters. In the appendix a short description is given, how to calculate the parameters. The same parameters as for LADI have been taken for NADI.

In some tests we compared the computer time for LADI and for direct methods, i.e. Gauß elimination. The results are that even in the case of 100 grid points and reasonable time steps the LADI-procedure was less time consuming. Theoretical considerations show that the effort grows with N^2 for direct methods and $N \cdot \log N$ for LADI, where N is the number of grid points. An other advantage of iterative methods is that a decrease of time step on a large scale causes only a small increase of computer time. This is because for smaller time steps the extrapolation yields a better first estimate of the solution v^{k+1}. Thus the iterative method is able to exploite information from former time steps.

3.4 Handling boundary conditions

The part Γ_I of the boundary, on which inequality boundary conditions are given, can be devided into two parts, that part, where $v = v_{min}$ and $q \leq q_{max}$ and the other, where $q = q_{max}$ and $v > v_{min}$. The point, which seperates the two parts, this is for instance the upper point of a seepage face, in general moves with time. For the numerical procedure this means that for every time step it must be found out, to which of the parts a grid point on Γ_I belongs. This is done in the following way:

If at time t^k for some point for instance we have $v^k = v^k_{min}$ and $q^k \leq q^k_{max}$, we try the same for t^{k+1}, this means that we prescribe $v^{k+1} = v^{k+1}_{min}$ and solve the flow equations for v^{k+1}. After this we test whether the inequalities on Γ_I are satisfied, that is if $q^{k+1} \leq q^{k+1}_{max}$. If this inequality is violated, we do the calculations again, prescribing $q^{k+1} = q^{k+1}_{max}$ instead of v^{k+1} at that point. Similar calculations are done, if at time t^k we have $q^k = q^k_{max}$ and $v^k < v^k_{min}$.

The advantage of ADI iteration is, that the control of inequality boundary conditions can already be done during the iteration procedure. For direct methods first the solution has to be calculated and then the decision has to be made, whether or not the solution fits the boundary conditions.

3.5 Time step control

The time step sizes Δt^k are chosen automatically and according to an idea of Edwards (1972) in dependence on some parameters Δt_{min}, Δt_{max} and Δv_{opt}. The time steps will be adjusted in such a way that from one time level to the next the maximal change of v is about Δv_{opt}, but not larger than $2 \cdot \Delta v_{opt}$. In detail:

At the beginning let be $\Delta t^o = \Delta t_{min}$. Then, if Δt^{k-1} was the time step for the last time interval and Δv the largest change of v during this period, we define $r = \Delta v_{opt}/\Delta v$.

If $r < 1/2$, the last time step is repeated with $\Delta t^{k-1} = \max\{\Delta t_{min}, \Delta t^{k-1}/100\}$ as step size. This is to assure that rapid changes in the boundary conditions are managed adequately.

Otherwise the new step size is calculated by

$$\Delta t^k = \begin{cases} r^2 \cdot \Delta t^{k-1} & \text{if } r < 1 \\ 0.5 \cdot (1+r) \cdot \Delta t^{k-1} & \text{if } r \geq 1 \end{cases}$$

under the restrictions

$0.5 \cdot \Delta t^{k-1} \leq \Delta t^k \leq 2 \cdot \Delta t^{k-1}$ and

$\Delta t_{min} \leq \Delta t^k \leq \Delta t_{max}$.

4 TEST EXAMPLE

4.1 Analytical solution

In order to test the two discretisations FID and LID an artifical flow problem was simulated. All types of boundary conditions have been posed in the problem as shown in figure 5. The region was chosen to be the unit square. The analytical solution of (3) and (4) resp. with

$$\Theta(v) = 10 \cdot T \cdot \begin{cases} 1 & \text{if } v \geq 0 \\ 1 - v^2/20 & \text{if } v < 0 \end{cases}$$

is given by the formula

$$v(t,x,y) = \begin{cases} -z(+\infty) & & \text{if } t \leq 0 \\ -\lambda \cdot \ln s & \text{for } 0 < s \leq 1 & \\ -z(s) & \text{for } s > 1 & \end{cases} \Big\} \text{if } t>0, \qquad (9)$$

where $s = r \cdot \sqrt{T/t}$ and $r = \sqrt{(x-x_o)^2 + (y-y_o)^2}$ is the distance to the source S. The function z is the solution of the ordinary differential equation

$$z''(s) = - [1/s + 0.5 \cdot s \cdot z(s)] \cdot z'(s),$$

$$z(1) = 0, \; z'(1) = \lambda ,$$

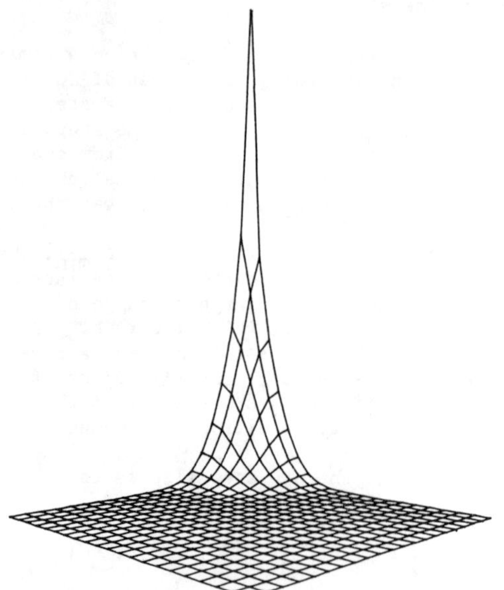

Fig. 2 Hydraulic head at time t = 3.2

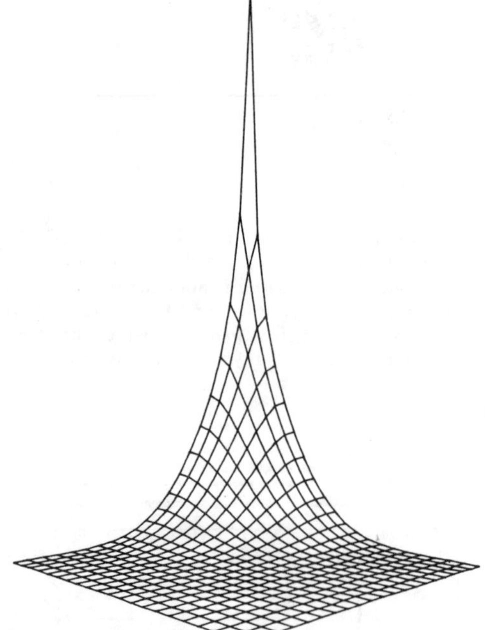

Fig. 3 Hydraulic head at time t = 10

which is solved by numerical methods once for all. The flow derived from (9) is radial and the circles in figure 5 show the front of saturation at different times ($\lambda = 5$, $T = 100$, $x_o = 0$, $y_o = -0.01$). At the beginning the region is totally unsaturated and then becomes wet by infiltration from the source S. Fig. 4 shows v as

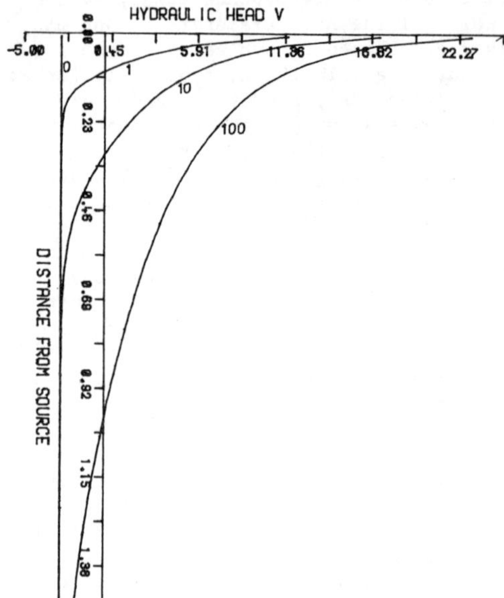

Fig. 4 The exact solution

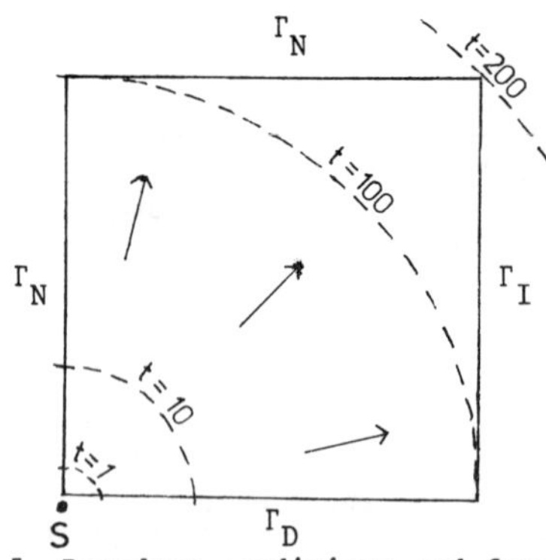

Fig. 5 Boundary conditions and front of saturation at different moments

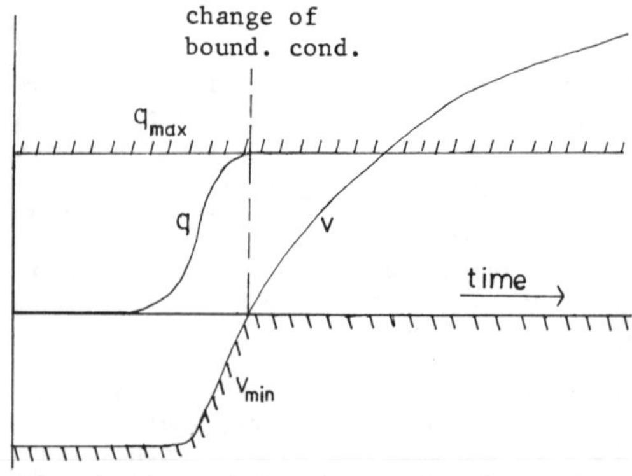

Fig. 6 Potential and actual values of q and v for some point on Γ_I

a function of distance to S at different times. Simulation of the flow given by (9) is a hard problem because of the sharp wetting front at the beginning of infiltration. Plots of such a front at different times are given in fig. 2 and 3.

The prescribed boundary values are the values of the analytical solution. Figure 6 shows the values of q, q_{max} and v, v_{min} for some point of the boundary Γ_I as functions of time. The change from a prescribed head to a prescribed flow occurs, when the saturation front passes this point.

4.2 Results

The numerical experiments had the following results:

1. The water balance of FID represented by eq. (6) is exact for all space and time discretisations; the computational time is about 20-30% larger than for LID.

2. Both discretisations do almost not differ with respect to the maximal or mean error, if compared with the exact solution. The difference between calculated and exact water content is smaller for FID.

3. For a given spatial discretisation there is a bound for Δt_{mean} such that further decrease of Δt only effects the water balance of LID, but does not reduce the error of the approximation.

In this context error means $err(t) = \max |v_{ij}^{num.} - v_{ij}^{analyt.}|$ and is a function of time. It is plotted in figure 7 for LID and FID. Discretisation parameters for this example are $\Delta x = \Delta y = 0.1$, $\Delta t_{min} = 0.1$, $\Delta t_{max} = 10$, $\Delta v_{opt} = 0.5$. The maximal error occurs at the beginning of the time interval (when the wetting front reaches the first internal grid point) and later on remains almost constant. Table 1 shows, that the maximal and the mean error, i.e.

$$\frac{1}{200}\int_0^{200} err(t)dt,$$

are hardly influenced by the time steps. Only a further refinement of the spatial descretisation reduces the error. A further decrease of time step size has an effect only in the water balance of LID. For FID the water balance is always exact. This fact is espacially important, if the problem becomes stationary. Then the approximation error is of minor importance but for LID time step size has to be controlled in order to yield a good water balance. From this point of view FID is less time consuming than LID.

Table 1. Error and water balance of LID for different time steps.

Δx	Δt_{mean}	max. error	mean error	error in wat. bal.
0.1	4.0	1.69	0.95	3.7 %
0.1	1.1	1.60	0.95	0.9 %
0.05	3.8	1.34	0.65	3.7 %
0.05	1.0	1.27	0.64	0.9 %

Table 2. Comparison of LID and FID

	Δx	Δt_{mean}	max. error	error in wat.cont.	CPU sec
LID	0.1	4.0	1.69	1.5 %	6
FID			1.53	0.8 %	7
LID	0.05	3.8	1.34	1.1 %	28
FID			1.13	0.5 %	32

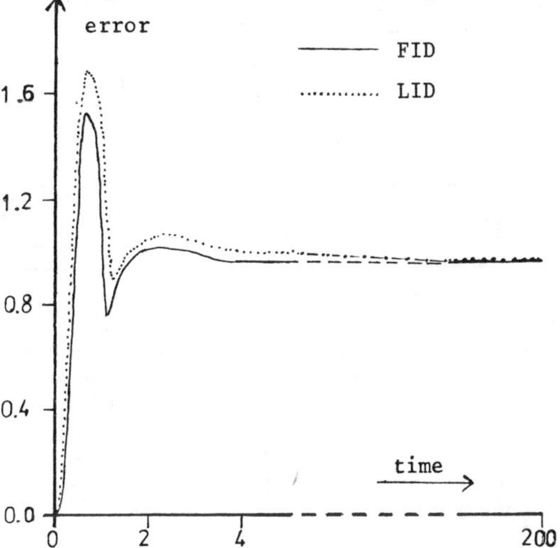

Fig. 7 Error as function of time

5 CONCLUSION

This study shows that with some additional effort an important improvement can be achieved. While for the numerical schemes commonly used the water balance can be improved only by a refinement of spatial and time discretisation, the FID scheme has an exact water balance for all discretisations.

6 REMARK

Further studies will be done to solve eq. (1) by the FID-NADI method directly without applying Kirchhoff's transformation. This has already been done successfully for Dirichlet boundary conditions. In addition to an exact water balance a better approximation than with LID was achieved, because the conductivity is adjusted during the iteration.

7 APPENDIX : ADI-parameters

For the spatial discretisation shown in fig. 1 optimal ADI-parameters can be found by using the following procedure: First the minimal and the maximal eigenvalue of the matrices belonging to the discretisation of $-\partial_x^2$ and $-\partial_y^2$ are estimated. These depend on the boundary conditions as shown in table 3.

Table 3. Eigenvalues for different boundary conditions.

boundary conditions D Dirichlet-c. N Neumann-c.	minimal eigenvalue a	maximal eigenvalue b
D ——————— D	$\beta\cdot\sin^2(\alpha)$	β
D ——————— N	$\beta\cdot\sin^2(\frac{\alpha}{2})$	$\beta\cdot\cos^2(\frac{\alpha}{2})$
N ——————— N	0	$\beta\cdot\sin^2(\frac{\alpha}{2})$

Here $\alpha=\pi/2m$, $\beta=4m^2$ and m is the number of interior grid points in one direction. These formulas are valid, if on every side the type of boundary conditions does not change. If, as in fig. 5, the type of boundary conditions changes with time, the best device must be found by experiment. For the boundary problem of fig. 5 we used

$$a_x = 4m^2\cdot\sin^2(\pi/4m),\quad b_x = 4m^2\cdot\cos^2(\pi/4m),$$

$$a_y = 4n^2\cdot\sin^2(\pi/4n),\quad b_y = 4n^2\cdot\cos^2(\pi/4n).$$

After a_x, b_x, a_y, b_y being fixed, the values

$$\xi = 2\cdot\frac{(b_y-a_y)\cdot(b_x-a_x)}{(b_y+b_x)\cdot(a_y+a_x)},\quad k' = 1/(1+\sqrt{\xi(\xi+2)}),$$

$$k = \sqrt{1-k'^2}\quad\text{and}\quad K=K(k) = \int_0^{\pi/2}\frac{dx}{\sqrt{1-k^2\cdot\sin^2 x}}$$

are calculated.

Then $\rho^\nu = \mathrm{dn}(\frac{2\nu-1}{2\cdot s}\cdot K,k)$ is calculated for $\nu=1,\ldots,s$, where dn is a Jacobian elliptic function, which, as well as K(k), can usually be found in a program library.

Then

$$\delta = 2\cdot\frac{k'\cdot(b_x+b_y)-(a_x+a_y)}{(a_x+a_y)\cdot(b_x-b_y)+k'\cdot(b_x+b_y)\cdot(a_y-a_x)}$$

$$(=0 \text{ if } a_x = a_y \text{ and } b_x = b_y)$$

$$\alpha = \frac{k'\cdot(a_y-a_x+2\cdot\delta\cdot a_x\cdot a_y)}{a_x+a_y}$$

$$\beta = \frac{2+\delta\cdot(b_x-b_y)}{b_x+b_y}$$

and

$$\omega_x^\nu = \frac{\rho^\nu+\alpha}{\beta+\delta\cdot\rho^\nu},\qquad \omega_y^\nu = \frac{\rho^\nu-\alpha}{\beta-\delta\cdot\rho^\nu}$$

for $\nu = 1,\ldots,s$ are determined.

8 REFERENCES

Edwards, A.L., 1972, TRUMP: A computer program for transient and steady state temperature distributions in multidimensional systems. Rep. UCRL-14754, Rev. 3, Lawrence Livermore Lab., Calif.

Hornung, U. & W. Messing 1980, A predictor-corrector alternating-direction implicit method for two-dimensional unsteady saturated-unsaturated flow in porous media, Journal of Hydrology 47: 317-323.

Neuman, S.P., R.A. Feddes & E. Bresler 1975, Galerkin method of simulating water uptake by plants. In G.C. Vansteenkiste (ed.), Computer Simulation of Water Resources System. Amsterdam, North Holland Publishing Company.

Remson, I., G.M. Hornberger & F.J. Molz 1971, Numerical methods in subsurfaces hydrology. New York, John Wiley & Sons.

Vauclin, M., G. Vachaud & J. Khanji 1975, Two dimensional numerical analysis of transient water transfer in saturated-unsaturated spoils. In G.C. Vansteenkiste (ed.), Computer Simulation of Water Resources Systems. Amsterdam, North Holland Publishing Company.

9 ACKNOWLEDGMENT

This study was supported by the Deutsche Forschungsgemeinschaft (Grant Ho 782/2).

Proceedings of Euromech 143 / Delft / 2-4 September 1981

Instability of a cylindrical foreign fluid substance in an aquifer

G.F.J.KRUIJTZER
University of Delft, Netherlands

SUMMARY

The instability of a downward penetrating foreign fluid substance in an aquifer is exemplified for the case in which the initial configuration of the substance is a cylinder.

1 INTRODUCTION

The potential theory of the flow of two inmiscible viscous fluids in a porous medium and in particular the motion of a foreign fluid substance of finite size in an aquifer, has been presented by o.a. Polubarinova-Kochina and Falkovich (1951), Taylor and Saffman (1959), Chia-Shun Yih (1963). The conditions at the interface are that the pressure p should be continuous and that the velocity components normal to the interface must be the same for both fluids. For the case in which the foreign fluid substance is of negligible viscosity, the substance moves as a rigid body and it can take infinitely many shapes. On the other hand, if the viscosity of the substance cannot be neglected, a rigid body motion of the foreign fluid substance will not be possible in general. This can be elucidated as follows. When the interface is assumed to move in a rigid manner, the potential flow outside the foreign substance is completely determined by the kinematic condition on the interface and the condition on the outer boundary of the aquifer. From the then known velocity potential the pressure in the aquifer can be calculated. The requirement that the pressure p should be continuous across the interface should give rise to a uniform velocity field inside the foreign fluid substance (see, for instance, Kruijtzer, 1980). In general this may not be expected. However it has been established that fluid substances with the configuration of an elliptic cylinder and an ellipsoid in an unbounded aquifer move as a solid body (Chia-Shun Yih, 1963). In this paper it is investigated whether or not such a rigid motion is stable. In particular a circular cylinder is considered.

2 FORMULATION OF THE PROBLEM

The problem to be investigated is the stability of a downward moving cylindrical foreign fluid mass in a water-bearing porous medium of infinite extent, see the figure 1.

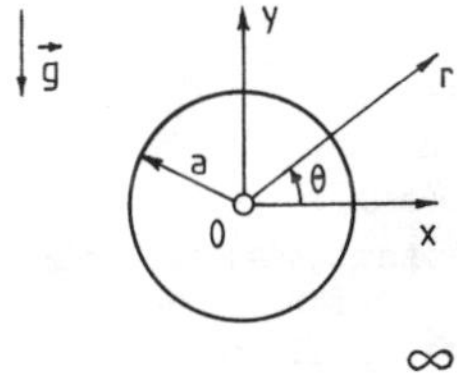

Fig. 1. Cylindrical foreign fluid substance in an aquifer of infinite extent.

The motion in the aquifer is assumed to be a plane motion in vertical planes perpendicular to the axis of the cylinder. In these planes a fixed coordinate system x,y is introduced. Its origin coincides instantaneously at time t with the center of the circular cross section of the cylinder. The fluid flows through the pores of the aquifer is governed by Darcy's law and the requirement of continuity (see, for instance, Verruijt, 1970). In Darcy's concept, the driving force represented by the specific weight of the fluid $\rho\bar{g}$ (density ρ, acceleration due to gravity $\bar{g}$) and the gradient of the pore pressure p, equalizes the resistance force of magnitude $(\mu/\kappa)\bar{q}$, encountered by the incompressible fluid with dynamic viscosity μ flowing with a specific discharge $\bar{q}$ through a porous medium with intrinsic isotropic permeability κ.

3 MOTION OF THE GROUNDWATER

In terms of the rectangular Cartesian components q_x, q_y of the specific discharge vector, Darcy's law takes the form

$$q_x = -K \frac{\partial}{\partial x}(p + \rho g y) = -\frac{\partial \phi}{\partial x} = -\frac{\partial \psi}{\partial y} \qquad (r > a)$$
$$q_y = -K \frac{\partial}{\partial y}(p + \rho g y) = -\frac{\partial \phi}{\partial y} = \frac{\partial \psi}{\partial x}$$

where $K = \kappa/\mu$ is a constant, $\phi = K(p + \rho g y)$ is the specific discharge potential and ψ is the stream function which exists by virtue of the equation of continuity

$$\frac{\partial q_x}{\partial x} + \frac{\partial q_y}{\partial y} = 0 \qquad (r > a)$$

4 MOTION OF THE FOREIGN SUBSTANCE

The motion of the foreign fluid substance is governed by similar equations, except that now the density and viscosity are different. Darcy's law can be written as

$$q'_x = -K' \frac{\partial}{\partial x}(p' + \rho' g y) = -\frac{\partial \phi'}{\partial x} = -\frac{\partial \psi'}{\partial y} \qquad (r < a)$$
$$q'_y = -K' \frac{\partial}{\partial y}(p' + \rho' g y) = -\frac{\partial \phi'}{\partial y} = \frac{\partial \psi'}{\partial x}$$

where $K' = \kappa/\mu'$ is a constant, $\phi' = K'(p' + \rho' g y)$, and where the prime distinguishes the substance motion from the water flow. The equation of continuity becomes

$$\frac{\partial q'_x}{\partial x} + \frac{\partial q'_y}{\partial y} = 0 \qquad (r < a)$$

5 BOUNDARY CONDITIONS

The fluids are assumed to be inmiscible. On the sharp interface between the extraneous fluid and the water the radial specific discharge and the pressure are continuous, so that

$$q_r = q'_r \qquad p = p' \qquad (r = a)$$

where $q_r = (q_x{}^2 + q_y{}^2)^{\frac{1}{2}}$. Further it is assumed that at infinity the groundwater is at a state of rest.

6 THE SOLUTION

The unique solution of the boundary problem under consideration is well-known, and may be presented as follows.
Let $\Omega = \phi + i\phi$ and $\Omega' = \phi' + i\phi'$ be the complex potentials for respectively $r > a$ and $r < a$, which are analytic functions of $z = x + iy$. Further, let $q = q_x + iq_y$, $q' = q'_x + iq'_y$ be the complex specific discharges, so that

$$\frac{d\Omega}{dz} = -(q_x - iq_y)$$

$$\frac{d\Omega'}{dz} = -(q'_x - iq'_y)$$

the solution is given by

$$\Omega = -A\frac{ia^2}{z}, \quad \Omega' = -iAz$$

where

$$A = \frac{(\rho' - \rho) g\, KK'}{K + K'}$$

so that

$$q_x - iq_y = -\frac{d\Omega}{dz} = -\frac{iAa^2}{z^2}$$
$$q'_x - iq'_y = -\frac{d\Omega'}{dz} = iA$$

7 INSTABILITY

Now it is assumed that at a certain instant of time a foreign fluid substance has the form of a cylinder whose circular cross section possesses a small irregularity as depicted in figure 2.

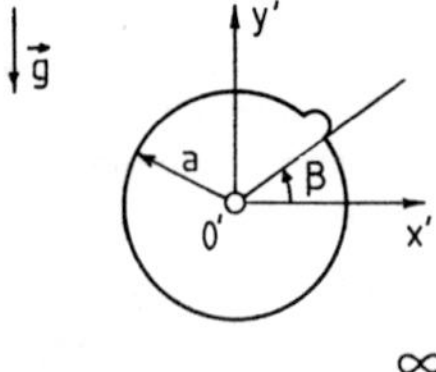

Fig. 2. Cylindrical foreign fluid substance whose circular cross-section shows an irregularity.

The (outer)boundary of the irregularity of the cross section is assumed to have the form of a circular arch. The fixed coordinate system being introduced in the section 2, is now designated by x',y'.

The influence of the small irregularity on the solution of the boundary value problem can be obtained with the method of conformal mapping. The circular cross section with small irregularity in the $z' = x + iy'$-plane of the figure 2 is mapped onto the purely circular cross section in the $z = (x + iy)$-plane of the figure 1. The conformal mapping is given approximately by (see Lawrentjew and Shabat, 1967),

$$z \simeq z' \{1 - \frac{\sigma}{2\pi} \frac{ae^{i\beta} + z'}{ae^{i\beta} - z'}\}$$

where σa^2 is the area of the irregularity. This approximation holds if both the condition $(\sigma/\pi) << 1$ as well as the condition $z' \neq ae^{i\beta}$ is satisfied. It follows

$$\frac{dz}{dz'} = \{1 - \frac{\sigma}{2\pi} \frac{ae^{i\beta} + z'}{ae^{i\beta} - z'} - \frac{\sigma}{2\pi} \frac{2\, ae^{i\beta}\, z'}{(ae^{i\beta} - z')^2}\}$$

It now is assumed that $e^{i\beta} = -i$ and $z' = (-i + \varepsilon)a$, so that the specific discharge at $z' = a(-i + \varepsilon)$ within the foreign fluid substance becomes

$$q_{x'} - iq_{y'} \simeq iA \{1 - \frac{\sigma}{2\pi} \frac{2i - \varepsilon}{\varepsilon} + \frac{\sigma}{2\pi} \frac{2(1 + i\varepsilon)}{\varepsilon^2}\}$$

where ε has been assumed to be small. It follows that at $z' = a(-i + \varepsilon)$

$$q_y' \simeq -A \{1 + \frac{\sigma}{2\pi} + \frac{\sigma}{\pi} \frac{1}{\varepsilon^2}\}$$

so that the configuration of the foreign fluid substance will not preserved. The motion is unstable.

REFERENCES

Polubarinova-Kochina, P.Y. and S.B. Falkovich, 1951, Theory of filtration of liquids in porous media, Advances Appl. Mech., vol.2, 154-217.

Taylor, Sir Geoffrey and P.G. Saffman, 1959, A note on the motion of bubbles in a Hele-Shaw cell and porous medium, Quart. Journ. Mech. and Appl. Math., vol.12, pt.3, 265-279.

Chia-Shun Yih, 1963, Velocity of a fluid mass imbedded in another fluid flowing in a porous medium, The Physics of fluids vol. 6, no.10, 1403-1407.

Lawrentjev, M.A. and B.W. Shabat, 1967, Methoden der Komplexen Funktionentheorie p.164, VEB Deutscher Verlag der Wissenschaften, Berlin.

Kruijtzer, G.F.J., 1980, The downward penetration of a spherical foreign fluid substance in an aquifer, L.G.M. mededelingen (Delft Soil Mechanics Laboratory) pt.21, no.2, 153-160.

Proceedings of Euromech 143 / Delft / 2-4 September 1981

Mathematical model for the evaluation of the recharge of aquifers in semiarid regions with lack of hydrogeological data

J.P.LOBO FERREIRA
Laboratório Nacional de Engenharia Civil, Lisbon, Portugal

SUMMARY

A mathematical model for calculating the daily sequential water balance is presented. This model, designed for regions with semiarid climates, was use with good results in the study of the water resources of Porto Santo island, Madeira archipelago, Portugal. It was later tested successfully in the portuguese region of Rio Maior wich has a mean annual precipitation of about 800 mm.

The object in the presentation of this theme is to exemplify how to obtain some of the hydrologic variables needed for the mathematical modelling of groundwater flows. The following hydrologic variables are updated daily by the model presented: rainfall, surface flow, surface infiltration, real evapotranspiration, soil moisture and deep recharge of aquifers.

1. INTRODUCTION

The need felt by LNEC to develop the model presented for calculating the daily sequential water balance of Porto Santo island was mainly due to three reasons. The first is concerned with the total absence of records for river discharges in Porto Santo, which makes it necessary to use mathematical models to simulate them. The second derives directly from the semiarid climate of the place, where the monthly potential evapotranspiration always exceeds the monthly rainfall, thus making it impossible to resort to the usual evalutions of the water balance for monthly periods. Finally, the small size of the island makes it necessary to calculate variables that are usually not considered, as for instance the quantification of the deep recharge of aquifers.

A brief reference to the evaluation of monthly water balances will be made followed by the daily sequential balance.

2. MONTHLY WATER BALANCE

The evaluation of the (monthly) water balance can be made chiefly by two methods. The first and easiest is the climatologic balance. To evaluate this balance the mean monthly values of rainfall and potential evapotranspiration are usually taken. To the mean monthly rainfall is subtracted the parcel that corresponds to the potential evapotranspiration. The remaining quantity – if the value of the rainfall is larger than that of the evapotranspiration – seeps through and increases the amount of water in the ground. When the limit of water that an unsaturated soil can hold is attained (specific retention) flow sets in. The application of this model to Porto Santo does not yield satisfactory results because, on a monthly scale and except in January, the value of evapotranspiration generally exceeds by far the value of the rainfall. The amount of water stored in the ground never reaches the limit considered and thus, according to the model, there would never be flow in Porto Santo, all the rainfall being consumed in evapotranspiration.

The second monthly water balance evaluation method is the sequential balance method. This is a more sophisticated method based on the month by month evaluation using as initial values for each month those calculated at the end of the previous month. To calculate flow, a given value for the limit capacity of the water that unsaturated soil is able to retain is also considered in this method. Results are generally given in tabular form after the series of monthly values for potential and real evapotranspiration, rainfall, soil moisture, lack and excess of water, have been treated statistically. Usually, the variable known as flow presupposes that half the excess water flows out in the month following that of the rainfall that originates it. For small basins this

supposition is, of course, not true and the excess of water variable should of course be assumed to be equal to the value of flow.

The method also failed to produce good results in Porto Santo island because with it we only get flow in the more rainy months of January and February, and this only happens in 20% of the years. This value drops down to 10% in December and March and to zero in the remaining months. Yearly, according to this method, there would only be flow in Porto Santo in 40% of the years.

Such values are not confirmed by reality since in Porto Santo the yearly storage of water in the dams has supplied water for irrigation, even in the driest years.

Neither of these models allows to separate and quantify the parcel of the flow that seeps in and recharges the aquifers from that which flows off superficially.

3. DAILY SEQUENTIAL WATER BALANCE

3.1 - Introduction

The main reason for the failure of the methods presented in section 2 to give plausible results in the case of Porto Santo is the the temporal increase they consider. In fact, the application of a monthly basis is not justified in the study of semiarid regions. In these regions the potential evapotranspiration shows values that are too high as compared with those of the rainfall and which will necessarily give values that are too high to the real evapotranspiration, if the time basis for the calculation is too long. Besides, if rainfall is of frontal and not orographic origin, like in Porto Santo, therefore usually involving strong sporadic rainfall at short intervals of time, the time basis for evaluating the water balance must be sufficiently small.

For this reason a quick analysis of the accuracy of the results of the water balance was carried out for 7, 3, 2 and 1 days. It seems that results are more realistic with the 1 day calculation basis. It is obvious that the amount of data to be treated will in this case be considerably larger.

3.2 - Development of an original computer programme for calculating the water balance

3.2.1 - General

This programme (called BALSEQ.FOR) calculates the total values recorded in the period under study as well as the mean values and monthly and annual standard deviations of the following hydrologic variables: rainfall, surface flow, real evapotranspiration, surface infiltration, soil moisture and deep aquifer recharge (see Table 1).

The initial data needed to carry out the programme are the limit amount of water that can be stored by unsaturated soil, the characteristic number of flow of the hydrographic basin, the value of the ground moisture on the first day of the balance, the daily rainfall (only necessary if above zero) and the monthly evapotranspiration.

The programme makes the daily updating of the values of the different variables of the hydrologic cycle referred to in the first paragraph.

The method for calculating the variables will be presented hereafter.

3.2.2 - Rainfall

An analysis of the udograph records of Porto Santo showed that the significant precipitation in the rainiest months take place within average time intervals of 7 days for January and February and 10 days for March. These heavy showers generally occur for a rather short period of time and have strong intensity, which means that, for the heaviest showers, the daily value of precipitation does not differ significantly from the precipitation recorded during the shower.

An auxiliary programme (CORBAL.FOR) was also developed to analyse the consistency of the time series of precipitation.

3.2.3 - Surface flow

Based on what has been said in the above sub-section, the evaluation of the surface flow according to the method developed by the United States Soil Conservation Service and proposed for Continental Portugal by LNEC 1976 seemed adequate. This method, which is described hereafter, seemes the more elaborate and suitable for the type of geologic and climatic data available. The model can be briefly described as follows:

a) Determination of the type or types of soil that exist in the basin under study and their identification according to the soil hydrologic classification of the United State Soil Conservation Service (type A, B, C or D soil).

b) Determination of the type or types of plant cover and of other surface conditions (for instance urban zones).

c) Weighted calculation of the characteristic coefficient of flow, N, with the help of Table 2.

d) Determination of the useful precipitation for flow P_u, corresponding to the total precipitations P_t, by equation:

$$P_u = \frac{25.4\left(\frac{P_t}{25.4} - \frac{200}{N} + 2\right)^2}{\frac{P_t}{25.4} + \frac{800}{N} - 8} \quad \text{(mm)}$$

The surface flow in a given section is obtained by multiplying the value obtained for the useful precipitation P_u by the area of the hydrographic basin defined by the section. This flow model is based on the construction of a triangular unitary hydrogramme. The duration of the showers should thus be of the same order of magnitude as the concentration time of the hydrographic basin under study. The model allows for initial losses of precipitation in order to permit saturation of the surface layer of the soil. For this reason, only the showers whose precipitation values exceed the minimum needed to saturate the surface layer of the soil are studied. The value of the total precipitation after which flow occurs depends on the type of soil and on its surface occupation and it may be expressed as a function of the characteristic number of flow N by equation:

$$P_t > \frac{5080}{N} + 50.8 \quad \text{(mm)}$$

The method also considers that the infiltration rate decreases exponentially after the beginning of the shower.

The hypotheses on which the model is based were deemed to satisfy for the conditions in which significant precipitations occur in Porto Santo and which were described in 3.2.2. As the spacing between the showers is wide enough (an average of 7 days at least) the moisture state considered for the soil at the beginning of each significant shower was that of moderately dry.

The programme memorizes for each precipitation that originates flow, the values of the total precipitation, of the infiltration, of the surface flow and the date of the shower.

3.2.4 - Real evapotranspiration

The potential evapotranspiration is calculated, for each month of the period under study, by one of the usual methods, according to the basic climatic data available, for instance, the Penman method for a given value of the characteristic albedo of the basin in question, or the Thornthwaite method.

If evaporimeter readings are available and if the monthly mean relation between these values and the potential evapotranspiration of the region is known, the monthly values observed may also be used after correction according to the referred mean relation.

These monthly values will later be divided in the programme by the number of days in the month, to obtain the mean daily potential evapotranspiration.

3.2.5 - Surface infiltration

If the daily precipitation exceeds the joint value of the daily surface flow and of the potential evapotranspiration the remaining precipitated volume will increase the moisture content of the soil. The value of the soil moisture should not be mistaken for that of aquifer recharge because the moisture in the soil may be evapotranspired partially or totally (up to the minimum limit of the hygroscopic water content) in the days following precipitation, without recharching underlying geologic formations.

3.2.6 - Soil moisture

This variable is closed related with the water that can be used for the evapotranspiration of a soil. This capacity is a content equal to the difference between specific retention and hygroscopic water content. It depends on the type of soil and on the relative humidity of the air of the region under study. The value of specific retention can be approximated in a laboratory by an other variable known as the equivalent moisture content of the soil. This content corresponds to the water retained in a sample that has been centrifugated during 40 minutes at 1000 atmopheres. The hygroscopic water content can also be obtained in the laboratory if the mean annual relative humidity of the air in the region under study is known. To determine the quantity of water used in evapotranspiration of the (unsaturated) soil per unit area it is necessary to know the depth of the soil subject to evapotranspiration. Direct evaporation of the soil decreases exponentially with the depth and, according to TODD 1959, it usually becomes negligible from one meter downward. The depth subjected to plant transpiration depends on the plant cover of the soil. For crops and other small plants the maximum depth usually attained by the roots does not exceed 50 or 70 cm. For trees the mean depth of the roots may attain two or three meters.

Assuming that the water table level lies deeper than that of the roots and that the soil is therefore unsaturated, it can be seen that the moisture content of the soil does neither drop below the hygroscopic water content, because it is then balanced with the atmospheric humidity, nor does it exceed that of specific retention because, as soon as this value is exceeded, (vertical) percolation sets in deep recharge of aquifers.

TABLE 1 - SEQUENTIAL DAILY WATER BALANCE:

HIDROGRAPHIC BASIN OF RIBEIRO COCHINO

AGUT: 70.0 MM

HIDROL. YEAR	OCT			NOV			DEC			JAN			FEB			MAR		
	PRC	EVP	ESC	PRC	EVP	ESC	PRC	EVP	ESC	PRC	EVP	ESC	PRC	EVP	ESC	PRC	EVP	ESC
1963-1964	17.3	80.0	0.2	26.7	60.0	0.0	91.4	43.0	0.0	83.8	44.0	3.5	8.7	44.0	0.0	34.5	52.0	0.0
1964-1965	11.3	82.0	0.0	11.1	63.0	0.0	38.4	43.0	0.0	79.5	41.0	2.8	44.9	41.0	2.6	16.2	55.0	0.0
1965-1966	139.1	71.0	11.4	42.5	50.0	0.0	13.6	45.0	0.0	6.5	50.0	0.0	33.1	43.0	0.0	92.2	51.0	4.6
1966-1967	118.7	80.0	21.7	41.2	50.0	0.0	45.5	44.0	0.0	14.2	45.0	0.0	22.3	40.0	0.0	2.6	57.0	0.0
1967-1968	76.7	84.0	1.6	146.0	54.0	3.5	30.9	45.0	0.0	13.2	42.0	0.0	78.9	43.0	5.1	96.8	43.0	2.2
1968-1969	49.7	88.0	1.0	110.2	60.0	6.2	56.3	45.0	0.2	137.9	50.0	26.1	88.4	44.0	0.0	39.1	53.0	0.0
1969-1970	54.6	77.0	0.3	36.1	58.0	0.0	25.8	42.0	0.0	151.6	46.0	26.1	53.6	38.0	2.1	94.9	47.0	20.0
1970-1971	11.9	83.0	0.0	62.6	60.0	0.2	99.1	43.0	3.4	37.8	46.0	0.0	62.0	38.0	2.3	35.3	44.0	0.0
1971-1972	1.7	91.0	0.0	38.5	59.0	0.0	64.8	46.0	15.5	48.1	43.0	0.0	62.3	37.0	2.8	47.3	44.0	0.0
1972-1973	65.4	73.0	0.0	30.7	56.0	0.0	58.7	43.0	0.0	46.9	43.0	1.2	29.8	40.0	0.0	26.1	53.0	0.1
1973-1974	36.9	79.0	0.0	39.3	59.0	0.0	22.2	45.0	0.0	7.8	44.0	0.0	15.5	42.0	0.0	85.1	48.0	0.1
1974-1975	33.7	76.0	0.0	6.2	57.0	0.0	12.1	50.0	0.0	59.9	46.0	0.4	49.0	42.0	3.8	15.0	52.0	0.0
1975-1976	17.9	77.0	0.0	14.2	60.0	0.0	108.5	40.0	1.3	20.5	41.0	0.0	57.6	39.0	0.0	26.8	48.0	0.0
1976-1977	24.2	80.0	0.0	28.4	57.0	0.0	103.5	46.0	0.9	48.4	42.0	0.0	38.7	41.0	0.0	9.8	56.0	0.0
1977-1978	62.5	77.0	4.1	67.8	58.0	5.8	116.7	49.0	8.2	61.9	39.0	0.0	25.0	44.0	0.0	2.9	50.0	0.0
TOTAL	722.	1198.	40.	702.	861.	16.	888.	669.	30.	818.	662.	60.	670.	616.	19.	625.	753.	27.
MEAN	48.1	79.9	2.7	46.8	57.4	1.0	59.2	44.6	2.0	54.5	44.1	4.0	44.7	41.1	1.2	41.6	50.2	1.8
STD.DEV.	39.8	5.3	6.1	37.6	3.7	2.2	36.4	2.6	4.3	44.2	3.1	9.0	23.0	2.3	1.7	34.1	4.4	5.2

PRC : PRECIPITATION (MM)
EVP : POTENTIAL EVAPOTRANSPIRATION (MM)
ESC : RUNOFF (MM)
AGUT: WATER CAPACITY AVAILABLE FOR EVAPOTRANSPIRATION
INITIAL VALUES: CARACT. N. = 79.6 AND SOIL MOISTURE = 0.0 MM

TABLE 1 - SEQUENTIAL DAILY WATER BALANCE:

HIDROGRAPHIC BASIN OF RIBEIRO COCHINO

AGUT: 70.0 MM

HIDROL. YEAR	OCT			NOV			DEC			JAN			FEB			MAR		
	RAQ	EVR	HUM	RAQ	EVR	HUM	RAQ	EVR	HUM	RAQ	EVR	HUM	RAQ	EVR	HUM	RAQ	EVR	HUM
1963-1964	0.0	10.3	6.8	0.0	32.6	0.9	0.0	43.0	49.3	24.5	44.0	61.1	0.0	44.0	25.8	0.0	52.0	8.3
1964-1965	0.0	11.3	0.0	0.0	11.1	0.0	0.0	31.4	7.0	0.0	36.8	46.9	0.8	41.0	47.4	0.0	55.0	8.6
1965-1966	0.0	67.4	60.3	11.1	50.0	41.7	0.0	45.0	10.3	0.0	16.8	0.0	0.0	26.0	7.1	0.0	45.0	49.7
1966-1967	0.0	40.4	56.5	4.5	50.0	43.2	0.0	44.0	44.7	0.0	45.0	13.9	0.0	36.2	0.0	0.0	2.6	0.0
1967-1968	0.0	54.1	21.0	44.9	54.0	64.6	2.3	45.0	48.2	0.0	42.0	19.4	0.0	43.0	50.2	32.9	43.0	68.8
1968-1969	0.0	20.8	27.9	6.3	60.0	65.6	30.5	45.0	46.2	62.9	50.0	45.2	19.7	44.0	69.8	9.6	53.0	46.3
1969-1970	0.0	51.1	3.3	0.0	34.1	5.2	0.0	31.0	0.0	18.4	46.0	61.1	23.1	38.0	51.5	26.3	47.0	53.1
1970-1971	0.0	11.9	0.0	0.0	38.2	24.2	7.1	43.0	69.8	2.4	46.0	59.3	16.8	38.0	64.1	0.0	44.0	55.4
1971-1972	0.0	1.7	0.0	0.0	38.5	0.0	0.0	16.1	33.2	0.0	43.0	38.3	0.0	37.0	60.9	6.0	44.0	58.1
1972-1973	0.0	73.0	30.0	0.0	51.8	8.9	0.0	43.0	24.6	0.0	43.0	27.3	0.0	40.0	17.1	0.0	42.1	1.0
1973-1974	0.0	26.5	10.4	0.0	49.7	0.0	0.0	22.2	0.0	0.0	7.8	0.0	0.0	12.8	2.7	0.0	42.8	44.9
1974-1975	0.0	33.7	0.0	0.0	6.2	0.0	0.0	12.1	0.0	0.0	36.8	22.7	0.0	42.0	25.9	0.0	39.0	1.9
1975-1976	0.0	17.9	0.0	0.0	14.2	0.0	0.0	39.1	68.1	0.0	41.0	47.6	2.4	39.0	63.7	2.4	48.0	40.1
1976-1977	0.0	24.3	0.0	0.0	28.4	0.0	0.0	37.2	65.3	4.3	42.0	67.4	0.2	41.0	64.9	0.0	56.0	18.7
1977-1978	0.0	50.1	8.3	0.0	29.8	40.6	31.1	49.0	68.9	33.4	39.0	58.4	0.0	44.0	39.4	0.0	42.3	0.0
TOTAL	0.0	494.6	224.5	66.8	548.5	295.0	71.0	546.2	535.7	145.8	579.3	568.6	63.0	566.0	590.6	77.4	655.8	455.0
MEAN	0.0	33.0	15.0	4.5	36.6	19.7	4.7	36.4	35.7	9.7	38.6	37.9	4.2	37.7	39.4	5.2	43.7	30.3
STD.DEV.	0.0	22.1	20.4	11.7	16.5	24.8	10.8	11.4	26.9	18.1	11.4	22.7	8.2	8.3	24.5	10.4	12.5	25.3

RAQ : AQUIFER RECHARGE (MM)
EVR : REAL EVAPOTRANSPIRATION (MM)
HUM : SOIL MOISTURE (MM)
AGUT: WATER CAPACITY AVAILABLE FOR EVAPOTRANSPIRATION
INITIAL VALUES: CARACT. N. = 79.6 AND SOIL MOISTURE = 0.0 MM

RECIP.,POTENTIAL EVAPOTRANS. AND RUNOFF

APR			MAY			JUN			JUL			AUG			SEP			ANNUAL VALUES		
[P]RC	EVP	ESC	PRC	EVP	ESC	PRC	EVP	ESC	PRC	EVP	ESC	PRC	EVP	ESC	PRC	EVP	ESC	PRC	EVP	ESC
7.0	54.0	0.0	0.0	81.0	0.0	16.3	90.0	0.0	0.6	100.0	0.0	6.1	105.0	0.0	43.8	95.0	3.0	356.2	848.0	6.7
8.7	58.0	0.0	0.4	83.0	0.0	13.6	96.0	0.0	6.0	108.0	0.0	8.1	108.0	0.0	5.2	92.0	0.0	243.4	870.0	5.4
4.9	68.0	0.0	8.1	73.0	0.0	9.3	89.0	0.0	5.4	96.0	0.0	11.7	111.0	0.0	26.0	92.0	0.0	392.4	839.0	16.0
2.8	54.0	0.0	58.9	69.0	1.5	0.3	85.0	0.0	2.7	104.0	0.0	0.3	114.0	0.0	13.7	100.0	0.0	363.2	842.0	23.2
9.0	55.0	0.0	6.2	72.0	0.0	0.5	91.0	0.0	0.0	112.0	0.0	2;9	117.0	0.0	5.5	104.0	0.0	486.6	862.0	12.4
5.0	53.0	0.0	19.8	70.0	0.0	14.6	81.0	0.0	0.0	104.0	0.0	2.6	113.0	0.0	21.9	91.0	0.0	555.5	852.0	33.5
8.3	61.0	0.0	7.1	74.0	0.0	13.6	87.0	0.0	2.8	108.0	0.0	1.8	111.0	0.0	14.0	98.0	0.0	464.2	847.0	48.4
6.2	54.0	0.0	18.5	66.0	0.0	9.3	83.0	0.0	0.0	103.0	0.0	8.2	103.0	0.0	8.0	92.0	0.0	378.9	815.0	5.9
4.7	57.0	0.0	3.2	68.0	0.0	7.8	82.0	0.0	7.5	105.0	0.0	3.1	110.0	0.0	73.1	95.0	2.7	372.1	837.0	20.9
9.2	62.0	0.0	24.5	75.0	0.0	2.1	86.0	0.0	5.3	103.0	0.0	3.2	107.0	0.0	14.1	99.0	0.0	326.0	840.0	1.4
8.0	52.0	0.0	20.9	76.0	0.1	5.9	87.0	0.0	1.6	99.0	0.0	2.5	102.0	0.0	16.4	89.0	0.0	272.1	822.0	0.2
6.3	57.0	0.0	5.5	73.0	0.0	9.9	88.0	0.0	1.4	111.0	0.0	2.1	116.0	0.0	4.2	93.0	0.0	225.3	861.0	4.2
9.3	53.0	0.0	15.5	65.0	0.0	1.6	90.0	0.0	2.4	115.0	0.0	9.1	131.0	0.0	23.4	104.0	0.0	316.8	863.0	1.3
3.0	58.0	0.0	3.9	69.0	0.0	4.4	85.0	0.0	2.9	97.0	0.0	8.6	104.0	0.0	7.6	111.0	0.0	283.4	846.0	0.9
9.3	60.0	0.0	8.2	70.0	0.0	7.3	81.0	0.0	0.5	105.0	0.0	0.6	111.0	0.0	5.5	107.0	0.0	368.2	851.0	18.1
72.	856.	0.	201.	1084.	2.	117.	1301.	0.	39.	1570.	0.	71.	1663.	0.	282.	1462.	6.	5404.	12695.	199.
8.1	57.1	0.0	13.4	72.3	0.1	7.8	86.7	0.0	2.6	104.7	0.0	4.7	110.9	0.0	18.8	97.5	0.4	360.3	846.3	13.2
0.8	4.3	0.0	14.8	5.1	0.4	5.3	4.1	0.0	2.4	5.5	0.0	3.6	7.2	0.0	18.4	6.6	1.0	90.7	15.0	13.8

UIFER RECHARGE, REAL EVAPOTRANS AND SOIL MOISTURE (Cont.)

APR			MAY			JUN			JUL			AUG			SEP			ANNUAL VALUES		
[RA]Q	EVR	HUM	RAQ	EVR	HUM	RAQ	EVR	HUM	RAQ	EVR	HUM	RAQ	EVR	HUM	RAQ	EVR	HUM	ANO	RAQ	EVR
.0	35.3	0.0	0.0	0.0	0.0	0.0	16.3	0.0	0.0	0.6	0.0	0.0	6.1	0.0	0.0	40.8	0.0	63-64	24.5	325.0
.0	17.3	0.0	0.0	0.4	0.0	0.0	13.6	0.0	0.0	6.0	0.0	0.0	8.1	0.0	0.0	5.2	0.0	64-65	0.8	237.3
.0	54.6	0.0	0.0	3.1	5.0	0.0	14.3	0.0	0.0	5.4	0.0	0.0	11.7	0.0	0.0	26.0	0.0	65-66	11.1	365.4
.0	36.0	6.8	0.0	64.2	0.0	0.0	0.3	0.0	0.0	2.7	0.0	0.0	0.3	0.0	0.0	13.7	0.0	66-67	4.5	335.4
.1	55.0	30.7	0.0	36.9	0.0	0.0	0.5	0.0	0.0	0.0	0.0	0.0	2.9	0.0	0.0	5.5	0.0	67-68	92.3	381.9
.0	53.0	8.3	0.0	28.1	0.0	0.0	14.6	0.0	0.0	0.0	0.0	0.0	2.6	0.0	0.0	21.9	0.0	68-69	129.0	393.1
.0	61.0	0.4	0.0	7.5	0.0	0.0	13.6	0.0	0.0	2.8	0.0	0.0	1.8	0.0	0.0	14.0	0.0	69-70	67.8	348.0
.0	54.0	27.6	0.0	46.1	0.0	0.0	9.3	0.0	0.0	0.0	0.0	0.0	8.2	0.0	0.0	8.0	0.0	70-71	26.3	346.7
.0	57.0	15.8	0.0	19.0	0.0	0.0	7.8	0.0	0.0	7.5	0.0	0.0	3.1	0.0	0.0	32.8	37.6	71-72	6.0	307.5
.0	20.2	0.0	0.0	24.5	0.0	0.0	2.1	0.0	0.0	5.3	0.0	0.0	3.2	0.0	0.0	14.1	0.0	72-73	0.0	362.3
.0	52.0	10.9	0.0	22.5	9.2	0.0	15.1	0.0	0.0	1.6	0.0	0.0	2.5	0.0	0.0	16.4	0.0	73-74	0.0	271.9
.0	23.1	0.0	0.0	5.5	0.0	0.0	9.9	0.0	0.0	1.4	0.0	0.0	2.1	0.0	0.0	4.2	0.0	74-75	0.0	221.1
.0	53.0	6.4	0.0	21.9	0.0	0.0	1.6	0.0	0.0	2.4	0.0	0.0	9.1	0.0	0.0	23.3	0.1	75-76	4.9	310.5
.0	21.7	0.0	0.0	3.9	0.0	0.0	4.4	0.0	0.0	2.9	0.0	0.0	8.6	0.0	0.0	7.6	0.0	76-77	4.5	278.1
.0	9.3	0.0	0.0	8.2	0.0	0.0	7.3	0.0	0.0	0.5	0.0	0.0	0.6	0.0	0.0	5.5	0.0	77-78	64.5	285.5
.1	607.6	106.9	0.0	291.8	14.2	0.0	130.7	0.0	0.0	39.1	0.0	0.0	70.9	0.0	0.0	239.0	37.8	TOTAL	436.2	4769.6
.8	40.5	7.1	0.0	19.5	0.9	0.0	8.7	0.0	0.0	2.6	0.0	0.0	4.7	0.0	0.0	15.9	2.5	MEAN	29.1	318.0
.1	17.3	10.2	0.0	18.6	2.6	0.0	5.8	0.0	0.0	2.4	0.0	0.0	3.6	0.0	0.0	11.0	9.7	S.DEV.	40.3	51.3

3.2.7 - Deep recharge of aquifers

Once the value of specific retention is reached, as said in the above subsection, percolation of the exceeding water sets in. For a saturated soil, the water velocity is directly proportional to the hydraulic gradient, the proportionality constant being the permeability of the soil. For an unsaturated soil the law that governs the velocities ceases to be linear. While permeability decreases, since the air in the pores hinders percolation, the hydraulic gradient increases, since "negative pressures" as regards the water table level (in which they are zero) accellerate the distribution of the moisture content.

The final infiltration capacity of an unsaturated soil varies according to REMENIERAS 1965 from 120 mm/hour for sands, to 20 mm//hour for clay alluvial soils.

Considering a mean infiltration coefficient of about 50 mm/hour it can be seen that the infiltration caused by concentrated precipitation exceeds the depht of one meter (of soil subject to evapotranspiration) after one day. It may thus be assumed that the excess soil moisture, as regards specific retention, will recharge the underlying aquifer, because after one day (the time basis considered in this sequential balance) this excess content will cease to be subject to evapotranspiration. Section 4 will deal with the results obtained for Santo Santo island.

4 - APPLICATION OF THE STUDY TO PORTO SANTO ISLAND WATER RESOURCES

The values obtained for the surface flow in Porto Santo island with the BALSEQ programme described in section 3 were compared with those obtained in other regions with similar climate as, for instance, the Algarve (South Portugal) coastal region. The values obtained for Porto Santo using this programme are of the same order of magnitude as that of the isolines of the surface flow of the Algarve coastal region given in QUINTELA 1967.

Next, a study of the surface water balance of the reservoirs of the island was made. Owing to their oversized dimensions these reservoirs do not discharge and all the water they receive is spent in irrigation or lost by evaporation or by leakage due to underground drainage (which must, nevertheless, be small as the reservoirs are quite sanded up with very fine materials). After integrating the annual volumes of the water distributed for irrigation, the other variables of the balance were calculated (except leakage):

PREC + FLOW = EVP + IRR + LEAK (1)
LEAK = (PREC + FLOW) - (EVP + IRR)

PREC - direct rainfall in the reservoir surface
FLOW - overland and surface flows calculated with the BALSEQ programme
EVP - direct evaporation from the reservoir (Penman with 5% of albedo)
IRR - water distributed for irrigation
LEAK - leakage through aquifer recharge or through underground drainage of the reservoir

The value obtained in (1) for the leakage in Porto Santo reservoirs was 20%. This value seems rather realistic and confirms the calculation model used for the surface flow.

The value of the deep recharge of aquifers was confirmed by the analysis of the exploitation and saturation rates of several groundwater catchment areas of Porto Santo island. The results obtained seem indeed realistic and confirm the exhaustion of the reserves of groundwater in several zones now being exploited and indicated the possibility of reserves in other zones that have proved suitable for new catchments. The results obtained confirm their hydrogeologic differeces observed in the zones that were studied.

The results obtained for the water balance of one of the hydrographic basins of Porto Santo are shown in Table 1.

5. CONCLUSIONS

It is thought that the model presented for the evaluation of the water balance may be used successfully in other regions with climatic conditions similar to those of Porto Santo.

The generalization of the programme to regions with different climates will have to be done more carefully and, whenever possible, results should be confirmed by those commonly accepted as, for instance the values shown by the isolines of the different hydrologic variables, given for those regions.

The programme was recently used to determine the deep rain recharge of a semi-confined aquifer in the portuguese Rio Maior region wich has a mean annual precipitation of about 800 mm. The results obtained show good agreement with the values calculated for the natural recharge of the aquifer, which were quantified by several pumping tests carried out in the region.

TABLE 2 — Values of the flow number N

Use of soil or soil cover	Surfaces conditions	Type of soil			
		A	B	C	D
Tilled soil		77	86	91	94
Culture of trees	along the steepest slope	64	76	84	88
	along contour lines	62	74	82	85
	along contour lines in terrace	60	71	79	82
Rotation of cultures	along the steepest slope	62	75	83	87
	along contour lines	60	72	81	84
	along contour lines in terrace	57	70	78	82
Pastures	poor	68	79	86	89
	average	49	69	79	84
	good	39	61	74	80
	poor, along contour lines	47	67	81	88
	average, along contour lines	25	59	75	83
	good, along contour lines	6	35	70	79
Permanent pasture	average	30	58	71	78
Rural social areas	average	59	74	82	86
Roads	pervious pavement	72	82	87	89
	impervious pavement	74	84	90	92
Forests	very open or low transpiration	56	75	86	91
	open or low transpiration	46	68	78	84
	average	36	60	70	76
	dense or high transpiration	26	52	62	69
	very dense or high transpiration	15	44	54	61
Impervious surface		100	100	100	100

BIBLIOGRAPHY

Todd, D. 1959, Ground Water Hydrology. New York, John Wiley & Sons, Inc.
Remenieras, 1965, L'Hydrologie de l'Ingénieur. Paris, Editions Eyrolles.
Quintela, A. 1967, Recursos de Águas Superficiais em Portugal Continental. Lisboa, Edição do autor.
LNEC, 1976, Determinação de Caudais de Ponta de Cheia em Pequenas Bacias Hidrográficas. Lisboa, Laboratório Nacional de Engenharia Civil.

Proceedings of Euromech 143 / Delft / 2-4 September 1981

Seawater intrusion in aquifers: The modeling of interface discontinuities

ANTÓNIO SÁ DA COSTA
Universidade Técnica de Lisboa, Lisboa, Portugal

1 INTRODUCTION

Seawater intrusion in coastal aquifers is a problem that affects coastal areas and islands subject to increasing groundwater withdrawal, to supply the demands due to the growth of population and industrial activities. This conduces to a change on the long term natural balance of freshwater flowing underground to the sea, and the heavier underlying seawater wedge. Since the freswater flow is reduced the seawater wedge advances, enroaching the freshwater threatning its availability with contamination by brackish water.

Numerical modeling has been, in recent years, a very usefull tool to simulate and analyse this seawater intrusion in aquifers with success. In many ot the models used it is assumed an immiscible interface between freshwater and seawater. Such procedure brings large savings on computer costs, and usually producing good results for the problems under study. However, a difficulty rises when the seawater wedge has a finite extent, due to a limited aquifer thickness. This produces a discontinuity of the interface and of the seawater phase. Usually is difficult, and often quite expensive, to model numerically discontinuities, specially when its location varies with time, as is the case of seawater intrusion in aquifers.

This paper presents a simple and inexpensive technique to model the interface discontinuity. This technique has been used in a finite element model called SWIM, an acronym for SeaWater Intrusion Model. This model was implemented at the Massachusetts Institute of Technology, USA, to simulate groundwater flow in coastal areas and under offshore islands. However, the technique discrebed here can be also implemented in other models using Gauss quadrature for the spatial numerical integration.

2 GOVERNING EQUATIONS

In most practical situations the transition zone between freshwater and seawater, where mixing and dispersion phenomena occur, is quite narrow when compared with the overall saturated thickness of the aquifer formation. In these cases an immiscible interface between freshwater and seawater is an appropriate and very convinient assumption (Bear, 1972 and 1979; Shamir and Dagan, 1971; Pinder and Page, 1976). This approximation leads to the definition of two types of freshwater aquifers (see fig. 1): lens when the freshwater "floats" on top of a seawater layer; toe when the immiscible interface intersects the bottom of the aquifer, defining a seawater wedge "toe". SWIM, which utilizes the immiscible interface approach, can simulate both types of aquifers.

Vertically integrated flow equations, also called Dupuit equations, are used to represent regional, essentially horizontal two-dimensional (2-D) flow. The derivation of the governing equations, leads to the following tensor notation form (Sá da Costa and Wilson, 1979):

FRESHWATER:
$$\frac{\partial}{\partial x_i}\left(K^f_{x_i x_j} b^f \frac{\partial \phi^f}{\partial x_j}\right) + \frac{K'^f_m}{b'^f_m}(\phi'^f_m - \phi^f) + q^f + N = \left(S^f + n\frac{\gamma^f}{\Delta\gamma}\right)\frac{\partial \phi^f}{\partial t} - n\frac{\gamma^s}{\Delta\gamma}\frac{\partial \phi^f}{\partial t} \quad (1)$$

$i,j,m=1,2$

SEAWATER:
$$\frac{\partial}{\partial x_i}\left(K^s_{x_i x_j} b^s \frac{\partial \phi^s}{\partial x_j}\right) + \frac{K'^s_m}{b'^s_m}(\phi'^s_m - \phi^s) +$$

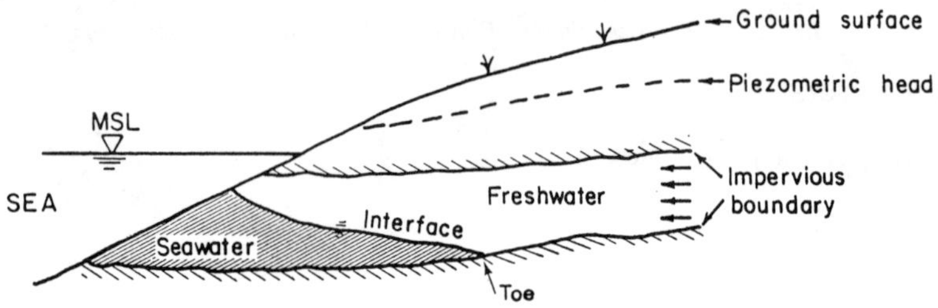

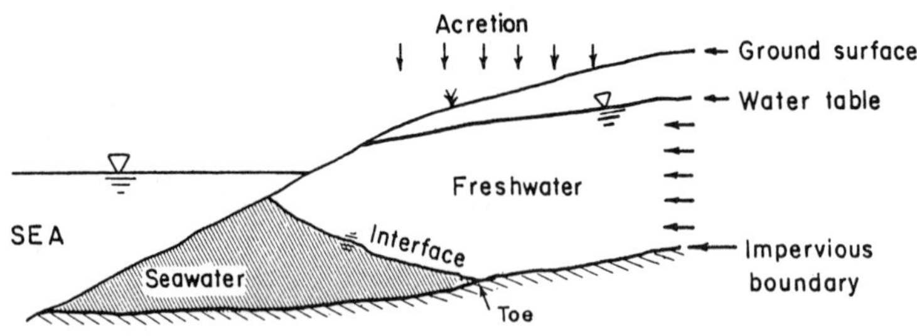

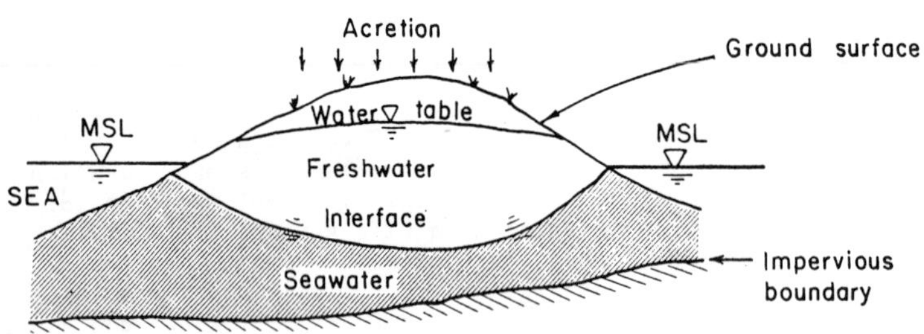

Fig.1 Seawater intrusion in different types of aquifers

$$q^s = (S^s + n\frac{\gamma^s}{\Delta\gamma})\frac{\partial\phi^s}{\partial t} - n\frac{\gamma^f}{\Delta\gamma}\frac{\partial\phi^f}{\partial t} \qquad (2)$$

$$i,j,m = 1,2$$

where $i,j = 1,2$ refer to ortogonal spatial coordinates, $x_i x_j$; and $m = 1,2$ refers to leakage from/to above or below. The interface depth, ζ, is defined by continuity in pressure at both sides of the interface. In terms of piezometric heads, this leads to the expression:

$$\zeta = \frac{\gamma^s}{\Delta\gamma}\phi^s - \frac{\gamma^f}{\Delta\gamma}\phi^f \qquad (3)$$

In these equations the superscripts ^{f} and ^{s} represent a freshwater and a seawater quantity, respectively , and

$K_{x_i x_j}$ are the components of the hidraulic conductivity tensor,

b is the saturated thickness,

ϕ is the piezometric head,

K'_m is the vertical hydraulic conductivity in the semi-pervious leaky layer (=0 if nonexistent) where m=1 top layer and m=2 bottom layer,

b'_m is the leaky layer thickness,

ϕ'_m is the piezometric head in the vertically adjacent aquifer on the other side of the leaky layer,

q represents the source/sink terms,

N is the natural accreation (=0 in confined aquifers),

S is the elastic storage for confined aquifers, or the specific yield for phreatic aquifers,

n is the effective porosity,

γ is the specific weight, $\Delta\gamma = \gamma^s - \gamma^f$,

t is the time.

The major assumptions behind the derivation of these equations are:
- immiscible interface separating freshwater and seawater,
- Darcy's law is applicable,
- vertical variations of the aquifer properties are neglected,
- Dupuit assumption is valid,
- vertical flow is only considered for leaky effects and for accreation,
- constant specific weight for both freshwater and seawater,
- freshwater and seawater are considered homogeneous, isotropic, and they completely fill all voids in the saturated zone of the porous media.

Using the Galerkin approximation to perform the space integration, the governing equations 1 and 2 can be reduced to a simpler matrix equation (Sá da Costa and Wilson, 1979):

$$\underline{\underline{C}}\frac{d\underline{X}}{dt} + \underline{\underline{K}}\underline{X} = \underline{F} \qquad (4)$$

where:

$\underline{\underline{C}}$ represents the storage matrix,

$\underline{\underline{K}}$ represents the transmissivity and leakage matrices,

$\underline{F}$ is the vector of independent variables,

$\underline{X}$ is the unknown piezometric heads vector.

SWIM uses mixed isoparametric elements, with functional coefficients for aquifer geomectric properties such as bottom and top elevations. Equation 4 is integrated in time using a fully implicit finite difference scheme and the resulting non-linear system of equations is solved iteratively using a modified Newton-Raphson procedure.

3 INTERFACE DISCONTINUITY. MODELING TECHNIQUES

As shown on the two top scheme of figure 1, quite often appear cases where the seawater wedge as a finite extent, defining a toe at the interface intersection with the bottom of the aquifer. This toe divides the aquifer in two. One, seaward of the toe, to the left of figure 1, where both freshwater and seawater exist; the other, inland of the toe, to the right of the same figure, where only freshwater exists. That is the toe also represents a discontinuity of the seawater phase.

For seawater intrusion in aquifers the term "toe" usually represents the intersection of the interface with the bottom of the aquifer. However, there are situations where the interface also intersects the top of a confined-leaky aquifer, and this "upper-toe" represents a discontinuity, now for the freshwater phase.

These upper and lower toe are moving boundaries in almost all the situations encountered in groundwater flow simulation, bringing some difficulties when one tries to model the exact position of the toe. Since both upper and lower toe will be treated in the same way, in this paper it will be refered only the lower toe, which is the more common case.

Some authors, to avoid the difficulty of the moving boundary at the interface discontinuity, ignored the existence of an impervious bedrock, or at least considered it very deep in order to avoid such discontiniuty. With such procedure they could model the aquifer as a lens situation, and later they determine the position of the toe by the intersection of the interface calculated ignoring the bottom with the bottom surface. The position found for both the interface and toe using this methodology is corrected only under steady state conditions, otherwise, it does not take in account a different transmissivity for both freshwater and seawater, due to the artificial deepning of the aquifer's bottom.

Since the steady state takes many years to attain, hundreds and sometimes thousands years, it can be stated that transient situations are the more common ones, and the ones that should be simulated. Steady state represents an ultimate situation, usually never achieved, and its only interest is the evaluation of the ultimate consequences of an exploitation policy.

Other technique to model the interface discontinuity, uses mesh regeneration or mesh displacement algorithms. This technique is quite difficult to implement and very costly to use, specially in cases with varying aquifer geometry and properties.

It is possible to represent the interface by the limits of the elements, when using finite elements models, or of cells, when using finite differences models. In other words,, the toe will never lie inside an element, or cell and will be always 'pushed' or 'pulled' to the nearest limit. This procedure can lead to over, or under, predict the toe's position depending on the seawater wedge movement and on its velocity.

A reliable and inexpensive alternative to track the toe movement, when using finite elements with quadratic elements, is to take advantage of the Gauss quadrature points used in the finite element numerical integration of the spatial matrices. This methodology is fully described in Sá da Costa and Wilson, 1979 and in Wilson and Sá da Costa 1980. This toe track algorithm is implemented in three steps. First the evaluation of the elements with a toe; this is achieved by calculating the interface elevation at the element's nodes and comparing it with the corresponding bottom's elevation. The second step consists in increasing the number of Gauss points, usually 2x2 to 4x4, to obtain more detail. In the third step the freshwater and seawater saturated thickness are evaluated at the Gauss points using the continuity in pressure at the interface (eq.3). This values are used to compute the transmissivity matrix in the usual way.

With this procedure two sets of Gauss points are implicitly defined: one inland of the toe where there is no seawater; the other seaward of the toe where there are two layers, one of freshwater the other of seawater (fig.2). The model recognizes, for computational purposes, that the toe lies between this two sets of Gauss points. The toe is exactly located only for output purposes.

This procedure alone does not lead to a great improvement, because the finite element method smooths out all sharp transitions such as the sharp corner exi ting at the toe, see

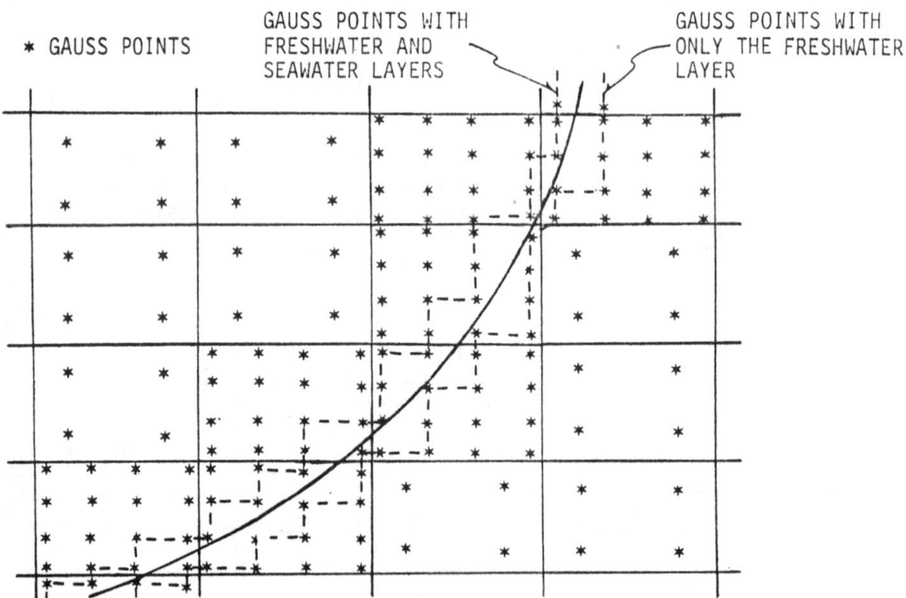

Fig. 2 Plan view of the toe location by a model using the Gauss quadrature points (SWIM)

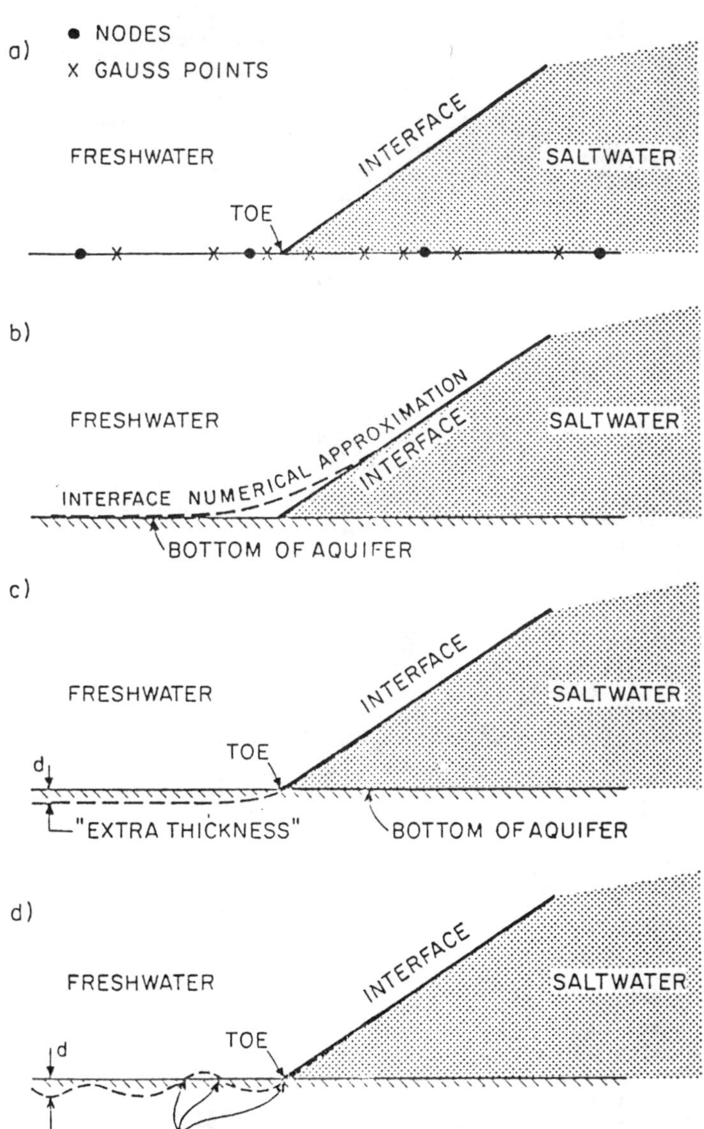

Fig.3 Representation of the interface's discontinuity at the toe: a) toe location within a finite element grid using Gauss points; b) interface smoothing at th e toe; c) interface numerical approximation using "extra-thickness" to define a pseudo-interface to the left of the toe; d) spatial numerical oscilations or "wrinkles" of the interface to the left of the toe.

fig.3. Therefore, the toe track algorithm must be conbined with the artifice of considering the interface beyond the toe to lie just below of the aquifer's bottom, defining the toe at the intersection of the interface with the bottom (fig.3c). This procedure eliminates the smoothing of the sharp corner at the toe, but the thickness of this "extra--thickness" should be kept small. A sensitivity analysis, Sá da Costa and Wilson,1979, shows that the ratio $d/b=10^{-3}$ is optimal (b is the aquifer's thickness). Larger values of d conduces to a decceleration of the toe movement, smaller values of d will not absorb the "wrinkles", spatial oscilations, of the interface just outside of the toe (fig.3d), due to the abrupt change of gradients.

Another advantage of adding the "extra--layer" is to avoid zeros on some rows of the transmissivity matrix, that could lead to a non-positive definite matrix. This because a finite saturated thickness for each phase is defined over the entire domain, or at least over the part of it where the two phases can coexist. This implies that where a phase is absent, its saturated thickness must be multiplied by some hydraulic conductivity K^d. For $10^{-5} \lesssim K^d/K \lesssim 1$ where $K=K^f$ (or K^s) the simulation results are almost unchanged. The "extra-layer" and the toe track algorithm only affect the toe region having no influence on the far field.

In Sá da Costa and Wilson, 1979 is presented a sensitivity analysis for the intrinsic factores of this technique. It is also described several applications of this methodology.

4 AQUIFER'S BOTTOM INFLUENCE ON THE INTERFACE SHAPE

The case of the development of a freshwater lens over seawater is presented here, because it illustrates quite well the point made in the previous section that the assumption of considering an infenitly deep aquifer will modify the interface shape.

An approximated analytical solution have been proposed by Hantush, 1968 for thick aquifers near the coast for the case of a coastal aquifer, with finite length L, receiving recharge from an infinitely long strip parallel to the coast (fig.4). Hantush, 1968 assumed Dupuit approximation, used Ghyben--Herzberg hypothesis, thus neglecting flow in the seawater phase, and employed a linearization at the resulting equation to arrive to an analytical expression for both interface depth and flow discharged to the sea for the situation schematized in fig.4.

SWIM was used to simulate this problem with the following characteristics:

L=100m

K^f=100m/day N=0.2m/day

γ^f=1000kg/m^3 γ^s=1025kg/m^3

The finite element grid used linear 4-node elements 5m long. The boundary conditions imposed were: at the coast a first type for seawater and a third type for freshwater; at the right limit a "no flow" condition. An arbitrary depth of the aquifer of 20m was selected for the simulation, the effect of the depth is discussed later. The initial condition for the interface and phreatic surface was specified as a horizontal plane with elevation equal to the meam sea level. In his work Hantush assumed a static seawater, this was simulated in SWIM by selecting an extremely large hydraulic conductivity for the seawater, $K^s/K^f = 10^4$.

Fig.5 presents the results from SWIM simulation and the analytical solution. At earlier stages Hantush solution does not conser-

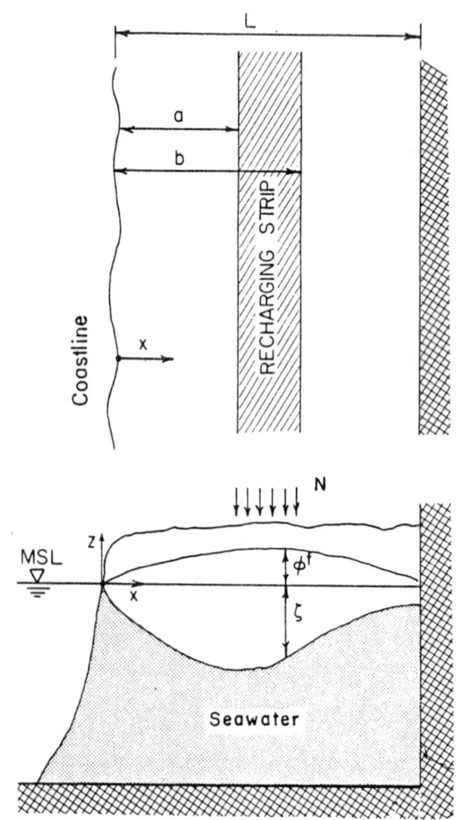

Fig.4 Schematic representation of the development of a freshwater lens over seawater. a) plan view; b) cross section.

ve mass. Later, $t \leq 100$ days, the solution becomes conservative. The explanation for this problem lies in the way the averaged interface depth is calculated. SWIM solution is always conservative. The final steady state solution is identical for both cases, because the Ghyben-Herzberg assumption is now applicable and the linearization for the analytical solution is now calculated correctly. The principal explanation for the observed differences between the two solutions can be attributed to these added approximations of the analytical solution: Ghyben-Herzberg and linearization. Fig.6a shows the drawdown at the midpoint of the recharge area, and both solutions are similar. The outflow to the sea per unit width is compared in fig.7a. The flow computed using Hantush's method is

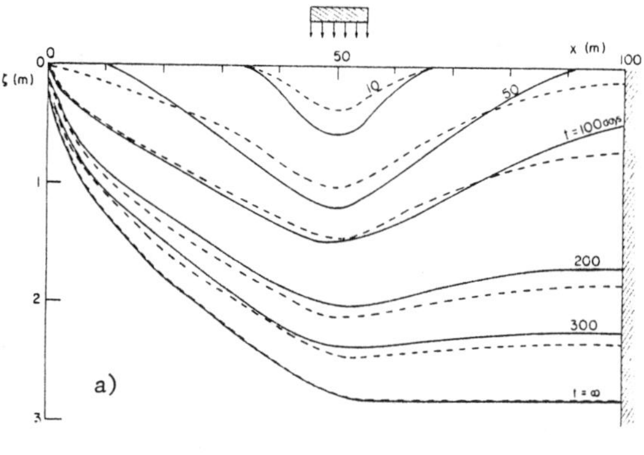

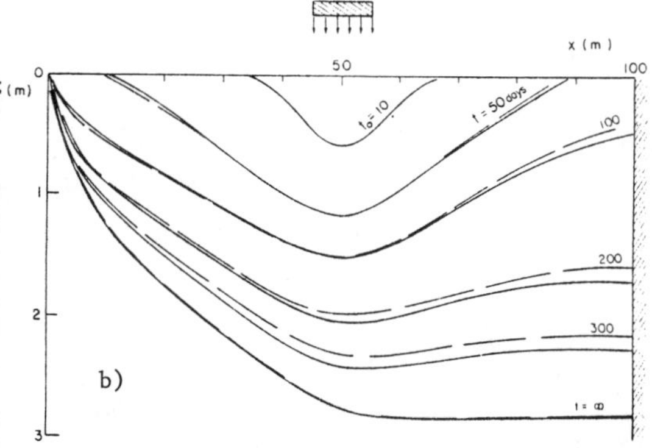

Fig.5 Interface movement in the development of a freshwater lens over seawater. a) Comparison between Hantush, 1968 (- - -) and SWIM, with Ghyben- Herzberg (——) and without (— — —); b) Effect of the Ghyben-Herzberg assumption.

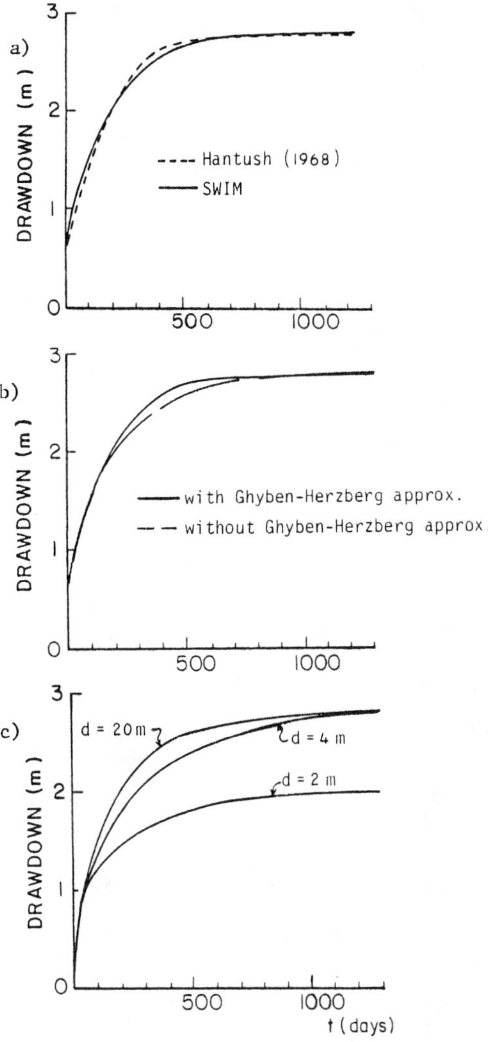

Fig.6 Comparison of the drawdown at the midpoint of the freshwater lens for different situations: a) Hantush, 1968 versus SWIM; b) effect of the Ghyben-Herzberg assumption; c) influence of the bottom elevation.

slightly smaller, consistent with its faster lens development.

To infer the influence of the Ghyben-Herzberg assumption a simulation with $K^s=K^f$ was performed. The results are compared with the previous simulation, in which $K^s/K^f=10^4$, in fig.5b. The final steady state solutions are identical. The transient interface that accounts for seawater dynamics moves slower than the seawater assumed static (fig.6b), although the discharge to the sea does not change much (fig.7b). In fact, during the first 40 to 50 days of lens development the

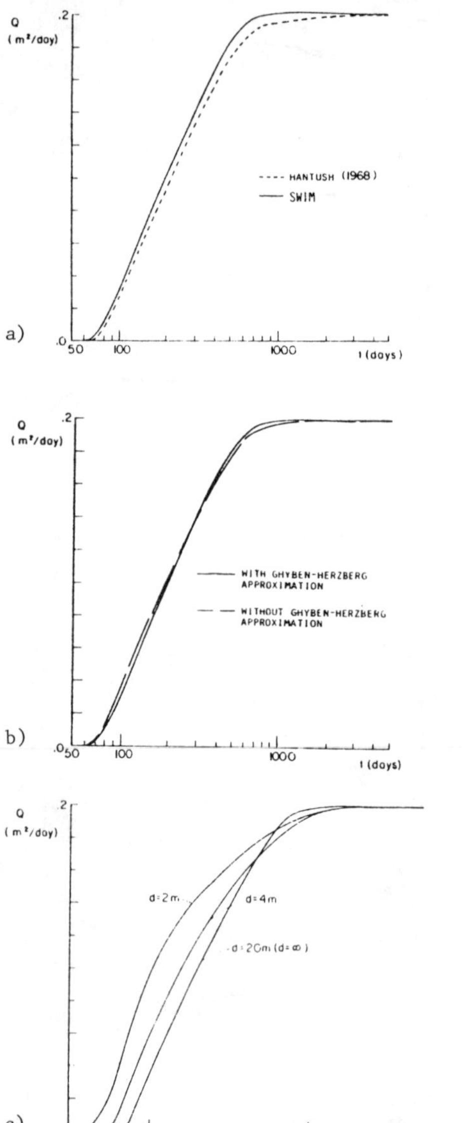

Fig.7 Comparison of the freshwater flow to the sea for different situations of the development of a freshwater lens over seawater: a) Hantush, 1968 versus SWIM; b) effect of the Ghyben-Herzberg assumption; c) influence of the bottom elevation.

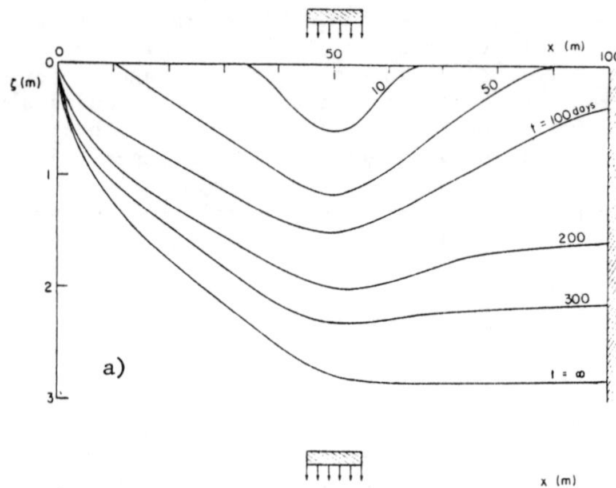

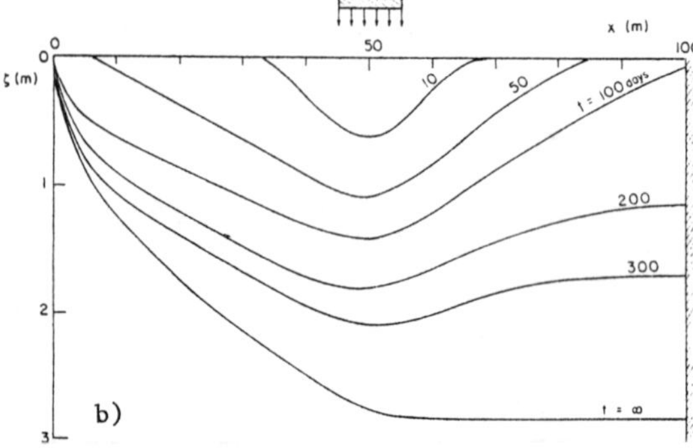

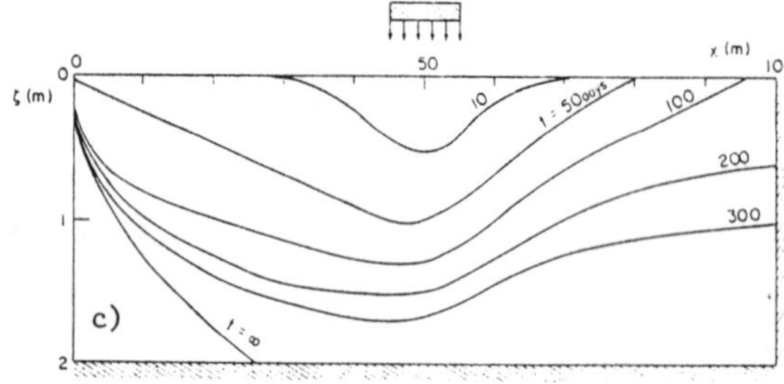

Fig.8 Interface movement in t he development of a freshwater lens over seawater. Influence of the bottom elevation. a) Depth of the bottom d=20m; b) d=4m; c) d=2m.

re is almost no difference in the solution. The lens is less than 1m thick and when this is compared to the overall aquifer thickness of 20m, the Ghyben-Herzberg assumption is reasonable. Later, however, the lens thickness grows and the approximation becomes less valid. Also note, for this example that the additional error of the linearization is as large as, or larger, than the error due to Ghyben -Herzberg approximation.

The appropriatness of this approximation is related to the aquifer's depth. In the solutions presented thus far the maximum depth of the int erface has been less than 3m, which is much smaller than the total depth of the aquifer, 20m. In these examples the aquifer can be considered a thick formation and the Ghyben-Herzberg approximation should apply. In the next set of simulations (fig.8) the seawater was assumed dynamic, while the elevation of the aquifer's bottom elevation was varied from depths of 20m, 4m and 2m. The interface positions for these three simulations are presented in fig.8, the drawdown at the midpoint of the recharge area is shown in fig.6c and the unit width freshwater discharge to the sea is given in fig.7c.

The influence of the bottom is felt in two important ways: a reduction of the rate of drawdown of the interface, and an increase of the freshwater discharge to the sea. These two phenomena are related and are due to the resistence the seawater encounters when it attempts to flow back from the aquifer to the sea. An interesting note is that for the depth of 2m no seawater is trapped in the aquifer by the freshwater lens, when it hits the bottom. As expected the steady state shape of the interface is the same for all three cases.

This example clearly demonstrates that in many cases the Ghyben-Herzberg assumption of a infinitely deep aquifer, to avoid the problem of modeling the toe and the seawater regime, is not always satisfactory, unless one is seeking a steady state solution.

5 CONCLUSIONS

The moving boundaries associated with the seawater wedge toe can be accuratly tracked using a fixed grid and a special algorithm (sá da Costa and Wilson, 1979; and Wilson and Sá da Costa, 1980). This algorithm locates the elements with a toe, then uses the interpolated piezometric heads and the interface pressure continuity boundary condition at the Gauss points to represent the discontinuity at the toe.

This methodology is quite easy to implement in existing codes using Gauss quadrature to evaluate the space integrals. The increase in CPU time due to this technique is minimal because the process described is only applied at elements with a toe, and because the calculations performed are quite simple. The accuracy associated with this methodology is comparable with that of grid regeneration schemes at a fraction of the cost (Wilson and Sá da Costa, 1980).

The results presented here confirmed that the Ghyben-Herzeberg approximation is not valid for transient problems, and that the aquifer's bottom depth influenciates the interface shape in transient problems.

6 ACKNOWLEDGMENTS

The author is gratefull to Prof. John Wilson of MIT who supervised him in the research presented here. This research was sponsered by the MIT Sea Grant Program, supported by NOAA' Office of Sea Grant, U.S. Department of Commerce, under Grant No. NA79AA-D-00101.

The U.S. government is authorized to produce and distribute reprints for governamental purposes, notwithstanding any copyright notation that may appear hereon.

7 REFERENCES

Bear, J. 1972, Dynamics of fluids in porous media, New York, American Elsevier Publishing Co. Inc.

Bear, J. 1979, Hydraulics of groundwater, New York, McGraw-Hill Book Co.

Hantush, M.S., 1968, Unsteady movement of fresh water in thick unconfined aquifers, Bulletin of International Association of Scientific Hydrology, XII(2), 40-60.

Pinder, G.F. and Page, R.H., 1976, Finite element simulation of salt-water intrusion in the South Fork of Long Island, Princeton, International Conference in Finite Elements in Water Resources.

Sá da Costa, A. and Wilson, J.L., 1979, A numerical model of seawater intrusion in aquifers, Cambridge (USA), R.M. Parsons Laboratory for Water Resources and Hydrodynamics, MIT.

Shamir, U. and Dagan, G., 1971, Motion of the seawater interface in coastal aquifers: a numerical sol ution, Water Resources Research, 7(3), 644-657.

Wilson, J.L. and Sá da Costa, A., 1980, Finite element simul ation of a saltwater/ /freshwater interface with indirect toe tracking, submited for publication in Water Resources Research.

Proceedings of Euromech 143 / Delft / 2-4 September 1981

The analysis of two-phase flow in the underground gas store considering memory effects

MARIUSZ R.SŁAWOMIRSKI
Polish Petroleum Institute, Kraków, Poland

1 INTRODUCTION

The underground gas store located in the depleted natural gas reservoir is alternately injected and exploited according to gas demand in the gas system.

Gas exploitation implies the progressive waterflooding of the store by contouring or underlaying water, whereas gas injection causes the displacement of water by the injected gas. Consequently, water-gas interphase surface moves back and forth.

Although water and gas are practically immiscible it is impossible to determine the interphase surface mentioned above because two-phase flow through porous media consists in increasing water saturation and decreasing gas saturation, or vice versa.

2 MULTIPHASE FLOW EQUATIONS

According to the wide-spread opinion the two-phase water-gas flow may be described by means of the generalized Darcy law which states that the velocity of each phase is a linear function of the pressure gradient, i.e.:

for water:

$$\vec{u}_w = \mathbb{K} \frac{k_w}{\mu_w} \operatorname{grad}(p_w - \varrho_w \gamma z) \tag{1}$$

for gas:

$$\vec{u}_g = \mathbb{K} \frac{k_g}{\mu_g} \operatorname{grad}(p_g - \varrho_g \gamma z) \tag{2}$$

$\vec{u}$ denotes here velocity vector, $\mathbb{K}$ is absolute permeability tensor, k is relative permeability, μ is fluid viscosity, p is pressure, ϱ is fluid density, γ is acceleration of gravity force, z is vertical component of position vector. Subscripts w, g refer correspondingly to water and gas.

The absolute permeability tensor $\mathbb{K}$ only depends on the properties of porous rock, whereas the relative permeabilities k_w, k_g vary in the interval [0 ; 1] and depend on saturations S_w, S_g respectively /cf. Fig.1/:

$$k_w = \chi_w(S_w) \tag{3}$$

$$k_g = \chi_g(S_g) \tag{4}$$

The saturation is understood here as the volume of pores occupied by each phase per total volume of all pores. Consequently:

$$S_w + S_g = 1 \tag{5}$$

3 MEMORY EFFECTS

Laboratory investigations as well as practical petroleum engineering experience have shown that for repeated alternate water and gas displacement in the porous medium the curves representing k vs. S dependence are different for decrease and increase of water saturation, i.e. the hysteresis loop for k vs. S relation must be taken into consideration /cf. Fig. 2/.

Consequently it has to be assumed that the relative permeability at given time moment depends not only on the saturation in this moment but also on all previous history of the exploitation of the reservoir.

Note, that the `time´ to which the history of the reservoir is referred may not be identified with the physical time $t$. For practical purposes it is better to define the `reservoir time´ θ in the following manner:

$$\theta = \underset{\tau \in (-\infty;\, t]}{\mathrm{Var}} \left\{ S_w(\tau) \right\} \tag{6}$$

where

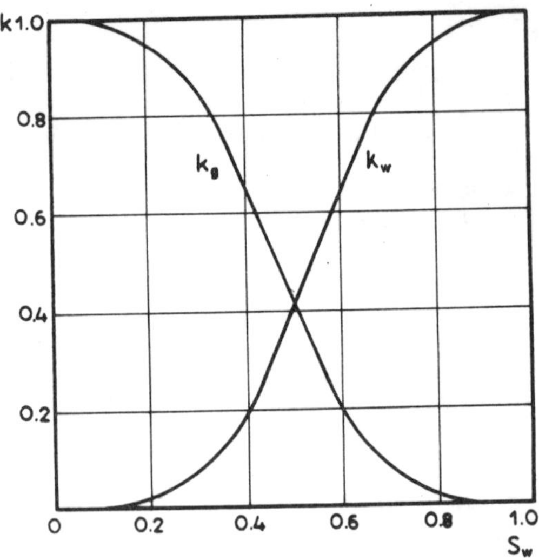

Fig. 1
Relative permeability curves for water $k_w(S_w)$ and gas $k_g(S_w)$

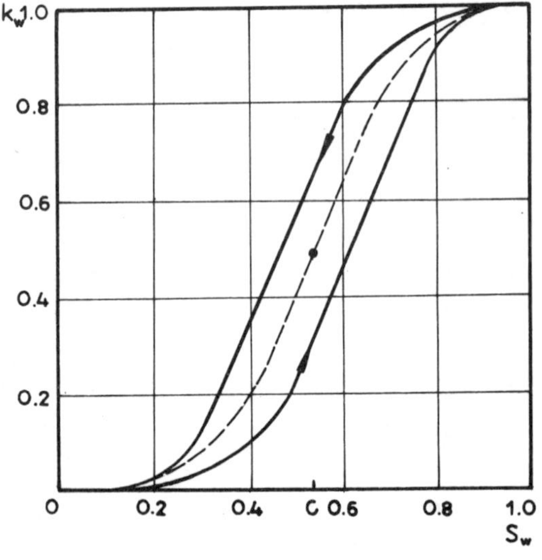

Fig. 2
Hysteresis loop for k_w vs. S_w dependence during increase and decrease of water saturation

$$Var\{S(\tau)\} = Sup\left\{\sum_j |S(\tau_{j-1}) - S(\tau_j)|\right\} \quad (7)$$

The history of repeated decrease and increase of water saturation $S_w(\theta)$ may be then presented in the `teeth of a saw´ form /cf. Fig.3/.

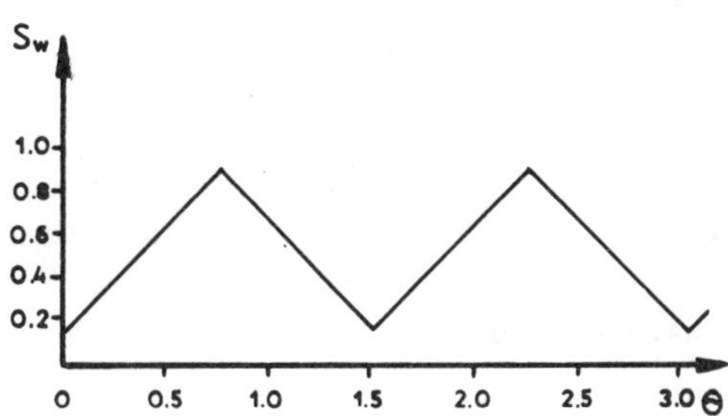

Fig. 3
The diagram of water saturation S_w vs. reservoir time θ dependence

Taking into account the remarks presented above the following equation instead of Eq. (3) may be assumed:

$$k_w(S_w,\theta) = \chi_w[S_w + M(\theta)] \quad (8)$$

The quantity M expresses the memory effects, and therefore it should be related to the previous history of the reservoir.

It may be easily seen that in case when memory effects are absent both branches of the hysteresis loop are the same, and then they are identical with light, uncontinuous line in Fig. 2.

4 FUNCTIONAL DESCRIPTION OF MEMORY EFFECTS

Empirical investigations have shown that the hysteresis loop for k_w vs. S_w dependence is approximately symmetric. The water saturation corresponding to the symmetry center of the loop will be called the `neutral water saturation´, and denoted by C.

For further considerations the following heuristic principle may be assumed:

"The value M describing memory effects in Eq. (8) at reservoir time θ depends on all the history of the deviation of saturation S_w from the neutral saturation C."

Consequently, the quantity M representing memory effects may be regarded as a functional of history of $S_w - C$ taken in the interval $(-\infty; \theta]$, i.e.:

$$M(\theta) = \mathop{\mathfrak{T}}_{\vartheta=-\infty}^{\vartheta=\theta}\left[S_w(\vartheta) - C\right] \quad (9)$$

Let us introduce a new reservoir time variable σ defined by:

$$\sigma = \theta - \vartheta \quad (10)$$

Eq. (9) may be then transformed to the following form:

$$M(\theta) = \mathop{\mathfrak{W}}_{\sigma=0}^{\sigma=\infty}\left[S(\sigma) - C\right] \quad (11)$$

where

$$S(\sigma) = S_w(\theta - \sigma) \quad (12)$$

5 FADING MEMORY

From the practical standpoint the entire history of the reservoir cannot be known.

On the other hand, the interpretation of experimental results indicates that the history of a reservoir before the begining of exploitation has no practical influence on the magnitude of the present memory effect.

In such a situation the following assumption called the principle of fading memory seems to be justified:

"Deviations from the neutral saturation which occurred in the distant past should have less influence in determining of $M(\theta)$ than deviations which occurred in the recent past".

In order to constitute the mathematical formulation of the principle of fading memory let us introduce the Hilbert space the elements of which are histories of deviation $S(\sigma)-C$.

The scalar product of two deviation histories $S(\sigma)-C$ and $S^*(\sigma)-C$ is defined by:

$$\langle S(\sigma)-C\,;\ S^*(\sigma)-C\rangle = \int_0^\infty \psi^2(\sigma)\,[S(\sigma)-C]\,[S^*(\sigma)-C]\,d\sigma \qquad (13)$$

where the influence function $\psi(\sigma)$ satisfies the following requirements:

$$\psi(\sigma) > 0 \quad \text{for all} \quad \sigma \geqslant 0 \qquad (14)$$

$$\psi(\sigma)\big|_{\sigma=0} = 1 \qquad (15)$$

$$\lim_{\sigma\to\infty}\{\sigma^\nu\psi(\sigma)\} = 0 \quad \text{for } \nu>\tfrac{1}{2} \qquad (16)$$

The norm of the deviation history is defined by:

$$\|S(\sigma)-C\| = \sqrt{\int_0^\infty \psi^2(\sigma)\,[S(\sigma)-C]^2 d\sigma} \qquad (17)$$

In the Hilbert space the norm of deviation $\|S(\sigma)-C\|$ may be interpreted as the `distance´ of history $S(\sigma)-C$ from the zero-history $O(\sigma)$. Since the influence function $\psi(\sigma)$ quickly decreases with increasing of σ, so the values of $S(\sigma)-C$ for small σ /recent past/ have greater influence in determining the norm $\|S(\sigma)-C\|$ than the values of $S(\sigma)-C$ for large σ /distant past/, according to Eq.(17).

The remarks cited above impose the following formulation of the principle of fading memory:

"There exists an influence function of order $\nu>\frac{1}{2}$ such that the functional $\mathfrak{W}[\ldots\ldots]$ in Eq. (11) is Fréchet-differentiable at the zero-history in the Hilbert space", i.e. the following requirement

$$\mathop{\mathfrak{W}}_{\sigma=0}^{\sigma=\infty}[S^*(\sigma)-C+S(\sigma)-C] - \mathop{\mathfrak{W}}_{\sigma=0}^{\sigma=\infty}[S^*(\sigma)-C]$$

$$= \delta\mathop{\mathfrak{W}}_{\sigma=0}^{\sigma=\infty}[S(\sigma)-C] + \mathop{\mathfrak{R}}_{\sigma=0}^{\sigma=\infty}[S(\sigma)-C] \qquad (18)$$

is satisfied when $S^*(\sigma)-C$

tends to the zero-history $O(\sigma)$.

The Fréchet-differential $\delta \mathfrak{W}[....]$ is the linear operator with respect to $S(\sigma) - C$, and the 'reminder' $\mathfrak{R}[......]$ decays to zero more quickly than $S(\sigma) - C$, i.e.:

$$\lim_{\|S(\sigma)-C\| \to 0} \frac{\left\| \underset{\sigma=0}{\overset{\sigma=\infty}{\mathfrak{R}}} [S(\sigma) - C] \right\|}{\| S(\sigma) - C \|} = 0 \quad (19)$$

6 APPROXIMATION OF THE MEMORY FUNCTIONAL

The concept of fading memory presented above makes it possible to determine the detailed form of memory functional $\mathfrak{T}[.....]$.

Taking into account that the Fréchet-differential is a linear operator, and applying the Riesz theorem according to which every continuous linear functional may be represented by a scalar product in the Hilbert space, one obtains:

$$M(\theta) = \int_0^\infty \Lambda(\theta, \sigma) [S(\sigma) - C] d\sigma + o\{\| S(\sigma) - C \|\} \quad (20)$$

Considering that the history of the reservoir for $\vartheta < 0$ has only small influence on memory effects at $\vartheta = \theta$, and neglecting a small value in the right hand side of Eq. (20), the following representation of $M(\theta)$ may be assumed:

$$M(\theta) = \int_0^\theta \Lambda_w(\theta, \vartheta) [S_w(\vartheta) - C] d\vartheta \quad (21)$$

It may be easily demonstrated that Eqs. (8), (21) describe correctly the memory effects related to the symmtric hysteresis loop.

7 NUMERICAL SIMULATION OF TWO -PHASE WATER-GAS FLOW IN THE RESERVOIR

The substitution of formulae (1), (2) into equations of continuity for multiphase flow

$$\operatorname{div}(\varrho_w h \vec{u}_w) = \varrho_w h q_w + \varrho_w h m \frac{\partial S_w}{\partial t} \quad (22)$$

$$\operatorname{div}(\varrho_g h \vec{u}_g) = \varrho_g h q_g + \varrho_g h m \frac{\partial S_g}{\partial t} \quad (23)$$

gives the following system of equations which describe two -phase flow through porous media:

for water:

$$\operatorname{div}(\mathbb{T}_w \operatorname{grad} \Phi_w) = h q_w + \frac{hm}{B_w} \frac{\partial S_w}{\partial t} \quad (24)$$

for gas:

$$\operatorname{div}(\mathbb{T}_g \operatorname{grad} \Phi_g) = h q_g + \frac{hm}{B_g} \frac{\partial S_g}{\partial t} \quad (25)$$

h denotes here porous strata thickness, q is magnitude of well production per unit of volume, m is porosity, B is bulk volume factor, Φ_w and Φ_g are water and gas velocity potentials related to water and gas pressure by means of the following formulae:

$$\Phi_w = p_w - \varrho_w \gamma z \qquad (26)$$

$$\Phi_g = p_g - \varrho_g \gamma z \qquad (27)$$

The transmissibility tensors for water $\mathbb{T}_w$ and gas $\mathbb{T}_g$ are defined by:

$$\mathbb{T}_w = \mathbb{K} \frac{k_w h}{\mu_w B_w} \qquad (28)$$

$$\mathbb{T}_g = \mathbb{K} \frac{k_g h}{\mu_g B_g} \qquad (29)$$

Considering that transmissibility tensors $\mathbb{T}_w$, $\mathbb{T}_g$ depend on the history of saturation, the system of equations (24), (25) is non-linear, and therefore it may be only solved by means of numerical methods.

The numerical finite difference method consists in the transformation of differential equations into difference equations. The difference equations mentioned above are then solved by means of the techniques applied in linear algebra.

Equations (24), (25) written in the finite difference terms are as follows:

for water:

$$\Delta(\mathbb{T}_w^{n+1} \Delta \Phi_w^{n+1}) = hq_w^{n+1} + \frac{hm}{B_w} \frac{S_w^{n+1} - S_w^n}{\Delta t} \qquad (30)$$

for gas:

$$\Delta(\mathbb{T}_g^{n+1} \Delta \Phi_g^{n+1}) = hq_g^{n+1} + \frac{hm}{B_g} \frac{S_g^{n+1} - S_g^n}{\Delta t} \qquad (31)$$

where Δ denotes the difference operator related to the \`nabla´ operator.

Transmissibilities $\mathbb{T}_w$, $\mathbb{T}_g$, potentials Φ_w, Φ_g and productions q_w, q_g have been taken here at the implicit $n+1$ time level in order to avoid the instability of the finite difference approximation.

The harmonic analysis associated with the Buchanan stability criterion confirms that the difference scheme (30), (31) is unconditionally stable, and therefore there is no restriction imposed on the magnitude of time step Δt.

After linearization the system of equations (30), (31) may be reduced to the following matrix form:

$$\mathbf{A}^{n+1} \mathbf{X}^{n+1} = \mathbf{R}^n \qquad (32)$$

where $\mathbf{A}$ is the penta- or septa-diagonal block matrix of coefficients, $\mathbf{X}$ is the \`numerical block vector´ of unknown values, and $\mathbf{R}$ is the \`numerical block vector´ of known values.

Considering that each element of block vector $\mathbf{X}$ consists of two-element matrix the components of which are water saturation S_w^{n+1} and water velocity potential

Φ_w^{n+1} taken at time step n+1 the solution of Eq. (32) performed by means of iterative techniques for successive time levels describes the water-gas /or gas-water/ displacement in blocks into which the reservoir is divided.

REFERENCES

Amyx J.W., Bass D.M., Whiting R. L. 1960, Petroleum Reservoir Engineering, McGraw-Hill, New York - Toronto - London.

Aziz K., Settari A. 1979, Petroleum Reservoir Simulation, Applied Science, London.

Bear J. 1972, Dynamics of Fluids in Porous Media, American Elsevier, New York - London - Amsterdam.

Buchanan M.L. 1963, Journal of the Society of Industrial and Applied Mathematics 11 : 474.

Coleman B.D., Noll W. 1961, Reviews of Modern Physics 33 : 239.

Evrenos A.I., Comer A.G. 1969, Sensitivity Studies of Gas-Water Relative Permeability and Capilarity in Reservoir Modeling, 44th Annual Fall Meeting of the Society of Petroleum Engineers of AIME, Denver.

Killough J.E. 1976, Society of Petroleum Engineers Journal 16: 37.

Paceman D.W. 1977, Fundamentals of Numerical Reservoir Simulation, Elsevier, Amsterdam - Oxford - New York.

Perzyna P. 1967, Rozprawy Inżynierskie 15 : 361.

Richtmyer R.D., Morton K.W. 1967, Difference Methods for Initial -Value Problems, Interscience, New York - London - Sydney.

Sławomirski M.R. 1981, The Modelling of Water-Gas Displacement in the Underground Gas Reservoir Applying Finite Difference Technique, in Proceedings of the 3rd International Conferrence on Applied Mathematical Modelling, Hamburg, 6 - 10 October, 1980, Springer.

Varga R.S. 1962, Matrix Iterative Analysis, Prentice-Hall, New York.

Proceedings of Euromech 143 / Delft / 2-4 September 1981

Infiltration into layered soils: Experiments and numerical simulation

F.STAUFFER
Federal Institute of Technology, Zurich, Switzerland

1 INTRODUCTION

The hydraulic processes between ground surface and ground water level are characterized by capillary phenomena. The following situation is considered:

1. Water infiltrates into the soil through the surface. Processes of this type occur during the natural or artificial recharge of ground water or during irrigation.
2. The flow may pass through saturated and unsaturated zones.
3. The soil is layered.
4. The flow is in a two dimensional vertical plane.
5. The process is unsteady and isothermal.

The physical processes are investigated by means of laboratory experiments with a sand packing.

In order to simulate such processes, a mathematical-numerical model is proposed and verified with the experimental data.

2 EXPERIMENTAL INSTALLATION

The experimental installation is used for investigating two dimensional vertical flows in porous sand packings. The sand packing is built up in a narrow prismatic trough (length 4.8 m, height 1.2 m, width 0.05 m, see fig. 1).

The inner walls of the trough consist of plexiglass. The bottom and the side walls are of nickel-plated steel. At one of the lateral ends of the trough, the water level can be adjusted to a pre-determined level. At this end, the sand packing is held by a sieve. A special technique, based on a concept of Wygal (1963) is used for the building up of the sand packing. According to this concept, thin strands of free falling sand hit a sequence of sieves in normal direction before reaching the surface of the sand body. With an optimal arrangement depending on the sand used, a relatively good homogeneity can be achieved (Stauffer, 1977).

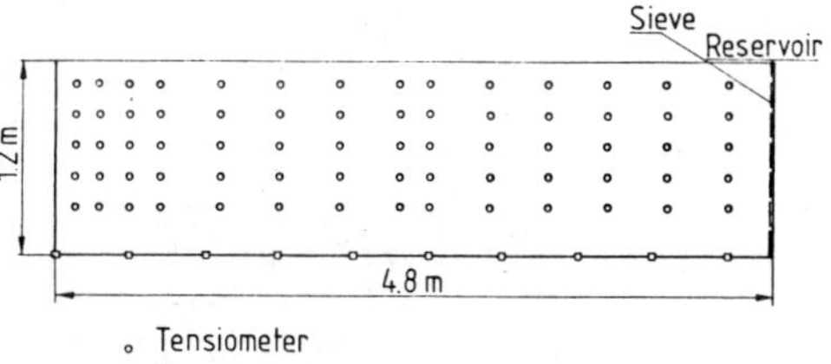

Fig. 1 Experimental trough

The total fluid saturation of the sand packing is achieved by a hydraulic recirculation system with air separation. The air is eliminated by continuous evacuation of the outflowing water before the reinjection into the sand packing.

During the experiments, the flow is monitored by measurements of the water content and of the capillary pressure.

The water content and the porosity are determined by the gamma-ray absorption method. The measurement chain consists of a gamma-ray source (Americium-241, activity 11/ns), two collimators, a Na J-scintillation-detector, an amplifier, a discriminator and a counter (Stauffer, 1977). With this device, the fluid saturation can be measured with a mean accuracy of about 2 %. Four probes, consisting of source, collimators and detector, are situated on two frames with a side length of 0.4 m (fig. 2). This arrangement allows a simultaneous measurement of the water content at four stations. The horizontal and vertical positioning of the frames carrying the probes, is carried out by step motors.

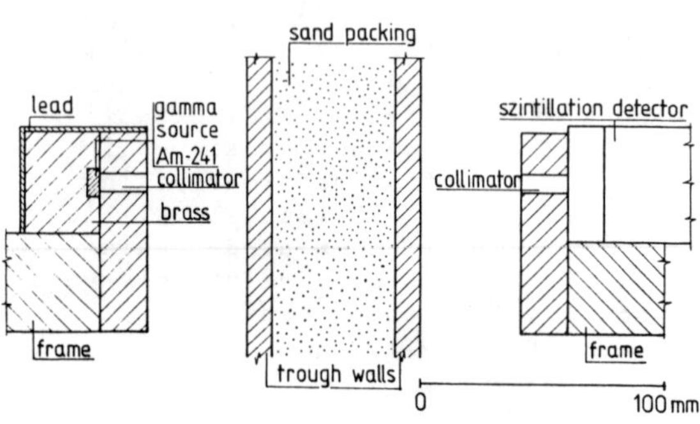

Fig. 2 Probe for the determination of water content and porosity

The capillary pressure or water pressure is determined at 80 stations (11 manometers and 69 tensiometers, see fig. 1). The measuring chain consists of a tensiometer (fig. 3) or a manometer, a scanivalve, a pressure transducer and an amplifier.

Four stations are simultaneously connected hydraulically to the corresponding pressure transducer. The differential pressure range is ± 0.2 bar and the mean error of a measurement is ± 0.15 mbar.

The data acquisition is controlled by a digital computer PDP 11/34. The pressure transducers are read by the AD-converter of the computer. The results of the gamma-ray counts are transmitted through a byte multiplexer to the digital input of the computer. Any sequence of the measurements can be programmed. During execution, the programme can be influenced by setting 6 bits of the digital input of the computer.

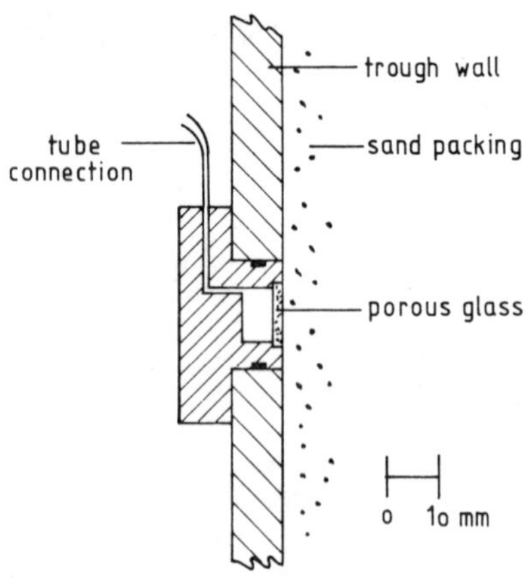

Fig. 3 Tensiometer

3 INFILTRATION EXPERIMENTS

The infiltration into layered soils is investigated by means of simple time-dependent experiments. They were carried out with a horizontally layered sand packing with the sequence fine / coarse / fine. The experimental situation is schematically shown in figure 4.

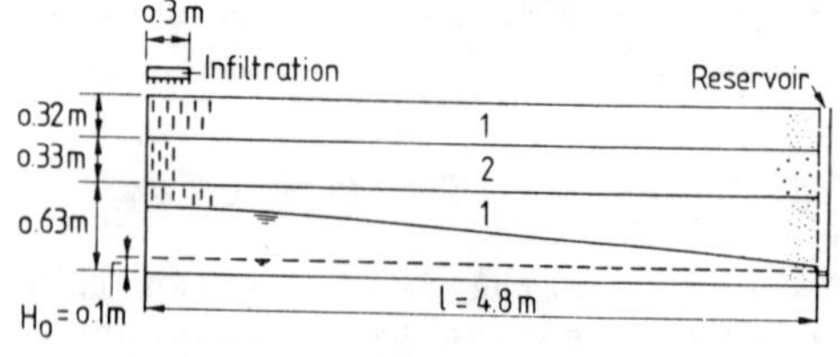

Fig. 4 Experimental situation, 1: fine layer, 2: coarse layer

The initial fluid pressure distribution in the sand packing is hydrostatic and determined by the position of the free surface level H_0. Above the free surface, a quasi static vertical water content profile is achieved. This initial state is the result of the lowering of the outflow level from the top of the saturated sand packing to the level H_0. In the unsaturated domain, the initial state corresponds to the first drainage loop in the relation between capillary pressure and water content. As an experimental boundary condition, the outflow level is held at a constant level. The infiltration is performed through a strip with a length of 0.3 m at the left end of the trough. The infiltration rate is held constant and is controlled by a volumetric pump. The water is distributed on the infiltration area by 28 capillary tubes and by a coarse filter which is placed on the wetted sand surface.

The grain size distribution of the sands is shown in figure 5.

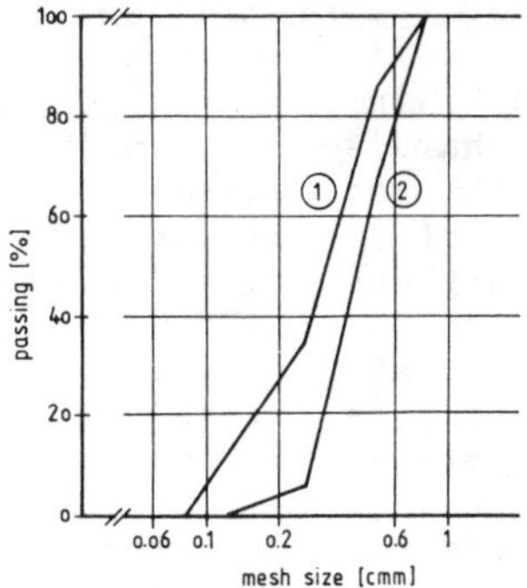

Fig. 5 Grain size distribution

The relations between capillary pressure p_c and saturation S at a representative station in the infiltration domain is shown in figure 6 (pressure head $h_c = p_c / \rho \cdot g$).

The relation between relative conductivity k_r and saturation S of the fine sand is shown in figure 7.

Both relations were determined for the imbibition loop (boundary wetting curve) starting from a residually saturated sand packing. The relation $p_c(S)$ was determined by a separate imbibition experiment with a raising water table. The relation $k_r(S)$ was determined by infiltration experiments in a homogeneous packing of the finer sand just below the infiltration strip during quasi vertical flow conditions.

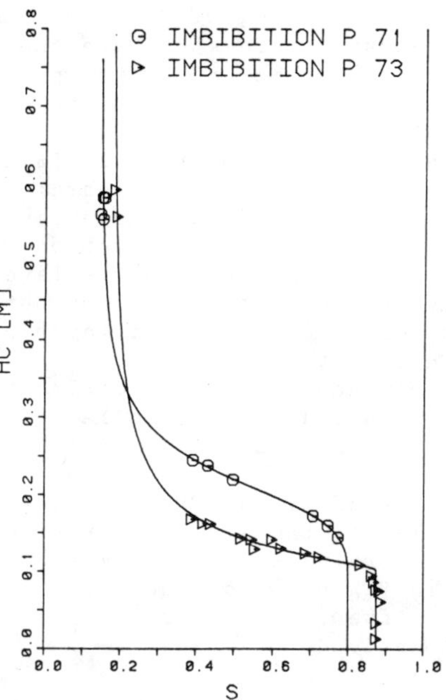

Fig. 6 $p_c(S)$

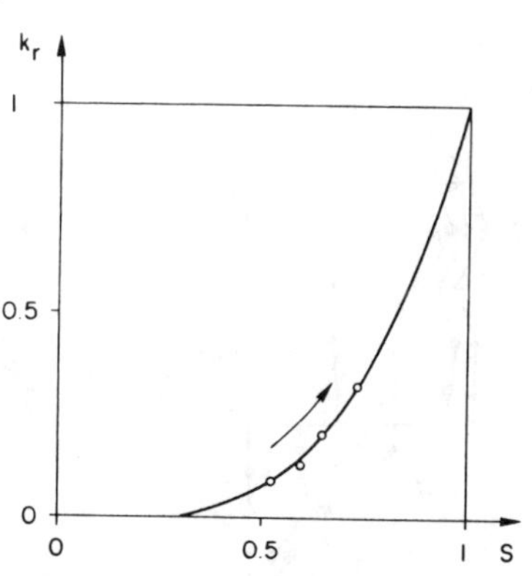

Fig. 7 $k_r(S)$

The mean hydraulic conductivity k at saturation and the porosity n are

$k = 0.25$ mm/s, $n = 0.36$

for the fine layers, and

$k = 0.73$ mm/s, $n = 0.39$

for the coarse layer. Both parameters were determined in the completely saturated packing. k was determined by simple experiments with constant inflow and outflow level.

Experiments were carried out with the following infiltration rates q:

q = 0.082, 0.048, 0.035, 0.021 mm/s

The measured parameters are the capillary pressure, the water pressure, the water content and the outflow rate.

During the experiments, the following stages can be distinguished:

1. Strip source in a homogeneous sand as long as the infiltration front does not reach yet the coarse layer.
2. Influence of the boundary between upper fine and coarse layer with a fluid retention effect.
3. Influence of the boundary between coarse and lower fine layer.
4. Transition into the saturated zone.
5. Reaction of the free surface.

Profiles of the fluid saturation at the coordinate x = 0.14 m are shown in figure 8 for the quasi steady state (10 h after begin of experiment).

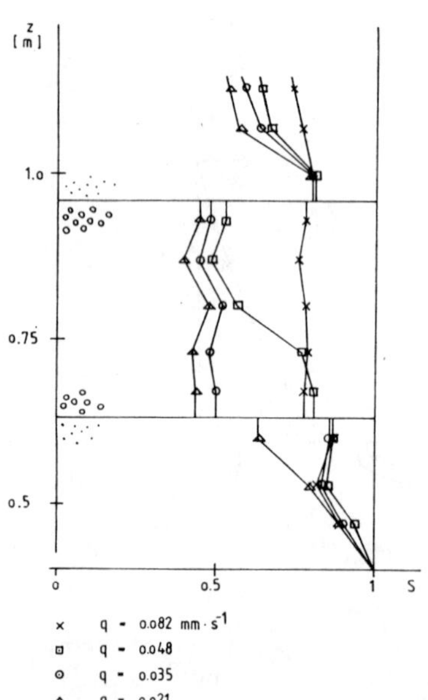

Fig. 8 Quasi steady state S-profiles

4 FINITE ELEMENT GALERKIN MODEL

The flow condition is assumed to satisfy Darcy's law for saturated and partially saturated flow, or

$$\underline{q} = -k \ (p/\rho g + z) \qquad (1)$$

Here, q is the specific flux, p is the fluid pressure, g is the gravitational constant and ρ is the density of the fluid which is assumed to be a constant. Frictional effects due to air flow in the unsaturated medium are neglected as well as evaporation.

Under these conditions and taking into consideration the flow continuity, the unsaturated flow is governed by the Richard's equation

$$n \cdot \frac{\partial S}{\partial t} = \nabla \cdot (k \cdot \nabla (p/\rho g + z)) \qquad (2)$$

where t is the time. For saturated conditions, where the saturation remains constant, the time derivative vanishes.

The problem of the simulation of two dimensional infiltration processes into soils involves the solution of the Richard's equation (2) together with the soil characteristics p(S) and k(S) including initial and boundary conditions. Since one is interested in simultaneous saturated - unsaturated flow and since the soil may be layered, the fluid pressure p(x,z) has to be used as the dependent variable. The resulting differential equation is highly nonlinear and, therefore, has to be integrated numerically.

The numerical procedure for the integration of the differential equation is based on the model of Neuman (1972). It uses a Galerkin-type finite element method. Quadrilateral elements are employed which are composed of four triangles with linear shape functions. The integration over the solution domain leads to an ordinary differential equation system

$$[A] \cdot \{p_i\} + [F] \cdot \{\frac{\partial p_i}{\partial t}\} + \{B\} = 0 \qquad (3)$$

with [ ] being m x m matrices and { } being vectors of length m. m is the number of nodes. p_i is the representation of the variable at the node i. For the integration over time, Neuman used a fully implicit backward difference scheme which results in a linear equation system

$$\left([A(t+\tfrac{\Delta t}{2})]+\tfrac{1}{\Delta t}\cdot[F(t+\tfrac{\Delta t}{2})]\right)\{p_i(t+\Delta t)\}$$
$$=\{B(t+\tfrac{\Delta t}{2})\}+\tfrac{1}{\Delta t}\cdot[F(t+\tfrac{\Delta t}{2})]\cdot\{p_i(t)\} \quad (4)$$

Δt is the length of the time step. Since the coefficients in [A], [B] and [F] are dependent on the variable p, which is unknown at the time level $t + \Delta t/2$, they have to be estimated and improved by an iterative process by

$$p_i\,(t + \Delta t/2) = 0.5\cdot(p_i(t)+p_i(t+\Delta t)) \quad (5)$$

and a reevaluation of the coefficients at each iteration.

Special attention had to be given to the stiffness behaviour of the system of ordinary differential equations (3) which may result in numerical stability problems (Mercer and Faust, 1976). This stiffness is mainly due to the existence of sharp fronts which are typical for infiltration processes into soils. The following measures are now proposed to alleviate these problems:

1. The soil characteristics p(S) and k(S) are represented by algebraic equations instead of linear interpolations of discrete pairs of values. For the description of the relation p(S), an approach similar to that of Su and Brooks (1976) is used. It is

$$p = -p_b\cdot\left(\frac{S-S_r}{1-S_r}\right)^{-1/\lambda}\cdot\left(\frac{S_m-S}{S_m-S_r}\right)^{\delta} \quad (6)$$

for $p \leq 0$ and $S = S_m$ for $p \geq 0$

S_r is the residual saturation and S_m is the maximum saturation ($S_m \leq 1$). p_b is approximately the minimum capillary pressure at which the non-wetting fluid (air) is continuous. λ and δ are dimensionless exponents which characterize the pore distribution. With this approach, the boundary wetting and drying curves of sandy soils can be well represented (see fig. 6, wetting curves). To describe the relation k(S), the approach of Brooks and Corey (1966) is used, which is

$$k = k_s \cdot S_e^{\varepsilon}$$

where $S_e = \dfrac{S-S_r}{1-S_r}$

k_s is the hydraulic conductivity at saturation. ε is a dimensionless exponent. By using these algebraic equations, the stiffness could be reduced.

2. The length of the time step is automatically adjusted during the programme execution. Δt is enlarged by a given factor, when the iterative solution process of the preceding time step has rapidly converged as measured by the number of iterations necessary for the fulfilment of a convergence criterion. Otherwise, Δt is reduced.

3. The entire equation system (4) is solved only for the first iteration. For the following iterations, the variables which show a small change during the time step in a given nodal point, are regarded as constants, provided that this condition is also satisfied for all points of the associated elements. This procedure reduces the computational effort, since small time steps are necessary to overcome the numerical stability problems. This procedure is justified for infiltration problems, because the rapid change of the variables is mainly restricted to the actual front region.

4. The initially hydrostatic profile of the fluid pressure in the unsaturated region with approximately residual saturation is replaced by a uniform pressure distribution. This is justified in infiltration problems, because the saturation behind the front is much higher than residual.

5 SIMULATION

The result of a simulation is presented for the first stage of the experiment with the highest flow rate, which is

$q = 0.082$ mm s^{-1}

This is the flow rate with which the sharpest fronts in the unsaturated domain are expected. The following experimentally determined characteristic parameters for the sand packing (see chapter 4) were

for the fine layer:	for the coarse layer:
$k_s = 2.5\cdot10^{-4}$ m s^{-1}	$k_s = 7.3\cdot10^{-4}$ m s^{-1}
$n = 0.37$	$n = 0.39$
$p_b/\rho g = 0.202$ m	$p_b/\rho g = 0.102$ m
$\lambda = 4.873$	$\lambda = 2.473$
$S_r = 0.145$	$S_r = 0.174$
$S_m = 0.801$	$S_m = 0.875$
$\delta = 0.124$	$\delta = 0.01$

The exponent ε, which has been determined for a homogeneous packing using the finer sand, has been found to be $\varepsilon = 3.0$. This value is taken also for the coarser sand, assuming a universal value of ε for sands.

The geometry of the region selected for the simulation is shown in fig. 9.

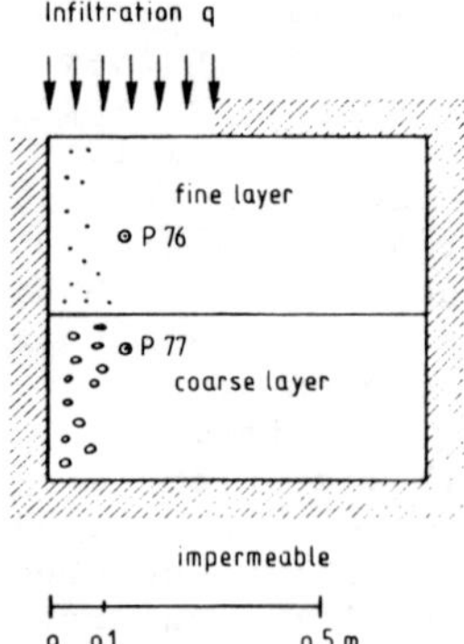

Fig. 9 Solution domain

Since only the first stage of the experiment was simulated, the region was restricted to the area influenced by the infiltration. As the initial condition, a uniform pressure head of -0.4 m was selected. Along the infiltration strip, the flux was prescribed. The other computational boundaries are assumed to be impermeable. The results can, therefore, be accepted, as long as the lower and the right vertical boundary are not influenced by the infiltration. The existence of an approximately residual saturation justifies the use of the boundary wetting curves in the soil characteristics. Hysteresis phenomena can, therefore, be neglected.

Figure 10 shows the pressure head distribution at the nodal points after 10 and 20 minutes.

Time series of the saturation at the tensiometer stations P 76 and P 77 are shown in figure 11, together with measured data. The values of the saturations at the two stations are determined from the pressure data by means of the relation p(S).

6 CONCLUSIONS

The hydraulic processes between ground surface and ground water level are characterized by capillary phenomena. Infiltration into soils results in more or less sharp fronts. In order to simulate such processes, a two dimensional finite element Galerkin model was formulated and tested by a laboratory experiment with infiltration into a layered sand packing. The process was simulated for the first stage of the infiltration into a soil region consisting of an upper fine layer and a lower coarse layer.

7 REFERENCES

Brooks, R. & A.T. Corey 1966, Properties of porous media afflecting fluid flow, J.Irrig.Drain.Div. 92: 61-88.

Mercer, J.W. & C.R. Faust 1976, The application of finite-element techniques to immiscible flow in porous media, in Finite Elements in Water Resources, Pentech Press, London.

Müller, A. & P. Ernst 1981, Hydraulic experiments: Data acquisition and control, in Proc. XIX. Congress IAHR, New Delhi.

Neuman, S.P. 1972, Finite element computer programs for flow in saturated-unsaturated porous media, Rep. NO. A10-SWC-77 Technion, Haifa.

Stauffer, F. 1977, Einfluss der kapillaren Zone auf instationäre Drainagevorgänge, Rep. R 13-77, IHW, ETH Zürich.

Su, C. & R.H. Brooks 1976, Hydraulic functions of soils from physical experiments and their application, Rep. WRRI-41 Water Res.Res.Inst. Corvallis, Oregon.

Wygal, R.J. 1963, Construction of models that simulate oil reservoirs, Soc. Petroleum Eng.J., Dec. 1963, pp 281-286.

TIME = .60000E+03 TIME STEP = 24

Z:	X: .0	.1	.2	.3	.4	.5	.6	.7
1.280	-.164	-.165	-.162	-.196	-.333	-.400	-.400	-.400
1.260	-.164	-.164	-.165	-.197	-.311	-.400	-.400	-.400
1.210	-.164	-.164	-.169	-.199	-.289	-.399	-.400	-.400
1.160	-.163	-.164	-.173	-.207	-.299	-.398	-.400	-.400
1.110	-.167	-.170	-.187	-.237	-.338	-.399	-.400	-.400
1.060	-.204	-.212	-.246	-.314	-.387	-.400	-.400	-.400
1.010	-.347	-.356	-.378	-.395	-.400	-.400	-.400	-.400
.960	-.400	-.400	-.400	-.400	-.400	-.400	-.400	-.400
.910	-.400	-.400	-.400	-.400	-.400	-.400	-.400	-.400
.860	-.400	-.400	-.400	-.400	-.400	-.400	-.400	-.400
.810	-.400	-.400	-.400	-.400	-.400	-.400	-.400	-.400
.760	-.400	-.400	-.400	-.400	-.400	-.400	-.400	-.400
.710	-.400	-.400	-.400	-.400	-.400	-.400	-.400	-.400
.660	-.400	-.400	-.400	-.400	-.400	-.400	-.400	-.400

TIME = .12000E+04 TIME STEP = 30

Z:	X: .0	.1	.2	.3	.4	.5	.6	.7
1.280	-.164	-.164	-.162	-.196	-.318	-.400	-.400	-.400
1.260	-.164	-.164	-.165	-.197	-.297	-.400	-.400	-.400
1.210	-.164	-.164	-.169	-.199	-.289	-.399	-.400	-.400
1.160	-.163	-.164	-.172	-.199	-.268	-.391	-.400	-.400
1.110	-.161	-.163	-.173	-.199	-.264	-.386	-.400	-.400
1.060	-.155	-.158	-.179	-.197	-.260	-.381	-.400	-.400
1.010	-.144	-.147	-.161	-.192	-.269	-.388	-.400	-.400
.960	-.135	-.136	-.143	-.173	-.263	-.394	-.400	-.400
.910	-.126	-.130	-.151	-.212	-.336	-.397	-.400	-.400
.860	-.139	-.152	-.199	-.299	-.388	-.400	-.400	-.400
.810	-.292	-.308	-.351	-.390	-.399	-.400	-.400	-.400
.760	-.400	-.400	-.400	-.400	-.400	-.400	-.400	-.400
.710	-.400	-.400	-.400	-.400	-.400	-.400	-.400	-.400
.660	-.400	-.400	-.400	-.400	-.400	-.400	-.400	-.400

Fig. 10 Simulated pressure head distribution for t = 10 and 20 min. (in m)

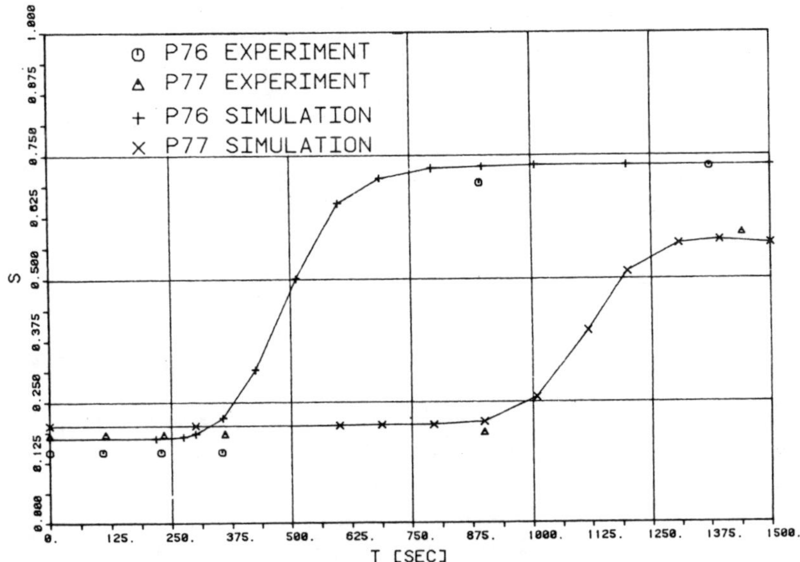

Fig. 11 Time series of S. Simulation and experimental data.

Proceedings of Euromech 143 / Delft / 2-4 September 1981

A numerical model for multiphase flow in porous media

A.VERRUIJT
University of Delft, Netherlands

1 INTRODUCTION

Multiphase flow in a porous medium is an important engineering problem in the Netherlands, because coastal aquifers, in which fresh and salt water may occur, are used for storage of drinking water. The prediction of the movement of the fresh water, and the behaviour of the interface with the salt water, which usually should remain outside the system, is an essential part of the design of a drinking water supply system. For these purposes numerical models are used extensively and one possible model is briefly presented here. The model was constructed with the aims that it should be fairly general, especially suitable for the prediction of the behaviour of an interface between fresh and salt groundwater, with a possible transition zone of brackish water. The model should especially be such that numerical dispersion (Bredehoeft, 1971) is avoided.

2 BASIC EQUATIONS

Let there be considered an isotropic incompressible porous medium of permeability κ, fully saturated with a fluid of viscosity μ, and density $\rho = \rho_o + c$, where ρ_o is the (constant) density of the pure liquid, and where c is the variable concentration of a certain substance, carried by the liquid, such as salt or a tracer. The flow is supposed to satisfy Darcy's law,

$$q_i = -\frac{\kappa}{\mu}\left(\frac{\partial p}{\partial x_i} - \rho g_i\right) \qquad (1)$$

where q_i is the specific discharge vector, p is the pressure in the fluid, x_i is the position vector and g_i is the gravity vector. If the porous medium is undeformable and fully saturated the flow must at all times satisfy the equation of continuity,

$$\frac{\partial q_i}{\partial x_i} = 0 \qquad (2)$$

where summation is implied by the repitition of the index i. Sunstitution of eq.(1) into (2) leads to the following basic differential equation

$$\frac{\partial}{\partial x_i}\left(\frac{\kappa}{\mu}\left(\frac{\partial p}{\partial x_i} - \rho g_i\right)\right) = 0 \qquad (3)$$

The boundary conditions on the pore fluid are in general that on part of the boundary the fluid pressure is prescribed, and on the remaining part of the boundary the flux normal to the boundary is given. The equations (1) - (3) together with the boundary conditions suffice to describe the flow phenomenon, provided that the fluid density ρ is a known function of the spatial coordinates. In order to describe the transport of the tracer, which determines the density distribution, the equation of motion of the tracer and a conservation equation should be formulated. For the motion of the tracer it is assumed that the dispersive transport can be disregarded with regard to the convective transport. In that case the differential equation for the tracer transport reduces to the hyperbolic equation

$$n\frac{\partial c}{\partial t} = -q_i\frac{\partial c}{\partial x_i} \qquad (4)$$

where n is the porosity of the porous medium. This means that the concentration c remains constant in the characteristic direction, which is defined by the equation,

$$\frac{dx_i}{dt} = \frac{q_i}{n} \tag{5}$$

This equation admits a simple numerical integration, such that a new distribution of the tracer concentration can be calculated from a previous distribution by noting that

$$c(x_i + \Delta x_i,\ t + \Delta t) = c(x_i, t) \tag{6}$$

where $\Delta x_i = q_i \Delta t/n$. Numerical integration is in general the only possibility because the specific discharge q_i can in general not be found analytically.

3 SOLUTION OF THE FLOW PROBLEM

The flow problem can conveniently be solved by the finite element method (Zienkiewicz, 1975). A simple and consistent numerical model for flow in a plane can be developed by using triangular elements, with the permeability, the viscosity and the density being defined and constant in the elements, and the pressure being defined in the nodes of the network. Details of the development of the computer program will not be given here, because this can be considered as a standard procedure.

For the modelling of multiphase flow a special problem arises along the interface between two fluids. A relatively simple approach is to consider a sharp interface as the limiting state of a thin transition zone in which the density changes linearly from one value to another. In the linear system of equations that is obtained in a finite element method this leads to a term in the right hand side proportional to the difference in density and vanishing for a vertical interface. The system of equations can be solved by a standard method of solution. The program described in this paper uses the Gauss-Seidel iterative method. The result is the pressure distribution compatible with a given distribution of densities. From this pressure distribution the new distribution of densities can be calculated, on the basis of the velocities. These velocities are calculated in each node, by a certain procedure of averaging over the elements surrounding that node. Several methods of averaging have been investigated, both analytically and experimentally. The best procedure seems to be to use a weight factor equal to the contribution of a certain element to the term on the main diagonal of the matrix of the system of equations. In this way an element having a sharp corner in a certain node contributes less to the velocity in that node than an element with a larger corner.

Once that the velocities in the nodal points have been calculated the displacement components Δx and Δy of the nodes during a small time step can be determined, in accordance with formula (6). The tracer material in each element, which causes the differences in density, is simply carried along, by keeping the density of each element unchanged. For the new configuration, with a modified network of elements one can now again calculate the new pressure distribution, make another time step, etcetera. In this way numerical dispersion, which is a disturbing phenomenon in many numerical models, is completely avoided, which seems to be a distinct advantage of the method used. The distortion of the element configuration is a disadvantage, which entails that the matrix coefficients have to be calculated again after each time step.

4 VALIDATION OF THE MODEL

In order to verify the numerical method a number of comparisons with existing analytical solutions have been made. All of these refer to the flow in a homogeneous layer, with a single interface. The most extensive series of verifications was done for the case of flow towards a single sink, above an interface between two fluids of different density, say fresh and salt water, see figure 1.

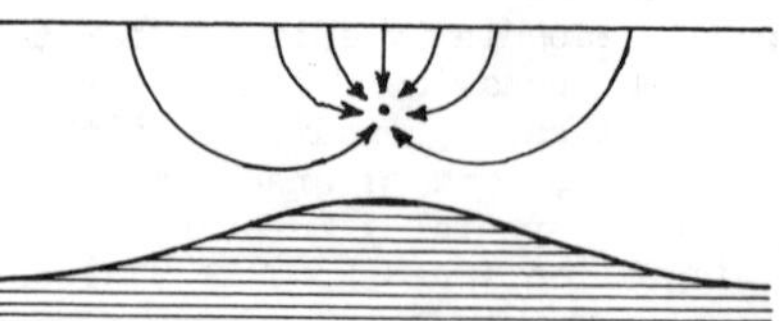

Fig. 1. Two-fluid flow.

The upper boundary is a potential line, and the left and right side boundaries are taken as impermeable. The lower boundary is either considered as impermeable (in which case the total amount of salt water is constant), or considered to be a potential line.

A first verification was to check whether an initially horizontal interface remains at rest when the discharge of the sink is zero, and whether the correct hydrostatic pressure distribution is obtained. This was indeed the case, with a relative accuracy depending upon the accuracy of the computer used. Although this seems to be a trivial verification, several other numerical models, for instance one using a definition of the concentration in the nodes, with linear interpolation in the elements, did not meet this requirement, and were therefore abandoned as possible alternatives.

The second verification concerned the steady position of the interface, obtained for a constant discharge of the sink. An analytical solution for this case has been derived by Strack (1973), by using the hodograph method. A comparison of one of Strack's stable steady state solutions with the position of a steady interface obtained numerically is given in fig. 2. Only one

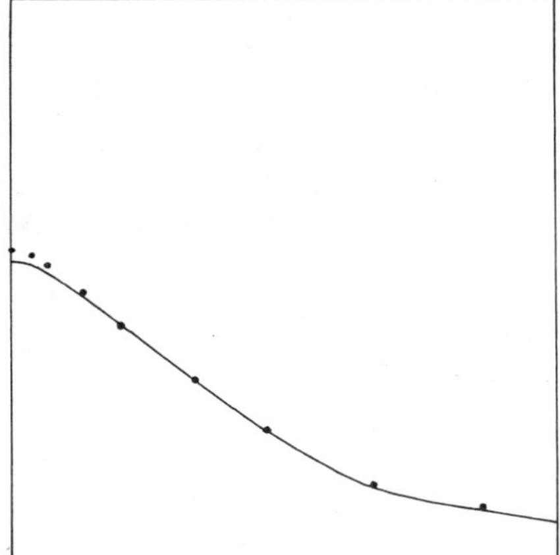

Fig. 2. Exact and numerical solution

half of the region was actually modelled, because of the symmetry of the problem. The network used consisted of 11 columns of 15 nodes, so that the total number of nodes was 165, and the number of elements was 280. This configuration could be handled by a BASIC program on a microcomputer with 32000 bytes memory (Verruijt, 1981).A somewhat finer network, with 600 elements, was analyzed by using a FORTRAN program and the computer of the University of Delft. In the numerical procedure the number of time steps needed to reach the steady state is very large, because the magnitude of the time step cannot be taken too large to prevent instabilities. Therefore an interface rather close to the steady state was used as an initial state. In the computer programs the nodes on the interface were constrained to move vertically only, in order to prevent a concentration of nodes (or even the disappearance of nodes) near the sink. The position of the fresh and the salt water were distributed uniformly over the height of each column.

In these calculations an essential difficulty was encountered, namely that the velocity is discontinuous across the interface. This is a well-known fact (De Josselin de Jong, 1960), which manifests itself by a non-zero velocity just above the interface, compatible with a zero velocity immediately below it. In the computer calculations the motion of the interface was calculated on the basis of the velocities in the elements below it only. If the average velocity of fresh and salt water was used no correct steady interface could be obtained, due to the fact that fresh water flows over the interface. Comparison of the numerical solution obtained by the procedure described above with the analytical solution (see fig. 2) shows that the agreement is satisfactory.

A third verification consisted of checking the time rate of change of the interface. This could conveniently be done for the velocity field in the first time step with an initially horizontal interface. For that case the velocity field does not depend upon the density distribution, and therefore admits a simple analytical solution. The largest vertical velocity occurs directly below the sink, and these values were compared with the numerical results, for various values of the parameters in the problem, notably the distance of the sink above the interface. In the comparison the lower boundary was considered as impermeable, and the analytical solution was determined by the complex variable method. A good agreement was obtained, with the numerical results usually about 5% smaller than the analytical ones.

A fourth verification also concerned the non-steady behaviour of an interface, in this case the rotation of an initially vertical front, in a layer of constant thickness. The initial velocities for this almost unstable case can be determined by using a distribution of singularities (De Josselin de Jong, 1960) or by Fourier analysis (Verruijt, 1980). In fig. 3 the horizontal velocities obtained numerically are

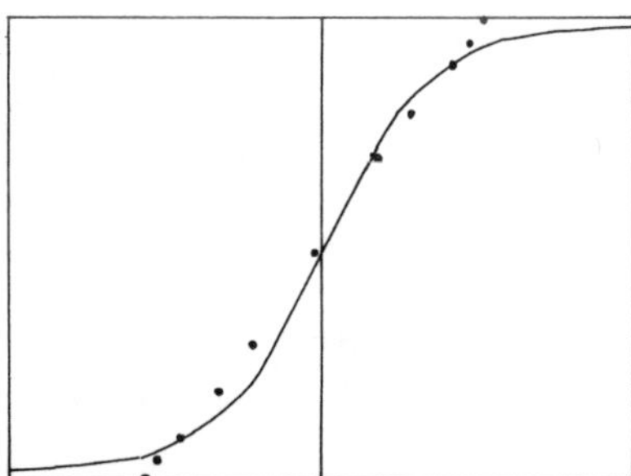

Fig. 3. Exact and numerical solution

compared with the analytical results. Again the agreement is reasonably good,as far as the general trend is concerned, but the numerical approach fails to predict the singular behaviour in the top and bottom points of the interface.

5 CONCLUSION

It may be concluded that the numerical method can be considered to give reasonably accurate results, at least if some care is taken in the distribution of the nodes over the domain, and provided that the time steps are taken small enough to ensure relatively small displacements in each step, say of the order of magnitude of $\frac{1}{4}$ of a typical element size. The method may seem costly, because in each time step a new network must be generated, but this may well be worth the effort, because numerical dispersion is completely avoided.

6 REFERENCES

Bredehoeft, J.D., Comment on numerical solution to the convective diffusion equation, Water Resources Res., Vol. 7, 755-756, 1971.

De Josselin de Jong, G., Singularity distributions for the analysis of multiple fluid through porous media, J. Geophys. Res., Vol. 65, 3739-3758, 1960.

Strack, O.D.L., Many-valuedness in groundwater flow, Doctor's thesis, Delft, 1973.

Verruijt, A., The rotation of a vertical interface in a porous medium, Water Resources Res., vol. 16, 239-240, 1980.

Verruijt, A., Some BASIC programs for finite element analysis, Adv. Engng. Software, vol. 3, 26-30, 1981.

Zienkiewicz, O.C., The finite element method, 3d ed., McGraw-Hill, London, 1977.

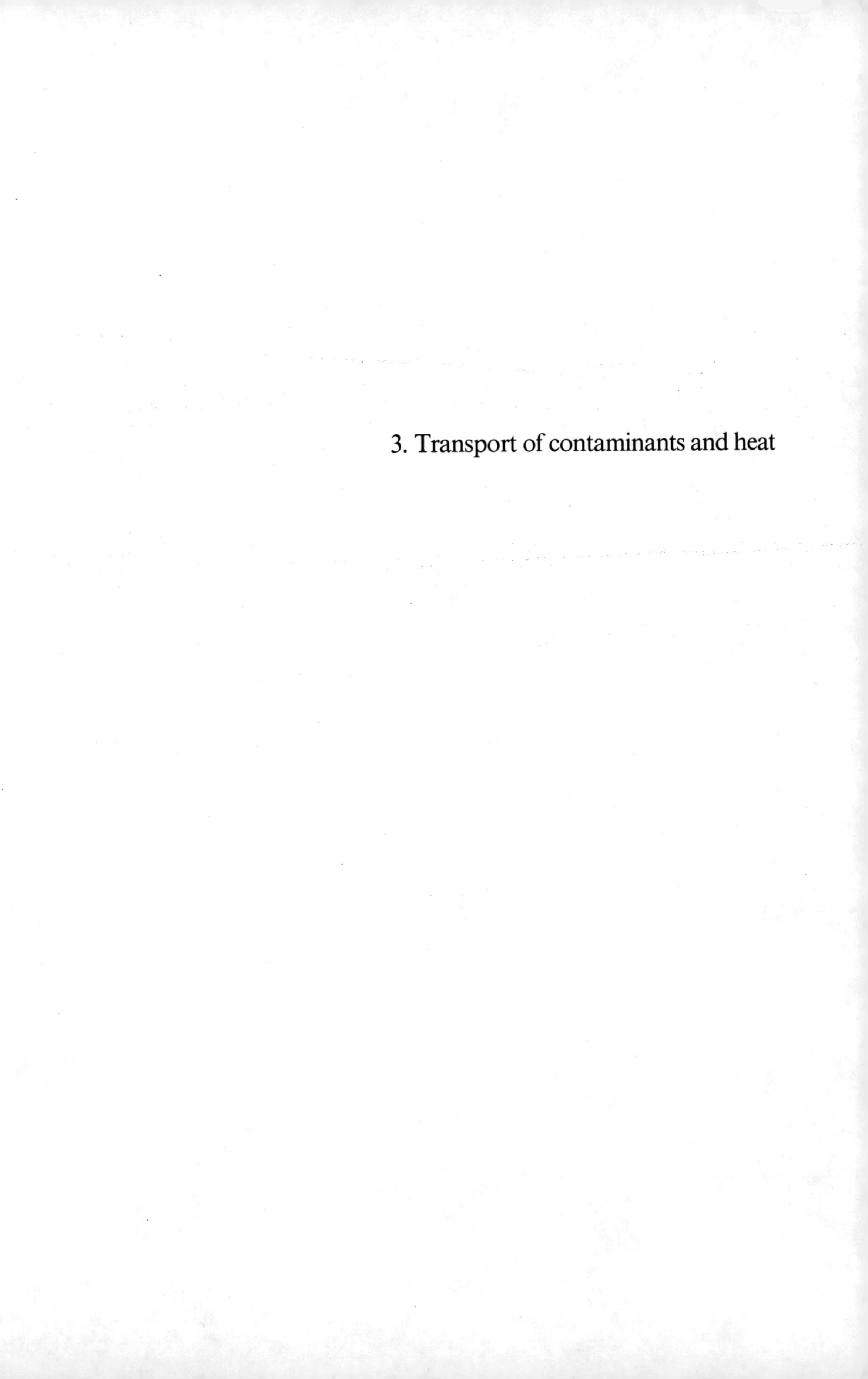

3. Transport of contaminants and heat

Proceedings of Euromech 143 / Delft / 2-4 September 1981

Annual variation of temperature in an array of wells adjacent to a river bank

M.BORELI & M.RADOJKOVIĆ
Belgrade University, Yugoslavia

1. INTRODUCTION

One of the essential advantages of water supply from groundwater resources is a fact that the annual temperature oscillation in a well adjacent to river bank is smaller (sometimes considerably) in comparison to the amplitude of temperature oscillations in the river.

Therefore, the prediction of the temperature regime in the well is important for the design of well systems.

The purpose of this work is to present a simple procedure for prediction of annual variation of the temperature in an array of wells, in a quasihomogeneous aquifer that can be easily handled by practical hydraulic engineers.

2. PROBLEM DEFINITION

Annual temperature variation in the river and adjacent systems of wells can be approximated by periodic functions where sinosoidal oscillations are usually good enough to fit observed data (Fig.1).

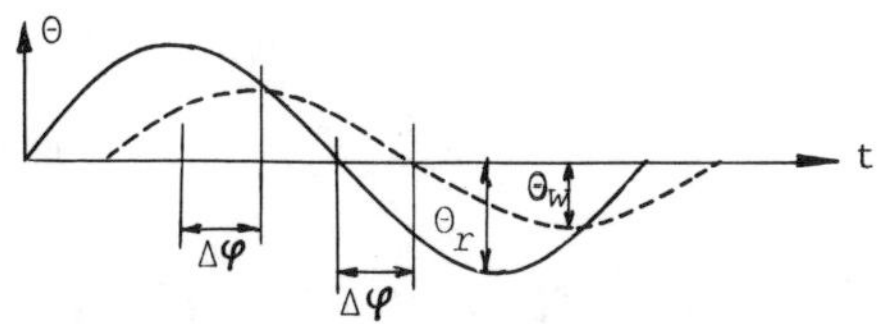

Fig. 1. Oscillations of temperature in river and wells

Therefore, temperature oscillations in the well are defined with two parameters:

- amlitude damping parameter (Θ^M_{*w}) where:

$$\Theta^M_{*w} = \frac{\Theta^{max}_w - \Theta^{min}_w}{\Theta^{max}_R - \Theta^{min}_R} \qquad (1)$$

- phase lag parameter ($\Delta\varphi_*$) where:

$$\Delta\varphi_* = \frac{\Delta\varphi}{T} \qquad (2)$$

For prediction of these two parameters one have to solve equations of heat transport and groundwater flow for relevant boundary conditions.

This problem can be solved either analyticaly or numericaly.

Numerical solutions are rather inacurate due to poor resolution of flow field in the vicinity of wells and due to dificulties for solving heat transport equation with dominant convective terms. For these reasons numerical approach for solving investigated problem is still rather a challenge to numerical analyst than a reliable tool for practical engineer.

Analytical solutions are complicated (except for a single well [1]), but, for simplified geometry, one can introduce universal dimensionless numbers and solve problems for all values which are of practical interest. Results than can be tabulated or approximated by elementary functions. Such approach to the problem solution is described in the sequel.

3. BASIC EQUATIONS

The heat transport problem is defined by aquifer thickness averaged equation, with supposition of equality of temperatures for the solid and fluid phase:

$$\frac{\partial \bar{D} \frac{\partial \bar{\Theta}}{\partial x_i}}{\partial x_i} - \frac{\partial \bar{\Theta} \bar{V}_i}{\partial x_i} = \frac{\partial \bar{\Theta}}{\partial t} \qquad i=1,2... \quad (3)$$

where:

$$\bar{V}_i = \bar{u}_i / \bar{n} \qquad (4)$$

$\bar{u}_i$ is local volume averaged velocity (Darcy's velocity) and $\bar{n}$ generalised porisity which takes account the heat exchange between solid and fluid phase defined as:

$$\bar{n} = n\left(1 - \frac{1-n}{n}\frac{C_s}{C_w}\right) \qquad (5)$$

$\bar{D}$ is dispersion coefficient which accounts for the following effects:

- heat conduction at the microscopic scale
- averaged convective transport at macroscopic (referent volume) scale (macrodispersion)
- averaged convective transport at megascopic [2] (aquifer thickness) scale (megadispersion)

In practice conduction is negligible in comparison to macrodispersion. Megadispersion occurs only in verticaly heterogeneous aquifers and than is usually much greater than macrodispersion. Both macro and megadispersion can be defined as [3]:

$$\bar{D} = \ell \, \bar{V} \qquad (6)$$

where ℓ is characteristics lenght (dispersivity) with order of milimeters for macrodispersion and order of meters for megadispersion [4]

The solution of problem can be simplified if one assumes that the oscillation of temperature in the well can be obtained by averaging the temperatures at the end of streamlines (i.e. by neglecting lateral dispersivity).

Due to different phase lags along each streamline, the amplitude of temperature oscillations is damped even with neglected dispersion term in eq. (3). Damping parameter (Θ^M_{*w}) obtained in this case has maximum value (for given horizontal geometry) and the procedure developed to determine it, together with phase lag parameter ($\Delta\varphi_*$) is derived in the next chapter. Practical applications of these parameters are given at the end of the paper.

4. DIMENSIONLESS SOLUTION PROCEDURE

The time of travel of a particle along the streamline Ψ (phase lag) is defined as:

$$t = \int_{\Psi_i} \frac{ds}{\bar{V}(a,b)} \qquad (7)$$

This equation is converted in dimensionless form:

$$t_* = \int_{\Psi_i} \frac{dS_*}{V_*(a_*)} \qquad (8)$$

where:

$$t_* = \frac{t_*}{t_o}; \quad S_* = \frac{S}{L_o}; \quad V_* = \frac{\bar{V}}{V_o}; \quad a_* = \frac{a}{L_o}$$

Reference values (subscript o) are defined as:

$$t_o = \frac{a}{b}\frac{2b^2 H\bar{n}}{Q}$$

$$L_o = b$$

$$V_o = \frac{Q}{2bH\bar{n}}$$

Evaluation of integral (7) is done numericaly with analytical solution for $V_*(a_*)$ derived by conformal mapping as given in the appendix 1.

Analytical solution for the heat transport equation along streamline (without dispersion) is:
dimensionless form:

$$\Theta_{*s} = \sin \frac{2\pi}{T}(t - t_s) \qquad (9)$$

This equation is converted into

$$\Theta_{*s} = \sin 2\pi\left(\frac{t}{T} - t_* a_* b_*\right) \qquad (10)$$

where:

$$b_* = \frac{t_o}{T} = \frac{2n^* b^2 H}{Q\,T}$$

Average temperature in a well is approximated as:

$$\Theta_{*w} = \frac{1}{N}\sum_{i=1}^{i=N} \Theta_{*s} \qquad (11)$$

$$\Theta^M_{*w} = e^{a_* \sqrt{b_*}} \qquad (14)$$

$$\Delta\varphi_* = a_* b_* \ \ln\left[\frac{3}{4}(e^{a_*} / \sqrt{b_*})^{2/3}\right] \qquad (15)$$

Correlation of the solutions with the approximations is given in figs. 4 and 5.

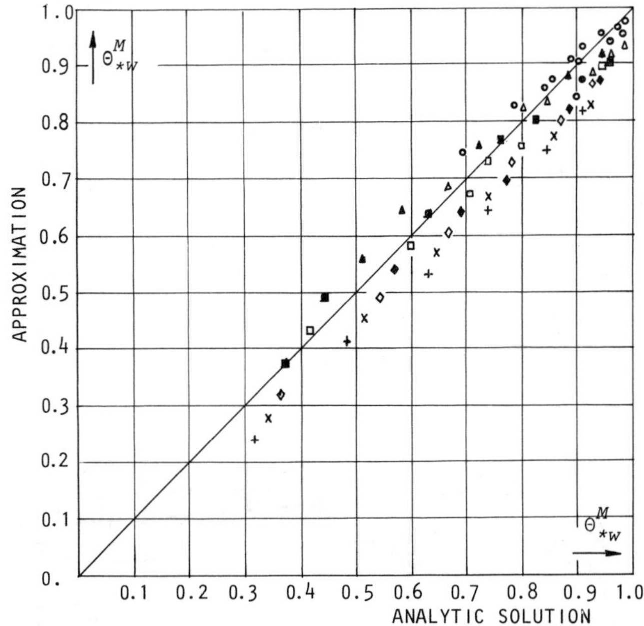

Fig.4. Correlation of analytical solution with approximation for temperature amplitude

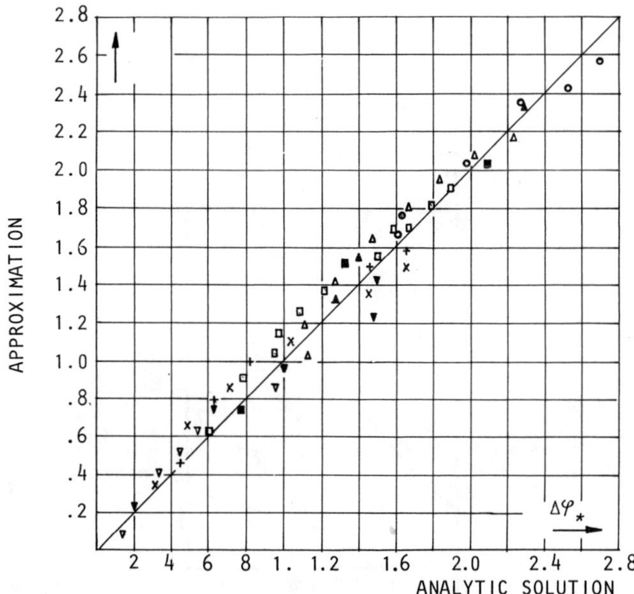

Fig.5. Correlation of analytical solution with approximation for phase lag

where N is a number of streamlines.

Amplitude damping parameter (Θ^M_{*w}) and phase lag parameter ($\Delta\varphi_*$) are easily computed from (11).

5. RESULTS

The foregoing procedure is applied to evaluate the functions:

$$\Theta^M_{*w} = f_1(a_*, b_*) \tag{12}$$

$$\Delta\varphi_* = f_2(a_*, b_*) \tag{13}$$

The computation of functions (2) and (3) was done within the range:

$0,2 \leqslant a_* \leqslant 2$

$0,01 \leqslant b_* \leqslant 0,5$

which includes the most of the conditions that might occur in practice.

By trial and error it was found that rather fine discretisation had to be used to obtain satisfactory accuracy. Discretisation factors were:

number of streamlines ($0 \leqslant \Psi \leqslant \frac{1}{2}$) ...250

number of segments along each streamline ...250

number of increments for one period ... 96

60 computations were done with the following values of a_* and b_*:

a_* = 0.2,0.4,0.6,0.8,1.0,1.2,1.4, 1.6,2.0

b_* = 0.01,0.02,0.05,0.1,0.2,0.5

Results of computations are given in figs. 2 and 3. With resonable accuracy, these results were approximated with the following functions:

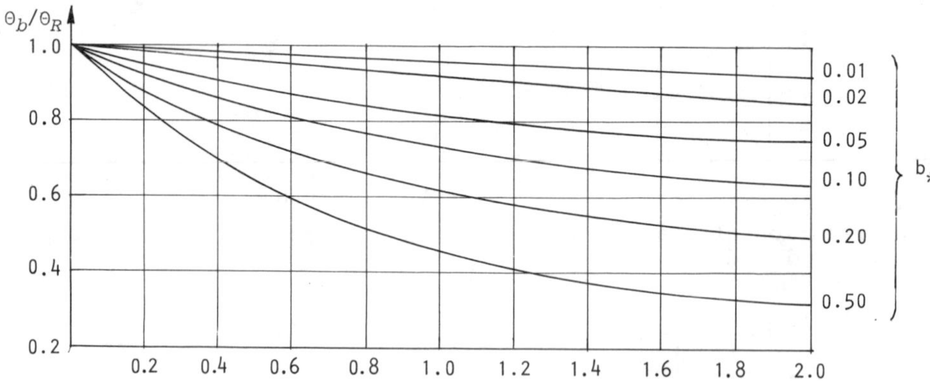

Fig.2. Analytical solution for temperature amplitude damping

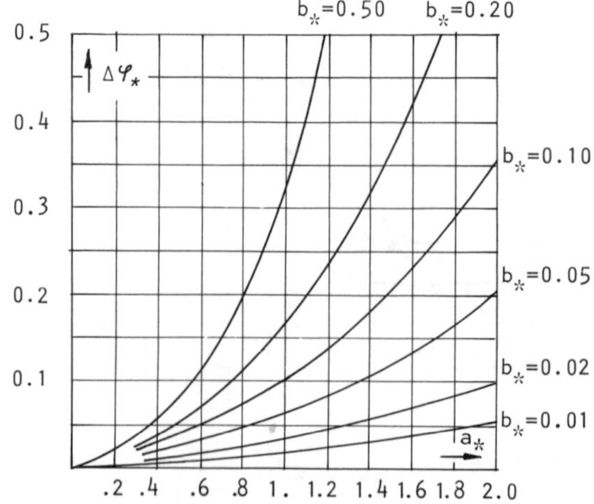

Fig.3. Analytical solution for phase lag

Equations (14) and (15) can be used for quick estimation of temperature oscillations in an array of wells adjacent to river bank if homogeneous aquifer with negligible dispersion is assumed.

6. FINAL REMARKS

The temperature damping and phase lag given by equations (14) and (15) are valid for a rather homogeneous aquifer where characteristic dispersion lenght is only a small fraction of the distance between array of wells and the river.

In the case of verticaly heterogeneous (multi layered) aquifer the damping of temperature oscillations can be much greater than one obtained by eq. (14) For instance, for an array of wells with horizontal drains located near the base of bed rock the authors have found that characteristic lenght and the geometric distance between well and river are of the some order (about 100 m).

The details on this problem will be given elsewhere. This paper is final answer to the problem of "hydraulic dispersion" (temperature damping and phase lag without macro or megadispersion).

7. ACKNOWLEDGEMENTS

Equations for description of flow field, given in appendix 1, were derived by Miss.R.Ivanović as partial fulfilment of her Batchelor of Science degree work under supervision of senior author of the paper.

8. LIST OF SYMBOLS

C_s - therewal capacity of solid phase
$\underline{C}_w$ - thermal capacity of water
$\underline{D}$ - dispersion coefficient
H - aquifer thickness
L_o - reference lenght
N - number od streamlines
$\underline{Q}$ - well discharge
u - Darcy's velocity averaged over H
$\bar{V}$ - efective velocity averaged over H
$\underline{V}_o$ - referent velocity
$\bar{V}$ - efective velocity along a streamline
V_* - dimensionless velocity
S - lenght along streamline
S_* - dimensionless lenght along streamline
T - period of oscillation (one year)
a - distance between array of wells and river
a_* - dimensionless parameter
b - distance between wells in an array
b_* - dimensionless parameter
i - index
ℓ - dispersion lenght (dispersivity)
n - porosity
$\bar{n}$ - generalized porosity
t - time
t_o - reference time
t_s - time along streamline
t_* - dimensionless time
x_i - cartesian coordinates
$\Delta\varphi$ - phase lag
$\Delta\varphi_*$ - dimensionless phase lag
Θ - temperature in aquifer averaged over H
$\Theta_w^{max}, \Theta_w^{min}$ - maximum and minimum temperatures in the well
$\Theta_r^{max}, \Theta_r^{max}$ - maximum and minimum temperatures in the river
Θ_{*w} - dimensionless temperature in the well
Θ_{*w}^{M} - maximum temperature in the well
Ψ_i - streamline

REFERENCES

[1] Ubell,K.1977, Time of travel at wells near a stream. Seventeenth Congress of IAHR (Seminar), Baden-Baden

[2] Bear,J. 1978, Relation between microscopic and macroscopic description in porous media. Symp. on Scale effect in orous Thessaloniki, pp 1-32.

[3] Bear,J.,Zaslavsky.D.,Irmay,S. 1968, Physical principles of water percolation and seepage, UNESCO, Paris.

[4] Gaillard,B.,Rousselot,D.,Sauty, J.P. 1977, Application d'une méthode de économique détermination sur le terrain des parametres de dispersion, Symp.on underground pollution and dispersion,Pavia.

10. APPENDIX 1

Velocity distribution in the flow field of an array of wells adjacent to river bank

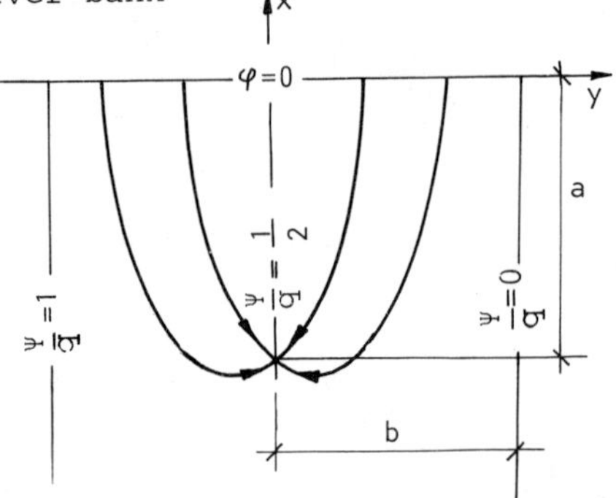

Fig. 1

Streamlines $\Psi = \text{const}$ are defined with the equation:

$$X^2 = \frac{2mY + K(1-m^2Y^2)}{K(m^2-Y^2) + 2mY} \qquad (1)$$

where:

$$X = \text{th}\,\frac{\pi x}{2b} \qquad (2)$$

$$Y = \text{tg}\,\frac{\pi y}{2b} \qquad (3)$$

$$K = \text{tg}\,\frac{2\pi\Psi}{q} \qquad (4)$$

$$m = \frac{1}{\text{th}\,\frac{a\pi}{2b}} \qquad (5)$$

Each streamline originates at point located at:

$$\frac{Y_0}{B} = \frac{2}{\pi}\,\text{arc tg}\left[\frac{\text{tg}\left(\frac{\pi}{2} - \frac{\pi\Psi}{q}\right)}{m}\right] \qquad (6)$$

Due to symmetry, problem is analised in the domain $0<\Psi<q/2$.

Velocity components are defined with the equations:

$$\frac{V_X}{q/2b} = \\ = \frac{-2m\left[(m^2-1)\,\text{ch}\frac{x}{b}\cos\frac{y}{b} + (m^2+1)\right]}{F} \qquad (7)$$

$$\frac{V_Y}{q/2b} = \\ = \frac{-2m\left[(m^2-1)\,\text{sh}\,\frac{x\pi}{b}\,\sin\,\frac{y\pi}{b}\right]}{F} \qquad (8)$$

where:

$$F = \left[(m^2-1)\,\text{ch}\,\frac{x\pi}{b}\,\cos\,\frac{y\pi}{b} - (m^2+1)\right]^2 + \\ + \left[(m^2-1)\,\text{sh}\,\frac{x\pi}{b}\,\sin\,\frac{y\pi}{b}\right]^2 \qquad (9)$$

Computation procedure is as follows:

For finite number of streamlines (N) each streamline has value:

$$\frac{\Psi}{q} = \frac{1}{N}\left(\frac{1}{4} + \frac{i}{2}\right) \qquad i = 0,1,2\ldots N-1 \qquad (10)$$

for which one finds y coordinate for x=0 from (6).

Streamline is then subdivided to a finite number of segments (Fig.1) with coordinates of segment points evaluated from (1)-(5).

The velocity components are computed for each point from equations (7) and (8) so that time of particle travel along streamline can be computed as:

$$t = \sum_{i=1}^{i=M} \frac{\sqrt{(x_{(i+1)}-x_{(i)})^2 + (y_{(i+1)}-y_{(i)})^2}}{\sqrt{v^2_{(x\,i)}+v^2_{(y\,i)}} + \sqrt{v^2_{(x\,i+1)}+v^2_{(y\,i+1)}}} \cdot 2 \qquad (11)$$

where M is the number of segment along streamline.

Proceedings of Euromech 143 / Delft / 2-4 September 1981

A note on the equation of transport of heat in solids

N.J.DAHL
Institute for Town and Country Planning, Copenhagen, Denmark

1 INTRODUCTORY REMARKS

A number of physical or technical problems - for instance non-steady one-dimensional ground water flow - leads, after a transcription into a dimensionless form, to the partial differential equation

$$(1) \quad \frac{\partial^2 y}{\partial x^2} - \frac{\partial y}{\partial \tau} = -z(x,\tau)$$

where y denotes the dependent variable in the (x,τ)-coordinate system giving the position (x) and the time (τ); in (1) $z(x,\tau)$ is an analytical function determined by the physical or technical conditions belonging to the actual problem. Thus we state that the solution $y(x,\tau)$ must fulfil equation (1) over a certain region in the (x,τ)-system together with the prescribed boundary and initial conditions.

Equation (1) is recognized as a special form of the equation of conduction of heat in solids and, as known, numerous solutions to this equation - corresponding to specified conditions - are found in the literature; reference is here made to Carslaw and Jaeger, 1978. Unfortunately the author has not succeeded in finding a systematic analytical method to solving (1) in the case of $z(x,\tau)$ being an arbitrary analytical function.

However, technicians have to solve their problems exactly - if possible - or - if not possible - to establish an approximative solution, applicable for specified cases. The problem from which equation (1) emerged is related to water supply from a ground water reservoir (N.J. Dahl, 1980 and 1981); the method used for the solving of (1) in these papers might, however, be useful in relation to other problems. Therefore, a concise description of the method is given in the following.

2 DEFINITION OF THE PROBLEM

We have to solve the equation

$$(2.1) \quad \frac{\partial^2 y}{\partial x^2} - \frac{\partial y}{\partial \tau} = -z(x,\tau)$$

over the region defined by, cf. Fig. 2.1,

$$(2.2) \quad 0 \leq x \leq 1 \qquad -\infty \leq \tau \leq \infty$$

with the boundary conditions

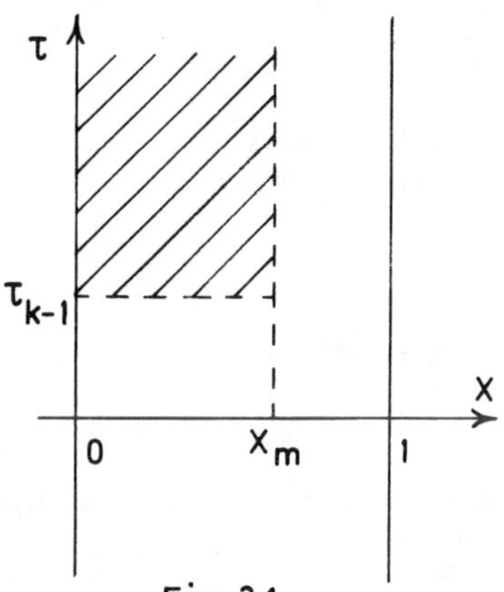

Fig. 2.1

$$(2.3) \quad \begin{cases} y(0,\tau) = 0 \\ y(1,\tau) = 0 \end{cases}$$

and the initial condition

$$(2.4) \quad y(x,\tau) = 0 \qquad \text{for } \tau \leq \tau_{k-1}$$

where τ_{k-1} is an arbitrary constant.

The function $z(x,\tau)$ is, in principle, an arbitrary function over the region of definition; at this moment we choose, however, the special function

$$(2.5) \quad z(x,\tau) = \begin{cases} 1 & \text{for } x \leqq x_m \text{ and } \tau \geqq \tau_{k-1} \\ 0 & \text{elsewhere} \end{cases}$$

i.e. we have $z(x,\tau) = 1$ over the hatched area in Fig. 2.1 and equal to zero over the rest of the region. Later on we shall see that it is possible to use a solution satisfying the conditions defined above as a toybrick to build up more complex solutions.

3 THE TOY-BRICK SOLUTION

We assume that this solution can be written in the form

$$(3.1) \quad y(x,\tau) = y_s(x) - y_n(x,\tau)$$

where $y_s(x)$ and $y_n(x,\tau)$ represent a steady, resp. a non - steady, part of the solution.

We now demand that $y_s(x)$ has to fulfil the equation, comp. (2.1),

$$(3.2) \quad \frac{\partial^2 y_s}{\partial x^2} = -z(x,\tau)$$

with the conditions

$$(3.3) \quad \begin{cases} y_s(0) = 0 \\ y_s(1) = 0 \end{cases}$$

and where $z(x,\tau)$ is defined by (2.5). It is easily proven that the complete solution to this problem is

$$(3.4) \quad \begin{cases} y_s(x) = \begin{cases} -\frac{1}{2}x^2+(x_m-\frac{1}{2}x_m{}^2)x & \text{for } 0 \leqq x \leqq x_m \\ \frac{1}{2}x_m{}^2(1-x) & \text{"} \quad x_m \leqq x \leqq 1 \end{cases} \\ y_s(x,\text{neg}) = 0 \end{cases}$$

where $y_s(x,\text{neg}) = 0$ symbolizes that we have $z(x,\tau) = 0$ if $\tau-\tau_{k-1} < 0$.

It is observed that $y_s(x)$ for $\tau-\tau_{k-1} \geqq 0$ is composed of a parabola over the interval $0 \leqq x \leqq x_m$ and a straight line over the interval $x_m \leqq x \leqq 1$, the line being a tangent to the parabola for $x = x_m$. We now interpret this function as an odd function over the interval $0 \leqq x \leqq 1$. To this function a Fourier - sine expansion exists, and it is easy to calculate the coefficients for this expansion, but for shortness we omit this work and content ourselves by quoting the result

$$(3.5) \quad \begin{cases} y_s(x) = \sum \frac{2}{(n\pi)^3} \cdot (1-\cos n\pi x_m) \sin n\pi x \\ y_s(x,\text{neg}) = 0 \end{cases}$$

where $n = 1, 2, 3, \ldots$. Finally we note that the solutions (3.4) and (3.5) are identical in every respect.

We now revert to (3.1); after differentiation and insertion in (2.1) we get, after simple reduction, that $y_n(x,\tau)$ has to fulfil the equation

$$(3.6) \quad \frac{\partial^2 y_n}{\partial x^2} - \frac{\partial y_n}{\partial \tau} = 0$$

and the boundary and initial conditions belonging to $y_n(x,\tau)$ are found as follows. The boundary conditions for $y(x,\tau)$ are given by (2.3), and since we have $y_s(x) = 0$ for $x = 0$ and $x = 1$, we conclude from (3.1) that the boundary conditions for $y_n(x,\tau)$ are

$$(3.7) \quad \begin{cases} y_n(0,\tau) = 0 \\ y_n(1,\tau) = 0 \end{cases}$$

The initial condition for $y(x,\tau)$ is defined by (2.4); since $y_s(x)$ is given by (3.4) or (3.5) we conclude from (3.1) that $y_n(x,\tau)$ must fulfil the conditions

$$(3.8) \quad \begin{cases} y_n(x,\tau) = y_s(x) & \text{for } \tau = \tau_{k-1} \\ y_n(x,\tau) = 0 & \text{"} \quad \tau < \tau_{k-1} \end{cases}$$

The solution to (3.6) fulfilling the conditions (3.7) and (3.8) can be found by using the theory of Laplace transforms including the unit function shown in Fig. 3.1. We

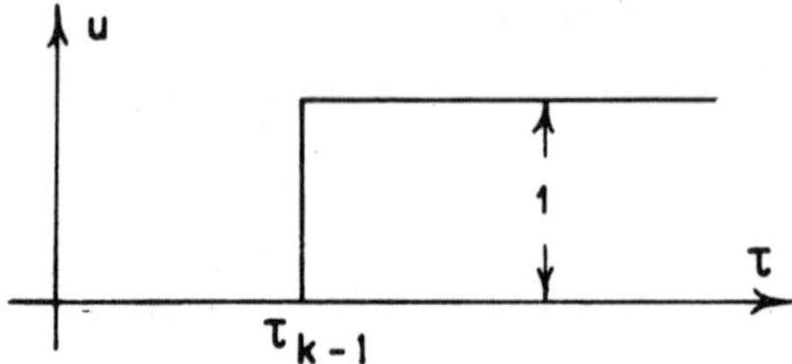

Fig. 3.1

omit the necessary calculations and quote the result which is

$$(3.9) \quad \begin{cases} y_n(x,\tau) = \sum \frac{2}{(n\pi)^3} (1-\cos n\pi x_m) \sin n\pi x \cdot \\ \quad \cdot \exp(-(n\pi)^2(\tau-\tau_{k-1})) \\ y_n(x,\text{neg}) = 0 \end{cases}$$

where $n = 1, 2, 3, \ldots$; the condition $y_n(x,\text{neg}) = 0$ arises from the use of the unit function.

Insertion of (3.5) and (3.9) in (3.1) now gives

$$(3.10)\quad \begin{cases} y(x,\tau) = \sum \frac{2}{(n\pi)^3}(1-\cos n\pi x_m)\cdot \sin n\pi x \\ \quad -\sum \frac{2}{(n\pi)^3}(1-\cos n\pi x_m)\cdot \sin n\pi x \cdot \\ \quad \cdot \exp(-(n\pi)^2(\tau-\tau_{k-1})) \\ y(x,neg) = 0 \end{cases}$$

where $n = 1, 2, 3, \ldots$. Since the two sums are identical term by term apart from the exponential factor we are entitled to rewrite (3.10) in the form

$$(3.11)\quad \begin{cases} y(x,\tau) = \sum_{n=1}^{\infty} \frac{2}{(n\pi)^3}(1-\cos n\pi x_m) \sin n\pi x \cdot \\ \quad \cdot [1-\exp(-(n\pi)^2(\tau-\tau_{k-1}))] \\ y(x,neg) = 0 \end{cases}$$

In the foregoing we have omitted some intermediate calculations and therefore it seems necessary to prove that (3.11) is a solution to (2.1) which fulfils all conditions attached to our problem, cf. Fig. 2.1. It is evident, because of the factor $\sin n\pi x$, that (3.11) fulfils the boundary condition (2.3) for $x = 0$ and $x = 1$. It is also evident, because of the condition $y(x,neg) = 0$, that the initial condition (2.4) is fulfilled for $\tau-\tau_{k-1} < 0$; the special case for $\tau-\tau_{k-1} = 0$ is also fulfilled since we have $\exp(0) = 1$.

By differentiation of (3.11) we now get

$$\frac{\partial^2 y}{\partial x^2} = -\sum_{n=1}^{\infty} \frac{2}{n\pi}(1-\cos n\pi x_m)\cdot \sin n\pi x \cdot$$
$$\cdot [1-\exp(-(n\pi)^2(\tau-\tau_{k-1}))]$$

$$\frac{\partial y}{\partial \tau} = \sum_{n=1}^{\infty} \frac{2}{n\pi}(1-\cos n\pi x_m)\cdot \sin n\pi x$$
$$\cdot \exp(-(n\pi)^2(\tau-\tau_{k-1}))$$

and insertion of these expressions in (2.1) gives

$$-\sum_{n=1}^{\infty} \frac{2}{n\pi}(1-\cos n\pi x_m) \sin n\pi x$$

$$+\sum_{n=1}^{\infty} \frac{2}{n}(1-\cos n\pi x_m) \sin n\pi x \cdot$$
$$\cdot \exp(-(n\pi)^2(\tau-\tau_{k-1}))$$

(cont.)

$$-\sum_{n=1}^{\infty} \frac{2}{n\pi}(1-\cos n\pi x_m) \sin n\pi x \cdot$$
$$\cdot \exp(-n\pi)^2(\tau-\tau_{k-1})$$
$$= -z(x,\tau)$$

or

$$\sum_{n=1}^{\infty} \frac{2}{n\pi}(1-\cos n\pi x_m) \sin n\pi x = z(x,\tau)$$

or, by insertion of (2.5),

$$(3.12)\quad \sum_{n=1}^{\infty} \frac{2}{n\pi}(1-\cos n\pi x_m) \sin n\pi x = \begin{cases} 1 & \text{for } 0 \leq x \leq x_m \\ 0 & \text{"} \quad x_m \leq x \leq 1 \end{cases}$$

for $\tau-\tau_{k-1} \geq 0$.

In fact, it is very easy to show, by differentiation of (3.4) combined with (3.5) or by direct calculation, that the left side of (3.12) is nothing but the Fourier - sine expansion of the right side.

For the sake of shortness in the following we now define the function

$$(3.13)\quad \begin{cases} \phi(x,x_m,(\tau-\tau_{k-1})) = \sum_{n=1}^{\infty} \frac{2}{(n\pi)^3} \cdot \\ \quad \cdot (1-\cos n\pi x_m) \sin n\pi x \cdot \\ \quad \cdot [1-\exp(-(n\pi^2)(\tau-\tau_{k-1}))] \\ \phi(x,x_m,neg) = 0 \end{cases}$$

and this function is precisely the toy -brick solution which we need for the construction of more complex solutions of (1) in case of $z(x,\tau)$ being a more complex function. It is observed that $\phi(x,x_m,(\tau-\tau_{k-1}))$ contains the two free parameters x_m and τ_{k-1}; it is the mere presence of these parameters which makes $\phi(\ )$ so flexible.

To complete the discussion of $\phi(\ )$ we note that $\phi(\ )$ is a monotone function of τ having the limit

$$(3.14)\quad \phi(x,x_m(\tau-\tau_{k-1})) \rightarrow$$
$$\rightarrow \sum_{n=1}^{\infty} \frac{2}{(n\pi)^3}(1-\cos n\pi x_m) \sin n\pi x$$
$$= y_s(x)$$

for $\tau \to \infty$. Fig. 3.2 shows the course of $\phi(\ )$ as a function of $\tau-\tau_{k-1}$ with x as a parameter; the curves correspond to $x_m = 1.0$; it is observed that the limit defined by (3.14) is reached with a fair exactness if $\tau-\tau_{k-1} > 0.5$.

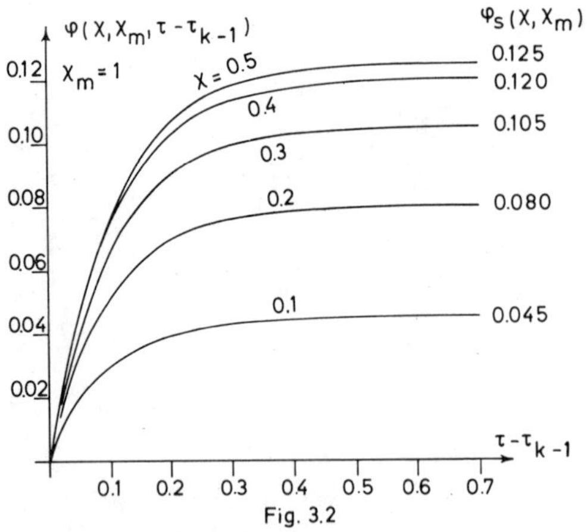

Fig. 3.2

4 $z(x,\tau)$ BEING A DOUBLE STAIRCASE FUNCTION

As mentioned above, x_m and τ_{k-1} are free parameters to which arbitrary, but fixed values could be ascribed. Nothing now prevents us from setting up the difference

$$(4.1) \quad y_k(x,\tau) = \phi(x,x_m,(\tau-\tau_{k-1})) - \phi(x,x_m,(\tau-\tau_k))$$

in which we have $\tau_k > \tau_{k-1}$. The 2nd ϕ-term on the right side corresponds to the definition of $z(x,\tau)$ shown in Fig. 4.1, i.e.

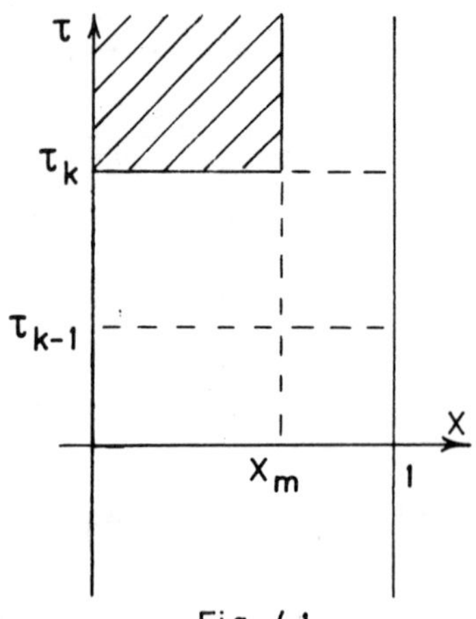

Fig. 4.1

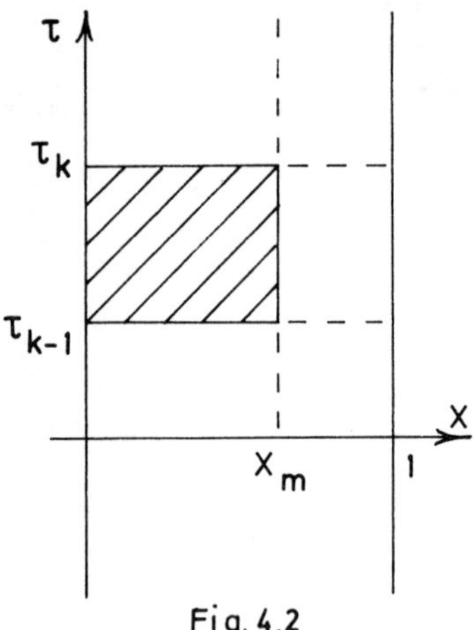

Fig. 4.2

$z(x,\tau) = 1$, over the hatched area and $z(x,\tau) = 0$ elsewhere. Consequently the solution (4.1) corresponds to the definition of $z(x,\tau)$ shown in Fig. 4.2, i.e. $z(x,\ ) = 1$ over the hatched area and $z(x,\tau) = 0$ elsewhere.

We now set up the double difference

$$(4.2) \quad y_{m,k}(x,\tau) = [\phi(x,x_m(\tau-\tau_{k-1})) - \phi(x,x_m,(\tau-\tau_k))] - [\phi(x,x_{m-1},(\tau-\tau_{k-1})) - \phi(x,x_{m-1},(\tau-\tau_k))]$$

in which we have $x_m > x_{m-1}$. It is now evident that (4.2) is the solution for which we have $z(x,\tau) = 1$ over the hatched rectangle shown in Fig. 4.3 and $z(x,\tau) = 0$ elsewhere. On this background we subdivide the

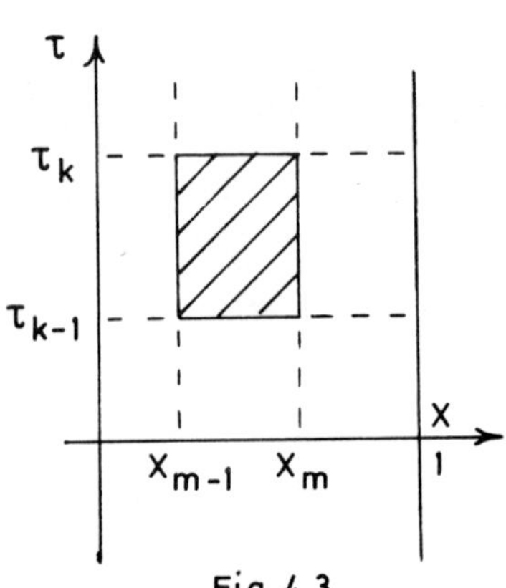

Fig. 4.3

region of definition into (small) rectangles using a finite number of lines $x = x_0$, $x = x_1$, ... over the interval $0 \leq x \leq 1$ and an infinite number of lines $\tau = \tau_0$, $\tau = \tau_1$, ...

Over each of these rectangles we ascribe a constant value $a_{m,k} \gtreqless 0$ to $z(x,\tau)$, and based on (4.2) we now set up the double sum

$$(4.3) \quad y(x,\tau) = \sum_m \sum_k a_{m,k}\{[\phi(x,x_m,(\tau-\tau_{k-1})) -\phi(x,x_m,(\tau-\tau_k))] -[\phi(x,x_{m-1},(\tau-\tau_{k-1}))- -\phi(x,x_{m-1},(\tau-\tau_k))]\}$$

which gives the solution corresponding to the case where $z(x,\tau)$ is a double staircase function covering the whole region of definition.

The symbols $\sum_m$ and $\sum_k$ indicate that we have to sum up over all actual rectangles; it should be emphasized that the number of addends is limited for finite values of τ since the condition $\phi(x,neg)$ is attached to the toy-brick function $\phi(\ )$.

5 $z(x,\tau)$ BEING A CONTINUOUS FUNCTION

The solution (4.3) is an exact solution to our problem if, for physical or technical reasons, the function $z(x,\tau)$ is defined as a double staircase function. On the other hand, if $z(x,\tau)$ is a continuous (analytical) function, the solution (4.3) will give an approximative solution if $z(x,\tau)$ is exchanged for a mean value $a_{m,k}$ inside the individual rectangles. The quality of the solution - in numerical respect - can be increased by a (very) fine subdivision of the region of definition but, of course, on the expense of numerical calculations.

Finally, we should add that the double summation in (4.3) gives rise to the question: is it possible to rewrite this double summation as a double integral? The author leaves this question open to mathematicians.

6 CLOSING REMARKS

It is not seldom found that a technician, working on theoretical problems, is facing an insoluble problem, for instance a partial differential equation combined with a number of conditions. In such cases he has nowadays, above all, recourse to methods of numerical calculation, f.inst. the method of finite differences. Such methods constitute a really strong tool when specified problems are to be solved; in the author's opinion they have, however, a grievous defect since they do not open up a prospect to analysing the physical/technical phenomena hidden in the actual problem.

In such cases we have to revert to the mathematical analysis, using various approximations; an example hereof is the linearizing of the actual equations. The aim of this paper has been to show that it will be possible, sometimes at least, to obtain solvable equations by exchanging a tricky condition for a simpler one, though in such a way that the obtained solution must coincide - with a reasonable exactness - with the unknown solution.

7 REFERENCES

Carslaw, H.S & Jaeger, J.C. 1978, Conduction of Heat in Solids, Oxford at the Clarendon Press.

Dahl, N.J. 1980, Base Flow and Water Supply for Irrigation, Series Paper 7, Inst. f. Town and Country Planning, Royal Vet. and Agricultural University, Denmark.

Dahl, N.J. 1981, Non-Steady Flow in Coned Layers, Inst. f. Town and Country Planning, Royal Vet. and Agricultural University, Denmark.

Proceedings of Euromech 143 / Delft / 2-4 September 1981

On heat transfer in porous media

H.I.ENE
INCREST, Bucharest, Romania

E.SANCHEZ-PALENCIA
Université Paris VI, France

1 INTRODUCTION

We consider the motion of a viscous fluid through a porous, rigid body. The "pores" have a periodic geometric structure, the period of which (associated with the small parameter $\varepsilon \to 0$) is small with respect to the other geometric magnitudes involved in the problem. In fact, we are in the general framework of the "homogeneization method", Bensoussan, Lions and Papanicolaou (1978) or Sanchez-Palencia (1980). It is known that the asymptotic process and the limit equations may have very different structure if several "small parameters" are involved in the problem.

In this paper we consider the problem of heat transfer in several cases where the density, the viscosity, the thermal conductivity or the thermal expansion coefficients are small (in fact functions of ε).

2 INCOMPRESSIBLE FLOW

We consider a parallelipipedic period Y of the space of the variables y_i (i=1,2,3) formed by a fluid and a solid part Y_f and Y_s, with smooth boundary Γ. We also denote by Y_f (resp.Y_s) the union of the Y_f(resp. Y_s) parts of all periods, and assume that Y_f (resp.Y_s) is connected. If Ω is the "porous body" in the space of the variables x_i, we introduce the small parameter ε and the fluid domain $\Omega_{\varepsilon f}$ (resp.the solid domain $\Omega_{\varepsilon s}$) defined by

$$\Omega_{\varepsilon f}=\{x;\ x\in\Omega,\ x\in\varepsilon Y_f\}$$

$$\Omega_{\varepsilon s}=\{x;\ x\in\Omega,\ x\in\varepsilon Y_s\}$$

If ρ, p^ε, T^ε and v^ε denote the density, pressure, temperature and velocity of the incompressible flow, they must satisfy the equations of conservation of momentum, mass and energy

$$\rho_f v_k^\varepsilon \frac{\partial v_i^\varepsilon}{\partial x_k} = -\frac{\partial p^\varepsilon}{\partial x_i} + \frac{\partial \tau_{ik}^\varepsilon}{\partial x_k} + \rho_f f_i \quad (1)$$

$$\frac{\partial v_i^\varepsilon}{\partial x_i} = 0 \quad (2)$$

$$\rho_f C_f v_k^\varepsilon \frac{\partial T^\varepsilon}{\partial x_k} = \tau_{jk}^\varepsilon \frac{\partial v_j^\varepsilon}{\partial x_k} + \frac{\partial}{\partial x_k}\left(\lambda_f' \frac{\partial T^\varepsilon}{\partial x_k}\right) \quad (3)$$

in $\Omega_{\varepsilon f}$, and

$$0 = \frac{\partial}{\partial x_k}\left(\lambda_s' \frac{\partial T^\varepsilon}{\partial x_k}\right) \quad (4)$$

in $\Omega_{\varepsilon s}$, where f_i are the components of the exterior body force by unit mass, τ_{ik}^ε are the components of the viscous stress tensor:

$$\tau_{ik}^\varepsilon = \mu'\left(\frac{\partial v_i^\varepsilon}{\partial x_k} + \frac{\partial v_k^\varepsilon}{\partial x_i}\right) \quad (5)$$

The boundary conditions on Γ are:

$$v_i^\varepsilon = 0 \quad (6)$$

$$T^\varepsilon|_f = T^\varepsilon|_s \quad (7)$$

$$\lambda'_f \frac{\partial T^\varepsilon}{\partial n}\Big|_f = \lambda'_s \frac{\partial T^\varepsilon}{\partial n}\Big|_s \tag{8}$$

In order to study the asymptotic process $\varepsilon \to 0$ we consider the classical expansions:

$$v_i^\varepsilon(x) = \varepsilon^n v_i^o(x,y) + \varepsilon^{n+1} v_i^1(x,y) + \dots \tag{9}$$

$$p^\varepsilon(x) = p^o(x,y) + \varepsilon p^1(x,y) + \dots \tag{10}$$

$$T^\varepsilon(x) = T^o(x,y) + \varepsilon T^1(x,y) + \dots \tag{11}$$

where $y = \frac{x}{\varepsilon}$ and all functions are considered to be Y periodic with respectic to the variable y and n is a positive parameter to be defined later (depending of the data). The two-scale asymptotic expansion is obtained by considering that the dependence in x is obtained directly and through the variable y. The derivatives must be considered as:

$$\frac{d}{dx_i} \to \frac{\partial}{\partial x_i} + \frac{1}{\varepsilon}\frac{\partial}{\partial y_i}$$

If we suppose that the viscosity is of the form $\mu' = \mu\varepsilon^m$ where μ is constant (independent of ε), it is well known Sanchez-Palencia (1980), Ene and Sanchez-Palencia (1975) that, for n+m=2, the asymptotic proces lead to the Darcy's law:

$$\tilde{v}_i^o = -\frac{K_{ij}}{\mu}\left(\frac{\partial p^o}{\partial x_j} - f_j\right) \tag{12}$$

$$\mathrm{div}_x \tilde{\underline{v}}^o = 0; \tag{13}$$

where $p^o = p^o(x)$ and $\sim$ is the mean operator:

$$\tilde{\cdot} = \frac{1}{|Y|}\int_Y \cdot\, dy \tag{14}$$

The matrix K_{ij}, named "permeability tensor", is defined by:

$$K_{ij} = \frac{1}{|Y|}\int_Y w_i^j dy \tag{15}$$

$$\underline{v}^o = \left(f_i - \frac{\partial p^o}{\partial x_i}\right)\underline{w}^i \tag{16}$$

where $\underline{w}^i$ denotes the Y-periodic flow corresponding to a mean pressure gradient equal to the unit vector in the direction of x_i, Ene and Sanchez (1975) and depend on the geometric structure of the period.

In order to obtain the "macroscopic equation" for the energy, we consider the case where $\lambda' = \lambda\varepsilon^p$ with constant λ in the two phases.

We shall see that, as it usually happens in homogeneization problems, T^o does not depend on y. Ussing (11) in (3) it is clear that the convective terms are significants if p=n. Moreover from the Darcy's law we have n+m=2, and equations (3) and (4) give with (11) at order ε^{n-2}:

$$\frac{\partial}{\partial y_i}\left(\lambda_{ij}(y)\frac{\partial T^o}{\partial y_j}\right) = 0 \tag{17}$$

where λ take the values λ_s, λ_f in Y_s and Y_f respectively. This equation holds in the hole Y in the sense of distributions and from the Y-periodicity we obtain $T^o = T^o(x)$.

Now, in the same way at order ε^{n-1} we obtain:

$$\frac{\partial}{\partial y_i}\left(\lambda_{ij}(y)\left(\frac{\partial T^o}{\partial x_j} + \frac{\partial T^1}{\partial y_j}\right)\right) = 0 \tag{18}$$

or

$$-\frac{\partial}{\partial y_i}\left(\lambda_{ij}(y)\frac{\partial T^1}{\partial y_j}\right) = \frac{\partial T^o}{\partial x_j}\frac{\partial \lambda_{ij}}{\partial y_i} \tag{18'}$$

This is the classical equation in homogeneization theory Bensoussan, Lions, Papanicolau (1978), Sanchez-Palencia (1980) and they give us:

$$\left(\lambda_{ij}(y)\left(\frac{\partial T^o}{\partial x_j} + \frac{\partial T^1}{\partial y_j}\right)\right)^\sim = \lambda^h_{ij}\frac{\partial T^o}{\partial x_j} \tag{19}$$

$$\lambda^h_{ij} = \left(\lambda_{ij}(y) + \lambda_{ik}(y)\frac{\partial \theta^j}{\partial y_k}\right)^\sim \tag{20}$$

$$T^1(x,y) = \theta^j(y)\frac{\partial T^o}{\partial x_j} + c(x) \tag{21}$$

where θ^j is the solution of the problem: find $\theta^j \in H^1_{per}(Y)$ with $\tilde{\theta}^j = 0$ satisfying

$$\int_Y \lambda_{ik}\frac{\partial \theta^j}{\partial y_k}\frac{\partial \phi}{\partial y_i} dy = \int_Y \lambda_{ik}\frac{\partial \phi}{\partial y_k} dy,$$

$$(\forall)\ \phi \in H^1_{per}(Y) \tag{22}$$

At order ε^n using (9)-(11), the equations (3) and (4) with the boundary conditions (7) (8) and the Y-periodicity give:

$$\rho_f C_f (v_k^o \frac{\partial T^o}{\partial x_k} + v_k^o \frac{\partial T^1}{\partial y_k}) = \mu (\frac{\partial v_j^o}{\partial y_k} + \frac{\partial v_k^o}{\partial y_j}) \frac{\partial v_j^o}{\partial y_k} + \frac{\partial}{\partial x_i} (\lambda_{ij}(y) (\frac{\partial T^o}{\partial x_j} + \frac{\partial T^1}{\partial Y_j})) + \frac{\partial}{\partial y_j} (\lambda_{ij}(y) (\frac{\partial T^1}{\partial x_j} + \frac{\partial T^2}{\partial y_j})) \quad (23)$$

in Y, where we admit that v_k^o take the value 0 on Y_s. If we take the mean value of the equation (23) we have the macroscopic energy equation:

$$\rho_f C_f (\delta_{sj} + \mu \alpha_{ij} (K^{-1})_{si}) \tilde{v}_s^o \frac{\partial T^o}{\partial x_j} = \mu (K^{-1})_{ji} \tilde{v}_i^o \tilde{v}_j^o + \frac{\partial}{\partial x_i} (\lambda_{ij}^h \frac{\partial T^o}{\partial x_j}) \quad (24)$$

where

$$\alpha_{ij} = (w_k^i \frac{\partial \theta^j}{\partial y_k})^{\sim} \quad (25)$$

The first term in the right hand side of the equation (24) is the viscous dissipation. In the problem of thermal combustion in porous media C.Marle (1981) gives a similar term in the energy equation.

The corrective term α_{ij} (25) in the convective coefficient was obtained by Ene and Sanchez-Palencia (1981 b). This term gives the influence of the difference of thermal conductivity $\lambda_f \neq \lambda_s$. If $\lambda_f = \lambda_s$ the homogeneization of the temperature is trivial and we have $T^1=0, \theta^j=0$ and consequently $\alpha_{ij}=0$.

3 NATURAL CONVECTION IN POROUS MEDIA

In order to obtain the Darcy's law used in the study of natural convection in porous media, we consider the equations of motion of a slightly compressible viscous fluid in the form:

$$\frac{\partial}{\partial x_i} (\rho^\varepsilon v_i^\varepsilon) = 0 \quad (26)$$

$$\rho^\varepsilon v_k^\varepsilon \frac{\partial v_i^\varepsilon}{\partial x_k} = - \frac{\partial p^\varepsilon}{\partial x_i} + \mu' \frac{\partial^2 v_i^\varepsilon}{\partial x_k \partial x_k} + (\rho^\varepsilon - \rho_o) g \delta_{i3} \quad (27)$$

$$\rho^\varepsilon = \rho_o (1 - \alpha' T^\varepsilon), \quad P^\varepsilon = p^\varepsilon + \rho_o g x_3 \quad (28)$$

$$\rho^\varepsilon C_f v_k^\varepsilon \frac{\partial T^\varepsilon}{\partial x_k} - \frac{P^\varepsilon}{\rho^\varepsilon} v_k^\varepsilon \frac{\partial \rho^\varepsilon}{\partial x_k} = \tau_{jk}^\varepsilon \frac{\partial v_j^\varepsilon}{\partial x_k} + \frac{\partial}{\partial x_k} (\lambda_f' \frac{\partial T^\varepsilon}{\partial x_k}) \quad (29)$$

where P^ε is the pressure and p^ε is the difference between P^ε and the Archimede's pressure for the reference temperature. Moreover, the coefficients are $\mu' = \mu \varepsilon^m$, $\lambda' = \lambda \varepsilon^p$ and $\alpha' = \alpha \varepsilon^r$ $(0<r<1)$.

The state equation (28) shows that temperature, but not pressure, is taken into account for density. In addition, the compresibility is small. Equations (27), (28) and $0<r<1$ amounts to the Boussinesq's approximation.

We now consider expansions (9), (11) for the velocity and the temperature; oppositely, according to the Boussinesq's approximation, we take for pressure:

$$p^\varepsilon(x) = \varepsilon^r p_o^o(x,y) + \varepsilon^{r+1} p^1(x,y) + \ldots \quad (30)$$

From (28) we have:

$$\rho^\varepsilon = \rho_o (1 - \alpha \varepsilon^r T^o - \alpha \varepsilon^{r+1} T^1 \ldots) \quad (31)$$

As in the section 2, we are obliged to take $n+m-2=r$ and from (26) (27) we have the Darcy's law:

$$\tilde{v}_i^o = - \frac{K_{ij}}{\mu} (\frac{\partial p^o}{\partial x_j} + \alpha \rho_o T^o g \delta_{i3}) \quad (32)$$

$$\operatorname{div}_x \tilde{v}^o = 0 \quad (33)$$

In the equation (29) the untrivial convective terms are obtained for $n=p$. Then at order ε^{n-2} and ε^{n-1} we obtain $T^o = T^o(x)$ and (21) with the homogeneized coefficient (20). But the terms of viscosity and compressibility are of order ε^{n+r} and are negligible with respect to convective terms (order ε^n). Consequently, instead of (24), we obtain:

$$\rho_f C_f (\delta_{sj} + \mu \alpha_{ij} (K^{-1})_{si}) \tilde{v}_s^o \frac{\partial T^o}{\partial x_j} =$$

$$\frac{\partial}{\partial x_i}(\lambda^h_{ij}\frac{\partial T^o}{\partial x_j}) \qquad (34)$$

where the coefficients α_{ij} are given by (25).

The system (32)-(34) is the classical system of equations for natural convection in porous media.

Remark 1
All considerations concerning the Darcy's law holds for the non-steady case, using a slow scale of time $\tau=\varepsilon^n t$. In equation (24) or (34) it appears a new term of the form:

$$(\rho c)^{\sim}\frac{\partial T^o}{\partial \tau}$$

or

$$(m\rho_f c_f + (1-m)\rho_s c_s)\frac{\partial T^o}{\partial \tau}$$

where "m" is the porosity of the medium defined by $m = \frac{|Y_f|}{|Y|}$.

Remark 2
If we consider p=0, then we have n=0 and m-r=2. In this case it is clear that for large viscosity, the problem is uncoupled, and consequently the temperature equation is the conduction one. The Darcy's law holds, but the macroscopic continuity equation contain an additional term in T^o, which gives the influence of the compressibility, Ene and Sanchez-Palencia (1981a).

4 NON-DIMENSIONAL NUMBERS

In order to obtain the physical meaning of the news terms which appears in the equations (24) and (34) we take a characteristic lenght ℓ of the pores, a characteristic lenght L of the domain Ω and a characteristic velocity Q of the filtration velocity $\tilde{v}^o$. Now, the small parameter is $\varepsilon = \frac{\ell}{L}$. We introduce the Reynolds number R_ε, the Prandtl number P, the Grashof number G_ε, the Rayleigh number Ra_ε and a new non-dimensional number S_ε defined by:

$$R_\varepsilon = \frac{Q\rho\ell}{\mu}, \quad P = \frac{\mu c}{\lambda}, \quad S_\varepsilon = \frac{\mu Q^2}{\lambda T}$$

$$G_\varepsilon = \frac{g\alpha\rho^2\ell^3 T}{\mu^2}, \quad Ra_\varepsilon = G_\varepsilon . P$$

where T is the difference between the temperature and the reference temperature. Then the equation (24) makes sense for:

$$PR_\varepsilon \sim \varepsilon^{-1} S_\varepsilon \qquad (35)$$

and (34) for:

$$Ra_\varepsilon \sim \varepsilon \qquad (36)$$

If $P \sim 1$, (35) show that:

$$\frac{\lambda\rho\ell^2 T}{\mu^2 L Q} \sim 1 \qquad (37)$$

If, instead of ℓ^3, we define the Rayleigh number with two-seales

$$Ra = \frac{\rho\alpha g\ell^2 TL}{\mu\chi} \; ; \; \chi = \frac{\lambda}{\rho c}$$

it is clear that this number if of order 1.

5 REFERENCES

Bensoussan, A. Lions, J.L. Papanicolaou 1978, Asymptotic analysis for periodic structures, North-Holland, Amsterdam.

Ene, H.I. Sanchez-Palencia, E. 1975, Equations et phénomènes de surface pour l'écoulement dans un modèle de milieu poreux, J.de Mécanique, vol.14, pp.73-108.

Ene, H.I. Sanchez-Palencia 1981a, Some thermal problems in flow through a periodic model of porous media, Int.J.Enging.Sci., vol.19, pp.117-127.

Ene, H.I. Sanchez-Palencia 1981b, On thermal equation for flow in porous media, Preprint INCREST nr.46.

Marle, C-M 1981, On macroscopic equations governing multiphase flow with diffusion and chemical reaction in porous media, Int. J.Enging.Sci (to appear).

Sanchez-Palencia, E 1980, Topics in non-homogeneous media and vibration theory, Lecture Notes in Physics 127, Springer-Berlin.

Proceedings of Euromech 143 / Delft / 2-4 September 1981

Effects of dilute polymer solutions on porous media flows
Part I: Basic concepts and experimental results

W.INTERTHAL
Hoechst AG, Frankfurt, Germany

R.HAAS
University of Karlsruhe, Germany

1 INTRODUCTION

Wall bounded flows of fluids are influenced largely by the flow geometry. This becomes pronounced particularly in the flow of dilute high-molecular polymer solutions, due to flow regions with elongation gradients such as occur in porous media flows. A drastic increase in the resistance is observed there, whilst in the more familiar shear flow the same diluted polymer solutions do not exhibit any extraordinary effects. This different flow behaviour must therefore be ascribable to differing motions and deformations of a dissolved macromolecule in shear flow and elongational flow.

2 MOTION AND DEFORMATION OF A PARTICLE IN SHEAR- AND ELONGATIONAL FLOW FIELDS

2.1 Simple shear flows

"Simple shear flow" occurs in all boundary controlled flows. A classic experimental design for realizing steady simple shear flows is shown in Fig. 1. The fluid is

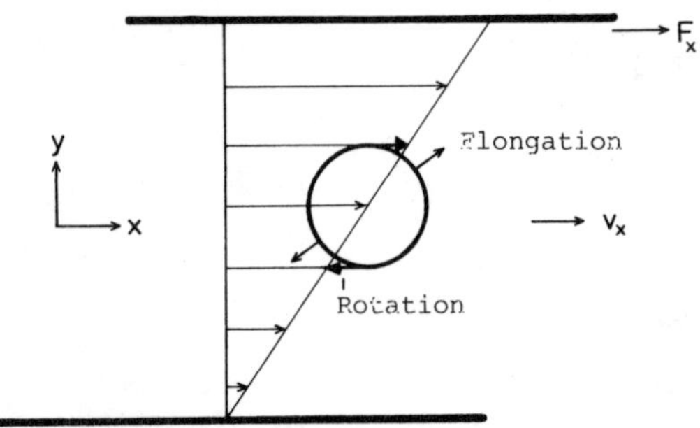

Fig. 1 Rotation and Elongation of a particle in simple shear flow

enclosed between two parallel plates, the one on top being moved at constant force F_x and constant velocity. In Newtonian fluids, like water, a linear velocity gradient dv_x/dy forms transversely to the flow direction. If a particle, which is large compared to the molecules of the fluid, is placed in the shear flow, it is caused to rotate driven by the velocity gradient (fig. 1). A detailed, quantitative description has been provided by Kuhn and Kuhn (1932). As explained by Debye (1946), a flexible particle, in neglected Brownian molecular motion, is elongated in addition at an angle of 45° relative to the flow direction. However, a dissolved, elastic macromolecule can only follow the extension to a limited extent because it is disturbed by the continued rotation. The rotation of entrained particles is an important characteristic of simple shear flow.

The relative motions of particles in shear flow causes a general increase in viscosity, which under idealized conditions for spherical rigid particles leads to the well-known Einstein (1911) equation

$$\eta_p = \eta_s \, (1 + 2{,}5\,\varphi) \qquad (1)$$

where η_p is the viscosity of the suspension of spheres, η_s the viscosity of the suspension fluid and φ the volumetric concentration. For flexible molecules Kuhn and Kuhn (1932) have developed basic concepts and carried out calculations on the basis of the classic dumbbell model.

2.2 Elongational flows

The inlet flow such as is shown in fig. 2 and frequently occurs in porous media flows, contrasts with the uniform geometry of simple shear flow. In the tapered inlet an accelerated flow develops with a velocity

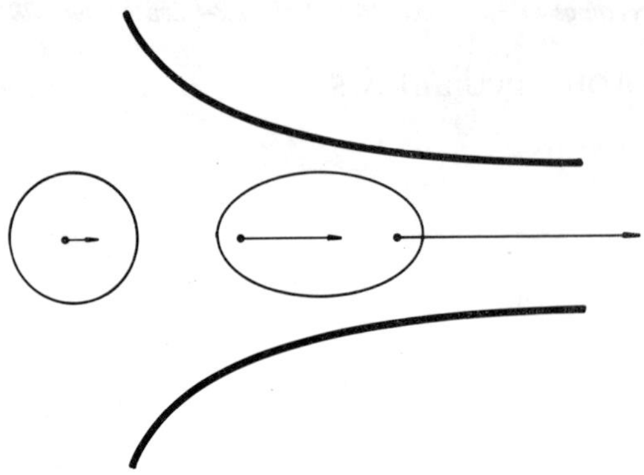

Fig. 2 Elongation of a particle in a converged flow section (inlet flow)

gradient dv_x/dx parallel to the direction of flow. A flexible particle (macromolecule) which is moving on the centre line is elongated uniaxially (fig. 2). The decisive difference compared to simple shear flow is that elongational flow is irrotational, that is to say the macromolecule can be considerably extended or elongated without being disturbed by rotation as in simple shear. Because of the undisturbed extension and deformation, dissolved macromolecules with high-molecular weight can cause a drastic increase in resistance, as has been described by Marshall and Metzner (1967), James and Mc Laren (1975), as well as Naudascher and Killen (1977).

The theoretical aspect of the viscosity-increasing influence of rigid, cylindrical particles in elongation gradients has been dealt with by Batchelor (1971).

Applying the theory developed by Batchelor for rigid particles to flexible long-chain polymer molecules, it was possible to describe some of the observed non-Newtonian effects at least qualitatively (see Naudascher and Killen (1977), Elata et al. (1977), James and Saringer (1980)). Unlike rigid particles the dissolved flexible macromolecules can, however, assume a large number of possible configurations which, as has been demonstrated, depend on the type of flow but also on the type of molecule and solvent. To obtain a basis for a quantitative theoretical description of the extensional effects as given by Haas et al. in Part II of this paper the essential thing to do is first of all to determine the interactions between molecule, solvent and flow experimentally.

Experimental setups for pure elongational flows have been described by Petrie (1979). However, they are either unsuitable for highly diluted polymer solutions or require an elaborate experimental setup. By contrast, the inlet flow shown in fig. 2 is not pure elongational flow, since shear flow, too, occurs at the boundary walls. It has been found, however, that the elongation and shear effects differ markedly and the former dominates. It is for this reason, that for low-viscosity polymer solutions wall bounded flows with elongation gradients are suitable for investigating the behaviour of dissolved elongated, macromolecules. A setup that can be handled particularly elegantly from the experimental aspect is the porous-media flow device described in the following.

3 ELONGATIONAL FLOW IN POROUS MEDIA

3.1 Porous-media flow device

A porous bed consisting of randomly packed spherical bodies is a system with continuesly changing flow cross-sections and constricted passages (inlet flow), in which considerable elongational flow gradients occurs locally. In addition, randomly packed spherical particles are suitable for describing flow behaviour in porous media under conditions approximating those in practice. A porous-media flow device with randomly packed particles is shown schematically in fig. 3. From a thermostatted storage container 1 the polymer solution under test is pumped through the test section 2 (diameter 3 cm, length 10 cm), which can be filled with glass spheres or sands of different particle diameters (> 50 µm). The pressure drop is measured after an initial length of 5 d (d = sphere diameter) using a differential pressure gage 4,6 and the flow rate by measuring the volume per time on a balance 7. The measured data are recorded and evaluated online by a computer 8. Measurement of these integral quantities is uncomplicated and makes it possible to cover a wide velocity range quickly.

3.2 Newtonian fluids in porous media flow

The flow of Newtonian fluids through a porous matrix is usually described in terms of modified Ergun coordinates (1952)

$$f = \Delta p \cdot d \cdot n^3 / \varrho \cdot \bar{v}^2 \cdot \Delta L \cdot (1-n) \quad \text{friction factor} \quad (2)$$

$$Re = \bar{v} \cdot d \cdot \varrho / \eta \cdot (1-n) \quad \text{Reynolds number} \quad (3)$$

where $\Delta p/\Delta L$ represents the pressure loss over the length, $\bar{v}$ the superficial velocity (i.e. the velocity related to the empty cross-section of the test column), d the mean particle diameter, η the dynamic

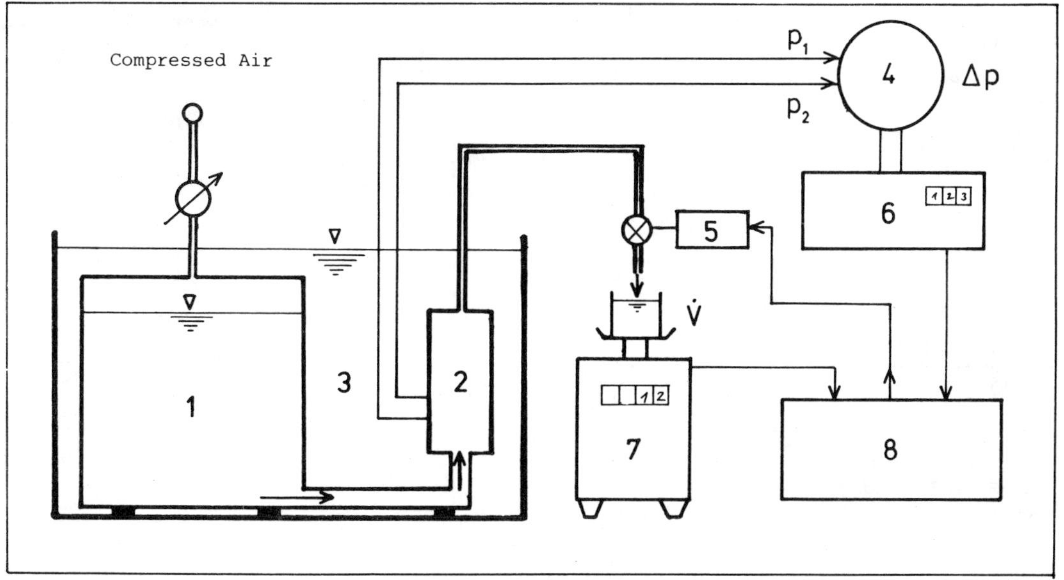

Fig. 3 Porous-media flow device
1 Storage container with polymer solution; 2 Test section; 3 Thermostatted bath;
4 Differential pressure gage; 5 Computer controlled,motor driven valve; 6 Amplifier;
7 Balance; 8 Computer

viscosity, ϱ the density of the fluid and n the porosity of the test specimen.

According to Ergun (1952) the flow behaviour of Newtonian fluids can be represented over a wide Reynolds number range by the empirical equation

$$\Lambda = 185 + 1.75 \cdot Re \quad (Re < 200) \qquad (4)$$

where $\Lambda = f \cdot Re$ is a normalized resistance coefficient.

In fig. 4 this correlation is confirmed for different particle diameters d, measured in the test section described in fig. 3. The full line corresponds to the

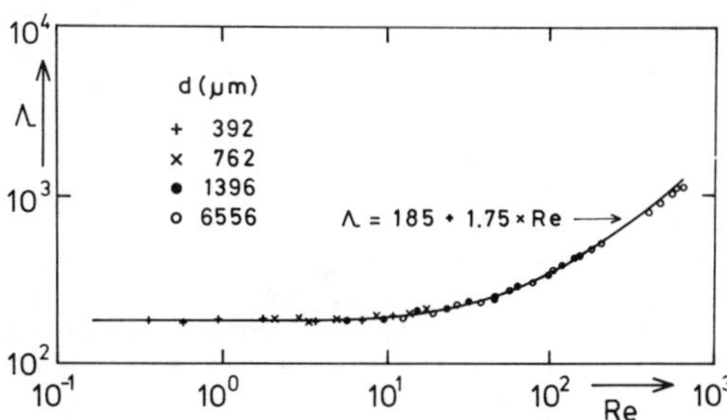

Fig. 4 Resistance coefficient Λ as a function of Reynolds number Re for several diameters d (Fluid: deionized water)

Ergun equation (4). In the case of Reynolds numbers greater than 5 the curve ascends, which is attributable to a separation of flow by inertia in the pore channel, as a result of which the elongational character of the flow slowly disappears.

3.3 Porous media flow of highly diluted polymer solutions

As stated above, highly diluted,high-molecular polymer solutions possess only a slightly higher viscosity than pure solvents in simple shear flow, whereas in porous-media flow they exhibit considerable non-Newtonian effects. These account for the drastic pressure loss for the Newtonian fluid and are caused by the interaction of polymer molecules and elongational flow. In fig. 5 the results measured in the porous-media flow device on dilute solutions of a high-molecular polyacrylamide PAM 2 (see table 1) at different concentrations are plotted in Ergun coordinates. For calculation of the Reynolds number the shear viscosity of the polymer solution is used according to the following formula

$$\eta_p = \eta_s \cdot (1 + [\eta] \cdot c) \qquad (5)$$

where η_s is the viscosity of the solvent, η_p the viscosity of the solution, $[\eta]$ the intrinsic viscosity and c the weight concentration.

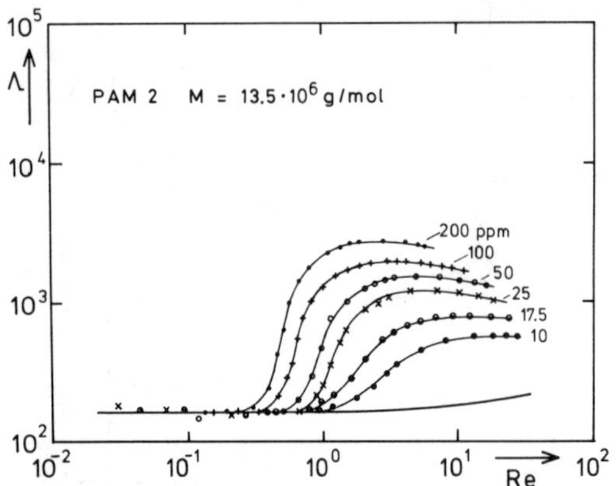

Fig. 5 Concentration dependence of the resistance coefficient Λ as a function of Reynolds number Re with dilute Polyacrylamide solutions (Solvent: 0.1 molar $MgCl_2$ in deionized water)

A very typical behaviour is the sudden and drastic increase in the resistance coefficient by more than one order of magnitude on a logarithmic scale starting from a certain onset Reynolds number. The onset point is shifted with increasing concentration towards smaller Reynolds numbers and the maximum resistance coefficient Λ_{max} increases, too. Before the onset point the macromolecules are only weakly influenced by elongation. They exhibit a flow behaviour which is equal to that of a Newtonian fluid.

3.4 Detailed experimental investigations with polyacrylamide solutions

Two high-molecular polyacrylamides (PAM) with different molecular weights and degrees of hydrolysis were used for the following investigations. Precise information about the polymers is given in table 1. The measurements were carried out

Table 1 Polyacrylamides used

Polymer	Hydrolysis (% by weight)	Intrinsic viscosity (cm^3/g)	Molecular weight (g/mol)
PAM 1	0	2830	$18.3 \cdot 10^6$
PAM 2	27	4570	$13.5 \cdot 10^6$

at 25° C in the porous media flow device. The mean sphere diameter of the packed spheres was 392 µm.

The flow of macromolecular solutions in a porous matrix is influenced by the particle diameter, the polymer type (electrical charge, linearity), polymer concentration, solvent quality, solvent viscosity and temperature.

The influence of the particle diameter is shown in fig. 6. The test section was

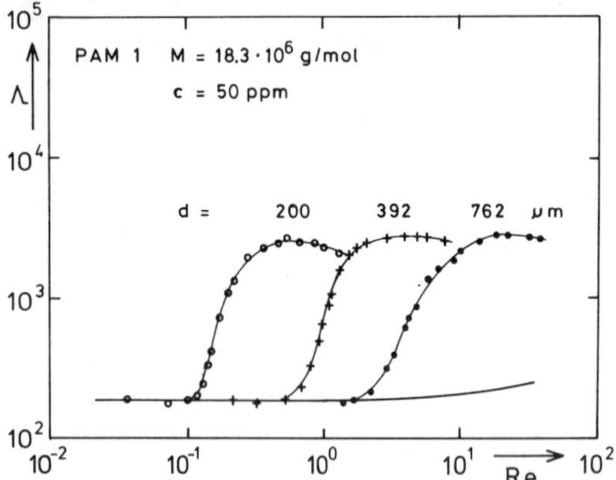

Fig. 6 Resistance coefficient Λ versus Reynolds number Re as a function of bead diameters. (Solvent : 0.5 molar NaCl)

filled with closely fractionated glass spheres with a mean sphere diameter of 200,392 and 762 µm, respectively. The polymer used was polyacrylamide PAM 1 at a concentration of 50 ppm in 0.5 m NaCl solution. As found by Durst et al (1979), the maximum resistance coefficient, Λ_{max} remains constant for all sphere diameters and only the onset point is shifted towards smaller Reynolds numbers with smaller sphere diameters. This result confirms that the following equation holds within certain limits for the elongation rate

$$\dot{\varepsilon} = k_1 \cdot (\bar{v}/d)$$

which has been proposed by Marshall and Metzner (1967), James and Mc Laren (1975), as well as by Tanner (1976). k_1 is a constant whose value, as shown by Haas et al. (Part II of this paper), can be assumed to be 4. Plotting the measured data in the form of an effective viscosity against a modified Deborah number De, which are defined in Part II, the obtained graph provides an invariant representation of the data. Based on these dependencies the following relationship for the onset elongation rate can be deduced:

$$\left(\frac{\bar{v}}{d}\right)_o = Re_o \cdot \eta_P \cdot (1-n)/d^2 \qquad (6)$$

The subscript o relates the parameters at onset conditions , which means that the Deborah number should have a value of 0.5.

The influence of electrical charges in the polymer chain of polyelectrolytes is shown in fig. 7. The more rodlike shape

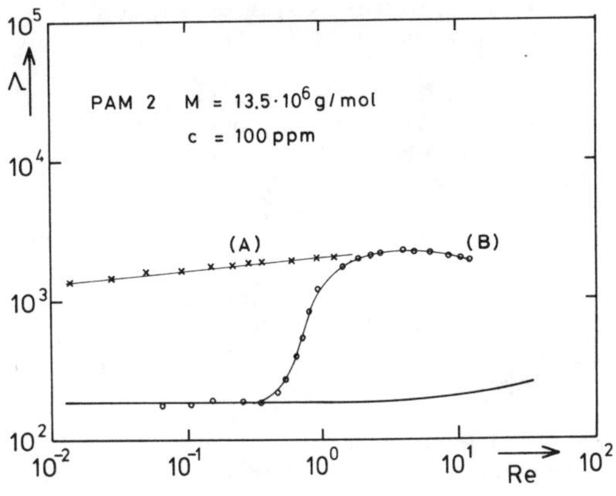

Fig. 7 Resistance coefficient versus Reynolds number Re of a polyelectrolyte solution in deionized water, curve (A), and in 0.1 molar $MgCl_2$ solution, curve (B)

of polyacrylamide PAM 2, (hydrolysed to a degree of 30 % by weight) such as forms in deionized water, has the effect that the onset behaviour, otherwise typical of non-charged polymers, curve (A), fig. 7, disappears. The rodlike polymer increases the resistance already at very much lower Reynolds numbers, this resistance increasing gradually to a maximum at higher Reynolds numbers. By contrast, the same polyelectrolyte with the same concentration in a 0.1 molar $MgCl_2$ solution behaves like a polymer without electrical charge (curve (B) fig. 7). Due to the screening effect of counterions the repellent forces, which have led to a widening of the polymer coil, are raised and the macromolecule assumes its classic unperturbed coiled shape. Then again a critical elongation rate is required in order to elongate the polymer, which is reflected in the onset behaviour. This sensitive effect can also be used to determine the degree of hydrolysis of a polyacrylamide as is stated by Durst et al. (1979). With increasing molecular weight, at a constant concentration, the onset point is shifted to smaller Reynolds numbers and the maximum resistance coefficient Λ_{max} becomes larger, as has been shown by Durst et al (1979), where it is also evaluated, that porous flow data can be inversely used to calculate molecular weights.

Solvent quality and temperature influence the molecule conformation. Enlarged, well solvatized macromolecules reinforce the increase in resistance and decrease the onset Reynolds number. With decreasing temperature (fig. 8) in the case of polyacrylamides, by contrast, the onset occurs earlier than at higher temperatures. The

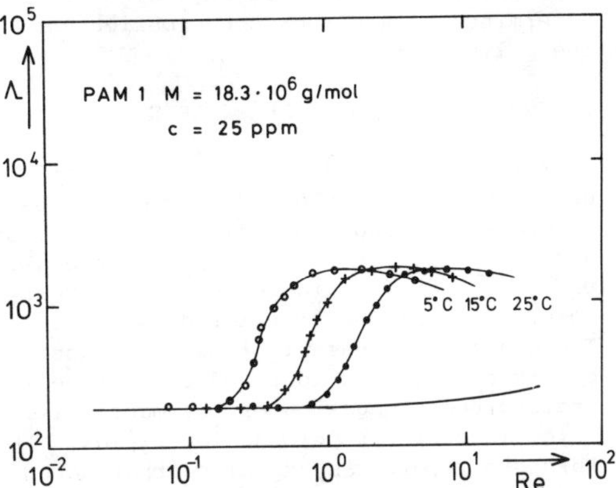

Fig. 8 Resistance coefficient versus Reynolds number as a function of temperature

maximum resistance coefficient Λ_{max} remains independent of the temperature. The onset shift is ascribable to the increase in the shear viscosity of the polymer solution at falling temperatures.

This influence of the viscosity outmanoeuvres the really better solvation of the polyacrylamide at elevated temperatures. Solvent quality, temperature and viscosity are interlinked variables which in part have an contrary effect to the onset Reynolds number and the resistance coefficient Λ.

4 POSSIBLE APPLICATIONS

In porous flow through sands and rock pores the particle diameters d are usually non-uniform and widely distributed.

However, the results obtained can be easily translated to widely distributed packed beds if, providing that the Ergun equation holds, an effective particle diameter d_{eff} is determined from the measured data obtained with water. Practical applications suggest themselves in all instances where the flow is to be slowed down and reduced in porous media. Thus, for example in tertiary oil recovery, the objective is to displace the oil located in the finer pores with the aid of a flooding agent. To achieve this, the viscosity of the flooding agent must be as high as possible and approximate the flow properties of the oil. Whilst in the past polymers have been added to the flooding agent (water) in order to adjust its shear viscosity to that of the oil, the oil might be displaced even more effectively by utilizing the aforementioned effects. In this case the flow parameters should be adjusted to appropriate flow conditions, which are determined by the rock conformation and the polymer used.

5 CONCLUSIONS AND FINAL REMARKS

Starting with the different motions and deformations of dissolved macromolecules in shear flow and elongational flow, first a qualitative interpretation of the drastic effects of dilute high-molecular polymer solutions in elongational flow was given. The drastic increase in flow resistance observed there was ascribed to a largely undisturbed elongation of the macrcmolecules in flow with elongation fields. Furthermore, it was explained that porous media flow in randomly packed beds exhibits considerable elongational behaviour and is suitable for investigating the behaviour of dissolved macromolecules experimentally under extensional stress.

A new measuring device was mentioned which permits the experiments to be carried out quickly. Detailed experimental investigations on dilute polyacrylamide solutions have shown that flow in porous material is suitable for drawing conclusions as to molecular parameters of high-molecular polymers, such as molecular weight and conformation as well as degree of hydrolysis.

The point of onset and the level of the increase in resistance coefficient have proved to be the main characteristic parameters.

Based on the concept of elongational flow in the porous medium and the experimental observations with dilute high-molecular polymer solutions a quantitative theory is developed by Haas et al. (Part II of this paper).

Practical application of the polymer effects in porous media flow might become important, for, among others, tertiary oil recovery.

6 REFERENCES

Batchelor, G.K.J. 1971, J. Fluid Mech. 46 : 813

Bird, R.B. Hassager, O., Armstrong, R.C., Curtiss, C.F. Dynamics of Polymeric Liquids New York 1977, J. Wiley & Sons

Debye, P. 1946, J.Chem.Phys. 14 (10):636

Durst, F., Haas, R., Naudascher, E., Schroeder, M. 1979, SFB 80-Report E/129, University of Karlsruhe

Einstein, A, 1911, Ann.Physik 19:289

Elata, C., Burger, J., Michlen, J., Taksermann, U. 1977, Phys. Fluids 20:280

Ergun, S. 1952, Chem.Eng.Prog. 48:89

James, D.F., Mc Laren, D.R. 1975, J.Fluid Mech. 70:733

James, D.F., Saringer, J.H. 1980, J.Fluid Mech. 97:655

Kuhn, W., Kuhn, H., 1932, Z.physikal.Ch. 161:1

Marshall, R.J, Metzner, A.B. 1967, Ind. Eng.Chem.Fundam. 6:393

Naudascher, E., Killen, J.M. 1977, Phys.Fluids 20:280

Petrie, C.J.S. 1979, Elongational Flows. London, Pitman

Tanner, R.I. 1976, Amer.Inst.Chem.Eng.J. 22:910

Wade, J.H.T., Kumar, P. 1972, J. Hydronautics 6 (1):40

Proceedings of Euromech 143 / Delft / 2-4 September 1981

Effects of dilute polymer solutions on porous media flows

Part II: Theory of molecule-flow-interaction and its verification

R.HAAS & F.DURST
University of Karlsruhe, Germany

W.INTERTHAL
Hoechst AG, Frankfurt, Germany

1 INTRODUCTION

Experimental investigations of flows of dilute polymer solutions through porous media yield large pressure-drop increases due to the local polymer action within the converging-diverging flow passages of the porous matrix. The industrial application of this "resistance-increasing effect" in tertiary oil recovery and in other problems of civil engineerings have been mentioned in several studies, e.g. see Savins (1969) and Durst et al. (1980). In addition to the application given by these authors, the non-Newtonian flow behaviour is also of interest in connection with the lay-out of technical filter systems.

The non-Newtonian flow properties of dilute polymer solutions are a consequence of flow-induced changes in the dimensions of the polymer molecules. One approach to the development of a constitutive equation to describe these polymer effects is therefore to first understand and model the macromolecular conformation changes as these occur by imposed shear- and elongational flow fields. An excellent review is given by Bird et al. (1977), in which the macromolecules are modeled as simple dumbbells suspended in a solvent. Major effects of these dumbbells are to be expected in extensional flow fields originating from alignment of the linear macromolecules in the direction of the flow paths and from their stretching in elongational flow fields. A detailed description of the differences between shear- and elongational flows is given by Interthal and Haas (Part I of this paper).

If one excepts the strong influence of elongational flow fields on the deformation of polymer molecules and their interaction with the flow field the porous media flow can be regarded as an excellent experimental device for testing theoretical equations derived by Bird et al. (1977) for dilute macromolecular solutions exposed to pure extensional flows. The observed pressure-drop increase is directly related to an increase in extensional viscosity, which occurs at a critical elongation rate of the flow. The aim of this paper is to compare modified theoretical viscosity equations with experimental results obtained in porous media flows with dilute polymer solutions. It can be shown, that all fluid mechanic and physico-chemical parameters that influence the measured, increased pressure losses, respond in the experiments as predicted by the theory.

2 THEORETICAL BACKGROUND

Important quantitative insight into the dynamic and steady-state responds of a polymer molecule to arbitrary flow can be obtained by simplifying the real macromolecule as a multi bead-spring model or as a simple dumbbell model as shown in Fig. 1.

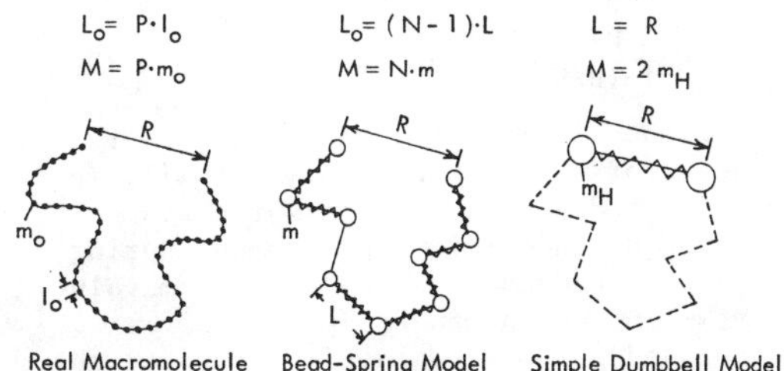

Fig. 1 Macromolecule models

According to the multi bead-spring model, the macromolecule is considered to be composed of (N-1) elastic connectors of length L and of N equivalent mass points each having the mass m. In contrast to this complicated model the simple dumbbell model consists only of one elastic connector and two mass points. In the latter case, the end-to-end distance R is determined by the motion of the free ends of the molecule. These are separated by Brownian motion against which an internal elastic force acts. Furthermore, the whole hydrodynamic resistance is assumed to act on the molecule on the two bead masses.

The probability density function $W(\vec{R})$ for the dumbbell with (N-1) subchains of lengths L was given by Kuhn (1934):

$$W(\vec{R}) = \left(\frac{3}{2\pi(N-1)L^2}\right)^{3/2} \exp(-3\vec{R}^2/2(N-1)L^2). \quad (1)$$

The vector $\vec{R}$ represents the instantaneous distance of the end points relatively to a fixed coordinate system in space. From eq. (1) the mean square end-to-end distance $\langle R^2\rangle$ of the polymer chain can be calculated in terms of the statistical values N and L:

$$\langle R^2\rangle = \int_{-\infty}^{+\infty} R^2 W(\vec{R})d\vec{R} = L^2(N-1). \quad (2)$$

The simplest dumbbell model takes the spring connector to be Hookeian, so that the intramolecular force $\vec{F}$ is linear in the molecular extension:

$$\vec{F}(\vec{R}) = \frac{3kT}{(N-1)L^2}\vec{R} = H\cdot\vec{R} \quad (3)$$

H is known as the Hookeian spring constant and can be obtained by energetic considerations. k represents the Boltzmann constant and T the absolute temperature.

It has been found in previous studies that the linear force-displacement law is not capable of describing a wide range of experimental observations of elongational flows of dilute polymer solutions. To overcome these shortcomings Warner (1972) proposed a FENE-(Finite Extensible, Non-linear Elastic) dumbbell model with the following relationship:

$$\vec{F}(\vec{R}) = \frac{H\cdot\vec{R}}{1-(R/R_o)^2} \quad R \leqq R_o. \quad (4)$$

The dumbbell with this force law reacts linear for small extensions, but will get stiffer and stiffer as the spring is further extended; furthermore the spring cannot be extended beyond a maximum value $R_o=(N-1)L$, which means a completely stretched polymer chain. If ζ denotes the friction coefficient for the bead masses, one can construct two times constants for the FENE-dumbbell. $\tau_H=\zeta/4H$ represents the Hookeian time constant whereas $\tau_R=\zeta R_o^2/12kT$ states the time constant for a rigid dumbbell with length R_o. Viscoelastic polymer effects are usually characterized in terms of the Hookeian time constant τ_H and a dimensionless ratio of time constants, $b=3\tau_R/\tau_H=3(N-1)$. Another time constant, $\tau_{H,e}$, is related to measurable properties of the polymer and the solvent by

$$\tau_{H,e} = \frac{[\eta]\,\eta_s M}{AkT}. \quad (5)$$

In this expression η_s represents the solvent viscosity, $[\eta]=(\eta_p-\eta_s)/\eta_s c$ the intrinsic viscosity of the polymer solution determined at zero shear rate and zero weight concentration c, M the molecular weight of the polymer and A the Avogadro number. For FENE-dumbbells τ_H is related to the experimental time constant by

$$\tau_H=(b+5)\tau_{H,e}/b. \quad (5a)$$

If dilute FENE-dumbbell solutions are subjected to pure elongational flow the extensional viscosity λ is increased above the Newtonian value $3\eta_s$ of the solvent by an amount which depends on the model parameters τ_H and b and on the elongation rate $\dot{\varepsilon}$ of the flow. Bird et al. (1977) have deduced theoretical extensional viscosity equations applicable at small and large elongation rates:

Pertubation expansion for small elongation rates:

$$\lambda_1 = 3\eta_s + \frac{3bn_o kT\cdot\tau_H}{(b+5)}\cdot\left[1 + \frac{b}{(b+7)}\cdot(\tau_H\cdot\dot{\varepsilon}) + \frac{3b^2(b+3)}{(b+5)(b+7)(b+9)}\cdot(\tau_H\cdot\dot{\varepsilon})^2 + \ldots\right] \quad (6a)$$

Asymptotic expansion for large elongation rates:

$$\lambda_2 = 3\eta_s + 2\cdot bn_o kT\tau_H\cdot\left[1 - \frac{(b+3)}{2\cdot b}\cdot\frac{1}{(\tau_H\cdot\dot{\varepsilon})} + \ldots\right] \quad (6b)$$

The number concentration n_o of dumbbells in the solvent can be replaced by the weight concentration $c=n_o M/A$. Defining a dimensionless Deborah number, $De=\dot{\varepsilon}\cdot\tau_{H,e}$, which characterizes the onset of viscoelastic polymer effects, one can simplify the extensional viscosity equations (6a) and (6b) for large values of the parameter b=3(N-1) in reduced form:

$$\lambda^* = \frac{\lambda-3\eta_s}{3\eta_s c[\eta]} \quad \text{Reduced extensional viscosity} \quad (7a)$$

$$\lambda_1^* = 1+De+3De^2\ldots \quad De \lessapprox 0.5 \quad (7b)$$

$$\lambda_2^* = 2N(1-1/(2De)+\ldots) \quad De > 0.5 \quad (7c)$$

An approximation for arbitrary values of

the Deborah number is obtained by "graphical matching" of the last two equations (7b) and (7c) as shown in Fig. 2 for different values of the parameter N.

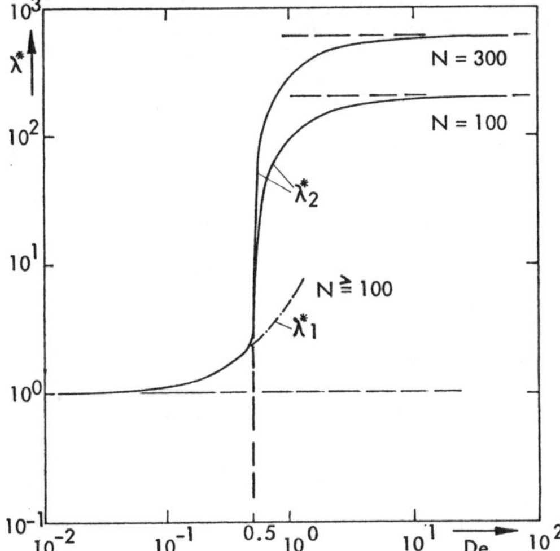

Fig. 2 Reduced extensional viscosity curves as function of the Deborah number

3 MODIFICATION OF THE DERIVED EXTENSIONAL VISCOSITY EQUATIONS TO POROUS MEDIA FLOWS

3.1 General description of laminar flow through porous media

Laminar flow of Newtonian fluids through porous media, such as regular packings of beads or randomly packed beds with narrow fractionated bead diameters (see Fig. 3) is usually described in terms of modified Ergun-coordinates (Ergun 1952) defined as follows:

$$f = \Delta p d n^3/\rho \bar{v}^2 \Delta L(1-n) \quad \text{friction factor} \tag{8}$$

$$Re = \bar{v} d\rho/\eta(1-n) \quad \text{Reynolds number} \tag{9}$$

$\Delta p/\Delta L$ represents the pressure drop per unit length of the porous bed, $\bar{v}$ the average flow velocity (flow rate per unit area of the empty bed), η the fluid viscosity, ρ the fluid density and n the porosity of the porous matrix.

The definition of a resistance coefficient $\Lambda = f \cdot Re$ leads to a useful representation of experimental data for Newtonian flow through beds with different bead diameter. When Λ is plotted versus the Reynolds number, a horizontal line is obtained for purely laminar flow with Re < 5 as indicated in Fig. 4. Taking in account the transition region to turbulent flow the experimental data obtained in randomly packed beds can be described empirically by a modified Ergun equation:

$$\Lambda = 185 + 1.75 \cdot Re \qquad Re < 200 \tag{10}$$

Similar results are obtained for packings with regular bead structures.

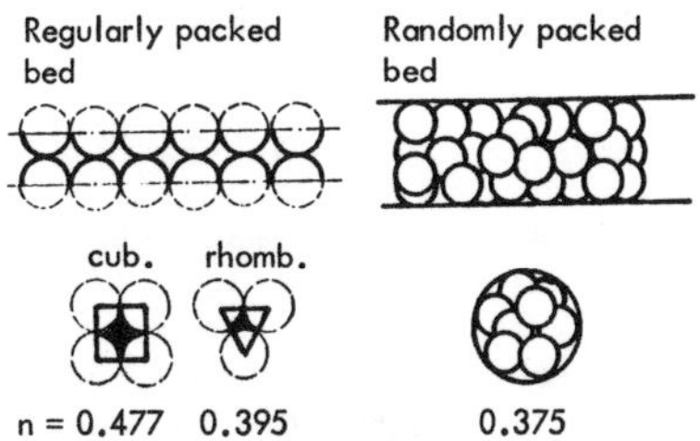

Fig. 3 Porous media structures

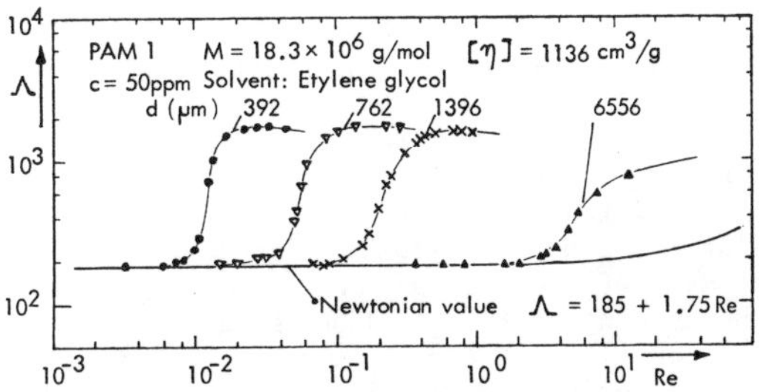

Fig. 4 Resistance coefficient/Reynolds number plots for Newtonian solvents and polymer solutions

3.2 Polymer effects on porous media flows

The influence of polymer additives on flows through porous media is best studied by quantifying deviations of the characteristic flow properties from those of Newtonian fluids. In Fig. 4 also the flow behaviour of dilute, high molecular weight polymer solutions is shown as the difference to the resistance curve for the pure solvent. It follows from this plot, that the polymer solution behaves Newtonian for very low Reynolds numbers, but will get more and more non-Newtonian as the Reynolds number is increased. Furthermore a maximum resistance-coefficient increase is observed at a certain Reynolds number. The sudden occurence of this viscoelastic polymer effect seems to be coupled with a dominant change in the "effective" viscosity of the polymer solution, when flowing through the converging-diverging passages of the porous matrix above a critical flow velocity. As shown by Interthal and Haas (Part I of this paper) quite a number of different parameters are influencing the onset and the maximum of the apparent polymer effects. The above considerations suggest, however, that those results can be unified by deducing modified dimensionless parameters η_e^* and De_e, to correlate all flow-induced and physico-chemical parameters and to describe the polymer effects in terms of an effective extensional viscosity dependend on an experimental Deborah number.

3.3 Measurable "effective" values η_e^* and De_e

For the comparison of the theoretically derived extensional viscosity equations with experimental results the porous media data have to be replotted in the following way. To calculate the apparent Reynolds number of the flow, the shear viscosity of the polymer solution is approximated by:

$$\eta_p = \eta_s(1 + c[\eta]). \qquad (11)$$

Eq. (11) follows from the definition of the intrinsic viscosity, which has to be known for each polymer-solvent-temperature system used in the experiments. The sharp increase in resistance coefficient for polymer solutions can be interpreted as a local increase of the extensional viscosity caused by the interaction of the elongational part of the flow and the polymer molecules. The relative increase of the resistance coefficient with respect to the Newtonian value at the same Reynolds number defines an "effective" reduced viscosity η_e^*.

$$\eta_e^* = \frac{\Lambda_p \cdot (1+[\eta]c) - \Lambda_s}{\Lambda_s [\eta] c} \qquad (12)$$

It follows from eq. (12) that for $\Lambda_p = \Lambda_s$ the reduced "effective" viscosity takes a value $\eta_e^* = 1$. This fact is consistent with the experimental observation, that no viscoelastic polymer effects are measured for low Reynolds numbers; the elongational flow field requires a certain onset-value to be "effective".

The elongation rate $\dot{\varepsilon}$ within the porous matrix can be approximated to be proportional to the average strain rate $(\bar{v}/d)$, which is determined by the ratio of the average superficial velocity $\bar{v}$ to the bead diameter d of the porous medium:

$$\dot{\varepsilon} = k_1(\bar{v}/d) = k_1(1-n)\eta Re/\rho d^2 \qquad (13)$$

The proportionality constant k_1 depends on the actual pore structure and takes for randomly packed beds a preferable value $2<k_1<20$. For the present investigations a value $k_1=4$ has been used to fit the data with the theoretical results. With eqs. (5) and (13) and the definition of the Deborah number, $De=\dot{\varepsilon}\cdot\tau_{H,e}$, one obtaines an experimentally determinable Deborah number:

$$De_e = \frac{4\eta_s^2(1-n)(1+[\eta]c)[\eta]M}{\rho d^2 AkT} Re \qquad (14)$$

The Deborah number De_e can be used to characterize the onset of viscoelastic polymer effects in porous media flows. From the theory it is known, that at a critical Deborah number De=0.5 dominant polymer effects should occur.

4 EXPERIMENTAL RESULTS

To verify the theoretical derivations experimental investigations were performed in a test section described by Interthal and Haas (Part I of this paper). The main parameters influencing the non-Newtonian flow behaviour of dilute polymer solutions in porous media flows are predicted from the theory to be:

a) the bead diameter d of the porous media
b) the type of polymer used in the experiments
c) the polymer weight concentration c
d) the polymer molecular weight M
e) the solvent viscosity η_s
f) the solvent quality
g) the temperature T.

A few examples are given in this chapter to show the consistency of experimental results with the applied theory. At first, the influence of the bead diameter d of the porous media on the resistance increase is discussed. The data presented in Fig. 4 show the non-Newtonian flow behaviour for dilute polyacrylamide solutions (PAM 1 in ethylene glycol, $M=18.3 \times 10^6$ g/mol, c=50ppm) for different bead diameters of the porous bed. As can be seen from the graphs the onset Reynolds number increases with increased bead diameter, whereas the maximum resistance coefficient remains constant in the laminar flow region. From eq. (14) it follows, that the experimentally determinable Deborah number should be proportional to Re/d^2. A plot in terms of the "effective" values η_e^* and De_e (see Fig. 5) reduces these different flow curves to a universal line as predicted by the theory.

The second example illustrates the concentration dependence of the viscoelastic polymer effects (see Fig. 6) for PAM 1-solutions with ethylene glycol as solvent. In thise case the data have been obtained for a constant bead diameter d = 392 µm. If one replots the experimental values of Fig. 6 in terms of the reduced parameters η_e^* and De_e the same correlation curve is found as for the bead diameter variation. Fig.7 when compared with Fig. 2 indicates this master curve. Similar results have been measured for polymer solutions with different molecular weights, solvent viscosities and temperatures and are shown by Durst and Haas (1981).

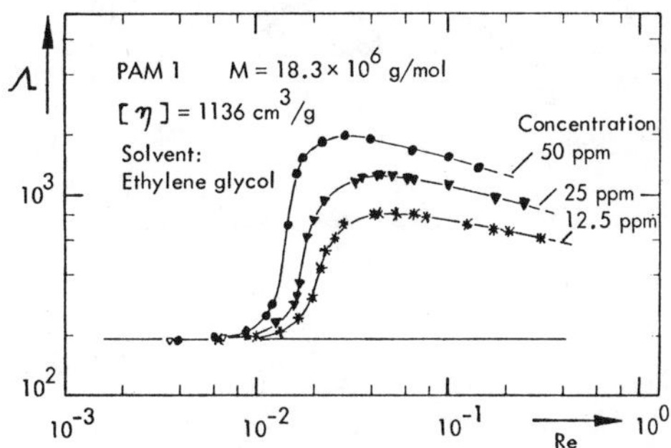

Fig. 6 Resistance coefficient/Reynolds number plots for PAM 1-solutions with different polymer concentrations

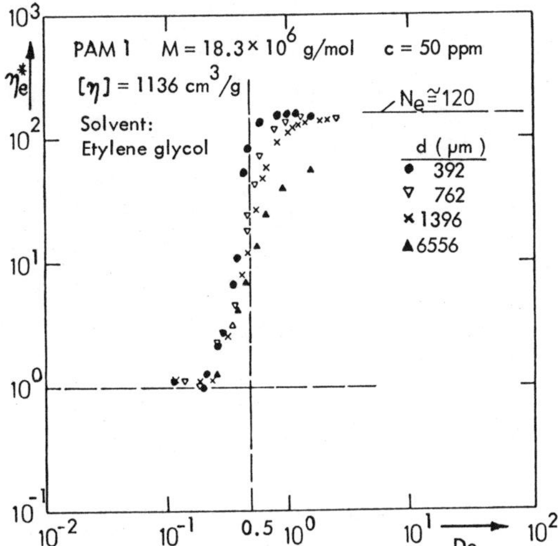

Fig. 5 Reduced "effective" viscosity η_e^* as function of the experimental Deborah number De_e for different bead diameters of the porous matrix

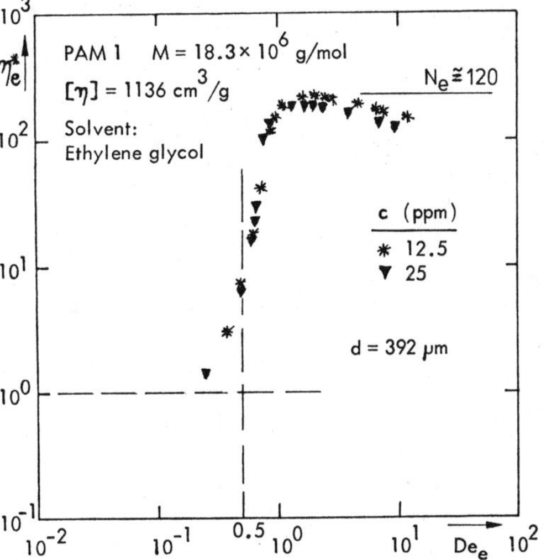

Fig. 7 Reduced "effective" viscosity η_e^* as function of the experimental Deborah number for PAM 1-solutions with different polymer concentrations

5 DISCUSSION AND FINAL REMARKS

The dumbbell model in its simplest form permits to deduce theoretical viscosity equations for pure elongational flow at small and large elongation rates. Two main parameters, the relaxation time τ_H, and a dimensionless time factor b, are necessary to describe the important features of the reduced extensional viscosity/Deborah number relationship for any kind of a polymer-solvent-temperature system. As has been shown by Durst and Haas (1981) the polymer effects in porous media flows mainly occur if the molecular weight of the used polymers exceeds a value of $M \approx 10^6$g/mol. For this condition the number of subchains $(N-1) \approx N$ can be estimated by means of the monomer units (m_o, 1_o), the molecular weight M and the intrinsic viscosity $[\eta]$ of the solution:

$$N = \left(\frac{1_o}{m_o}\right)^2 \left(\frac{\emptyset}{[\eta]}\right)^{2/3} M^{4/3} . \qquad (15)$$

$\emptyset$ is known by Flory & Fox (1951) to be a nearly universal constant $\emptyset = 2.5 \times 10^{23}$ mol^{-1}. The experimentally used system PAM 1/ethylene glycol at $T = 25^oC$ reveals a theoretically computed value $N \cong 23000$, which yields from eq. (15). The reduced graphs in Fig. (5) and (7) show experimental values $N_e \cong 120$ obtained with the assumption that porous media flows are pure extensional flows. Keeping in mind that only parts of the porous media flow possess pure elongational nature, one can assume that the macromolecules in certain flow parts are completely stretched. The lower value N_e with respect to the theoretical value N is due to the lower extension of the macromolecules in the shear parts of the porous media flow. The method outlined here to estimate resistance increases of porous media flows with dilute polymer solutions seems to be a useful tool in practical applications.

6 REFERENCES

Bird, R.B., O. Hassager, R.C. Armstrong & C.F. Curtiss 1977, Dynamics of polymeric liquids, Vol. 2, kinetic theory, New York, J. Wiley & Sons.

Durst, F. & R. Haas 1981, Dehnströmungen mit verdünnten Polymerlösungen: Ein theoretisches Modell und seine experimentelle Verifikation, Rheol. Acta, Vol. 20, pp. 179-192.

Durst, F., R. Haas & B.U. Kaczmar, 1980, Flow of dilute HPAM-solutions in porous media under various solvent conditions, SFB 80 report E/158, University of Karlsruhe.

Ergun, S. 1952, Fluid flow through packed columns, Chem. Eng. Prog., Vol. 48, pp. 89-94.

Flory, P.J. & T.G. Fox, Jr. 1951, Treatment of intrinsic viscosity, J. Amer. Chem. Sci., Vol. 73, pp. 1904-1908.

Kuhn, W. 1934, Über die Gestalt fadenförmiger Moleküle in Lösungen, Kolloidzeitschrift, Vol. 68, pp. 2-15.

Savins, J.G. 1969, Non-Newtonian flow through porous media, Ind. Eng. Chem., Vol. 61, pp. 18-47.

Warner, H.R. 1972, Kinetic theory and rheology of dilute suspensions of finitely extensible dumbells, Ind. Eng. Chem. Fundam., Vol. 11, pp. 379-387.

Proceedings of Euromech 143 / Delft / 2-4 September 1981

Coupled heat and moisture transfer in unsaturated porous bodies

C.TZIMOPOULOS & E.SIDIROPOULOS
Aristotle University of Thessaloniki, Greece

1.INTRODUCTION

Moisture movement inside unsaturated porous media is in most cases analyzed under the assumption of isothermal conditions. Systematic consideration of simultaneous heat and moisture movement dates back at least to Philip and de Vries (1957).However,actual computations of specific problems are being carried out only in recent years.

There are two reasons for this:
a)The inclusion of thermal effects introduces additional parameters and coefficients that are more difficult to measure or determine.
b)The problem is non-linear,coupling between the dependent variables is strong and,therefore,numerical methods and the use of powerful computational equipment must be resorted to for the solution of the governing partial differential equations.

The most intricate aspect of the thermal effects is the coupling between heat and moisture transport,since the pressure of heat is responsible for some of the mass transfer and,in turn,heat transfer is facilitated and carried along through the motion of moisture It is only necessary to consider linearized models in order to evaluate the significane of thermal effects.By taking into accout separately isotermal and non-isothermal conditions in a linearized model distinct moisture profiles are observed.

Due to the difficulty in obtaining some of the parameters that appear in the governing equations and the uncertainty about their values,it seems advisable to look into the order of magnitude of the various terms and/or the sensitivity of the solutions with respect to inaccuracies in the values of some coefficients.

In this paper the linear equations are solved both by analytical and numerical methods. Both finite difference and finite element schemes are employed and Fourier and Laplace transform techniques are implemented in order to derive analytic expression for the solutions.These analytic expressions open the way to sensitivity analyses with respect to parameters of interest such as the phase change number (denoted by ε in Luikov 1964).

A typical physical situation that requires the inclusion of thermal effects is the study of evapotranspiration.The influence of the groundwater level could be accounted for through a suitable boundary condition. The overall analysis of evapotranspiration should consist of a model based on simultaneous heat and moisture movement and including appropriate sources and sinks.

2.OVERVIEW OF VARIOUS MATHEMATICAL MODELS

There are two basic viewpoints from which problems of heat and moisture transfer are approached.The first one,mainly represented by Philip and de Vries is based on the microscopic interaction of liquid vapor and porous structure by invoking established physical laws (diffusion,Darcy,capillarity), equilibrium thermodynamics and conservation principles.The second viewpoint,mainly represented by Luikov and his collaborators, follows the formalism of linear Onsager relations as well as consrvation laws.

Both approaches lead to systems of differential equations with coefficients that have to be either determined empirically or measured experimetally.

The differential equations derived by Philip and de Vries (1957) and reiterated in many other studies are as follows:

$$\frac{\partial\theta}{\partial t} = \nabla(D_T\nabla T)+\nabla.(D_\theta\nabla\theta)+\frac{\partial K}{\partial z} \tag{1}$$

$$c\,\frac{\partial T}{\partial t} = \nabla(\lambda\nabla T)-L\nabla.(D_{\theta vap}\nabla\theta) \tag{2}$$

where t is the time,z the vertical coordinate,θ the volumetric liquid water content, T absolute temperature,D_T thermal moisture diffusivity, D_θ isothermal moisture diffusivity, K hydraulic conductivity, c heat capacity of porous body, λ thermal conductivity, L heat of vaporization, $D_{\theta vap}$ vapor diffusivity.

The first of the above equations is derived on the basis of continuity in combination with Darcy's conservation.

Analogous equations can be found in the review article by Luikov (1964). Upon transformation in terms of the volumetric moisture content θ they read as follows:

$$\frac{\partial\theta}{\partial t}=\nabla(D_\theta\nabla\theta+D_T\delta\nabla T)+\frac{\partial K}{\partial z} \quad (3)$$

$$\frac{\partial T}{\partial t}=\frac{1}{c^*}\nabla(\lambda\nabla T)+\frac{\varepsilon Q^*}{c^*}\frac{\partial\theta}{\partial t} \quad (4)$$

where $c^*=c\rho$ is the thermal capacity of the medium and Q^* is the enthalpy of evaporation, ε is the so called "phase change number". The respective equations of energy conservation in the two sets are not the same.

Within the Luikov model it is possible to derive a term similar to $L\nabla.(D_{\theta vap}\nabla\theta)$ if, instead of taking the energy flux equal to just $j_q=-\lambda\nabla T$ (Luikov 1964), another term is added attributed to a "Dufour action" as follows:

$j_q=\lambda\nabla T-L_D\nabla\theta$ (Luikov 1968). The original Philip and de Vries (1957) theory was extended by de Vries (1958) to include terms expressing evaporation. De Vries' analysis includes many terms, not all of which are equally important. Rose (1968), after order-of-magnitude estimates showed that the dominant term was of the form $\partial\theta/\theta t$, in agreement with the Luikov theory.

3. BASIC EQUATIONS AND THEIR NON-DIMENSIONAL FORM. ONE-DIMENSIONAL FLOW

Coupled heat and mass flow equations are derived in Luikov (1964). Luikov's equations, after suitable transformation, read as follows with respect to the volumetric moisture content: (Tzimopoulos and Sidiropoulos, 1981)

$$\frac{\partial\theta^*}{\partial t^*}=D\left(\frac{\rho_w}{\rho_o}\right)\frac{\partial^2\theta^*}{\partial z^{*2}}+\delta D\frac{\partial^2 T^*}{\partial z^{*2}} \quad (5)$$

$$c\rho_o\frac{\partial T^*}{\partial t^*}=k\frac{\partial^2 T^*}{\partial z^{*2}}+\varepsilon Q^*\frac{\partial\theta^*}{\partial t^*} \quad (6)$$

Let $\theta^*=\theta\theta_o$, $T^*=T\,T_o$, where θ_o and T_o are the initial values of moisture and temperature.

Also, $\alpha_m=D(\rho_w/\rho_o)$, $\alpha=k/c\rho_o$, $\varepsilon Q^*/c\rho_o=(L\varepsilon/c)(\rho_w/\rho_o)$, $t=(\alpha/\ell)t^*$, $z=z^*/\ell$, where ℓ is the length of the soil column.

Finally,

$\frac{\alpha_m}{\alpha}=L_u$, $\frac{\delta T_o}{\theta_o}\frac{\rho_o}{\rho_w}=P_n$, $\frac{\theta_o}{T_o}\frac{\rho_w}{\rho_o}\frac{L}{c}=K_o$, where, after Luikov, L_u, P_n and K_o denote the Luikov, Posnov and Kossovich numbers respectively.

The basic equations are as follows in non-dimensional form:

$$\frac{\partial\theta}{\partial t}=L_u\frac{\partial^2\theta}{\partial z^2}+L_uP_n\frac{\partial^2 T}{\partial z^2} \quad (7)$$

$$\frac{\partial T}{\partial t}=\frac{\partial^2 T}{\partial z^2}+\varepsilon K_o\frac{\partial\theta}{\partial t} \quad (8)$$

Initial conditions: $\theta(0,z)=1, T(0,z)=1$.

Boundary conditions:

$z=0$: $\theta=\theta_1$, $T=T_1$

$z=1$: $\theta=\theta_2$, $T=T_2$

4. METHODS OF SOLUTION

4.1 Finite-difference Crank-Nickolson Scheme

$$\frac{\theta_i^{n+1}-\theta_i^n}{\Delta t}=L_u(\Delta_\theta^{(n+1)}+\Delta_\theta^{(n)})+L_uP_n(\Delta_T^{(n+1)}+\Delta_T^{(n)}) \quad (9)$$

$$\frac{T_i^{n+1}-T_i^n}{\Delta t}=(\Delta_T^{(n+1)}+\Delta_T^{(n)})+\varepsilon K_o\frac{\theta_i^{n+1}-\theta_i^n}{\Delta t} \quad (10)$$

where

$$\Delta_\theta^{(n)}=(\theta_{i-1}^n-2\theta_i^n+\theta_{i+1}^n)/(2\Delta z^2)$$

$$\Delta_T^{(n)}=(T_{i-1}^n-2T_i^n+T_{i+1}^n)/(2\Delta z^2)$$

The above scheme leads to a system with a bi-tridiagonal coefficient matrix.

4.2. Exact solution as a Fourier series

The following finite Fourier transform is used:

$$\bar\theta(n,t)=\int_0^1\theta(z,t)\sin n\pi z\,dz,$$

$$\bar T(n,t)=\int_0^1 T(z,t)\sin n\pi z\,dz.$$

The solution, after inversion, is the following:

$$\theta(z,t)=2\sum_{n=1}^{\infty}\bar\theta(n,t)\sin n\pi z.$$

After some manipulation:

$$\theta(z,t)=\theta_2 z+\theta_1(1-z)+2\sum_{n=1}^{\infty}\left[A_1(n)e^{-n^2\pi^2\alpha_1 t}+A_2(n)e^{-n^2\pi^2\alpha_2 t}\right]\sin n\pi z$$

where

$\alpha_1=(b-\sqrt{b^2-4c})/2, \alpha_2=(b+\sqrt{b^2-4c})/2$

with $b=1+L_u+\varepsilon L_u P_n K_o$, $c=L_u$

$A_1(n)= K_1(-1)^n/n\pi +\lambda_1/n\pi,$

$A_2(n)= K_2(-1)^n/n\pi +\lambda_2/n\pi,$

with

$K_1=[c_2(-1+\theta_2)-(-1+T_2)]/(c_2-c_1),$

$\lambda_1=[c_2(1-\theta_1)-(1-T_1)]/(c_2-c_1),$

$K_2=[-c_1(-1+\theta_2)+(-1+T_2)]/(c_2-c_1),$

$\lambda_2=[-c_1(1-\theta_1)+(1-T_1)]/(c_2-c_1)$, where

$c_1=(\alpha_1-L_u)/(L_u P_n),$

$c_2=(\alpha_2-L_u)/(L_u P_n).$

4.3. Exact solution by means of Laplace transform

For small values of t the Fourier series does not converge.We can obtain the following series from Laplace transform,which does converge:

$$\theta(z,t)=1+2\ell_1\sum_{n=0}^{\infty}[\operatorname{erfc} B_n^{(1)}-\operatorname{erfc} C_n^{(1)}] +2\ell_2\sum_{n=0}^{\infty}[\operatorname{erfc} D_n^{(1)}-\operatorname{erfc} E_n^{(1)}] +2m_1\sum_{n=0}^{\infty}[\operatorname{erfc} B_n^{(2)}-\operatorname{erfc} C_n^{(2)}] +2m_2\sum_{n=0}^{\infty}[\operatorname{erfc} D_n^{(2)}-\operatorname{erfc} E_n^{(2)}]$$

where

$B_n^{(1)}= (B_1 z+2nB_1)/2\sqrt{t}$,

$C_n^{(1)}= (B_1(2-z)+2nB_1)/2\sqrt{t},$

$D_n^{(1)}= (B_1(1-z)+2nB_1)/2\sqrt{t},$

$E_n^{(1)}= (B_1(1+z)+2nB_1)/2\sqrt{t},$

$B_n^{(2)}= (B_2(2-z)+2nB_2)/2\sqrt{t},$

$C_n^{(2)}= (B_2 z+2nB_2)/2\sqrt{t},$

$D_n^{(2)}= (B_2(1+z)+2nB_2)/2\sqrt{t},$

$E_n^{(2)}= (B_2(1-z)+2nB_2)/2\sqrt{t}.$

5.CONCLUSIONS AND DISCUSSION

a.It is evident from Fig.1 that for the case of non-isothermal flow the moisture profile advances more rapidly,which was actually expected.

b.Figs.2 and 3 show the agreement between the analytical and numerical solutions.For

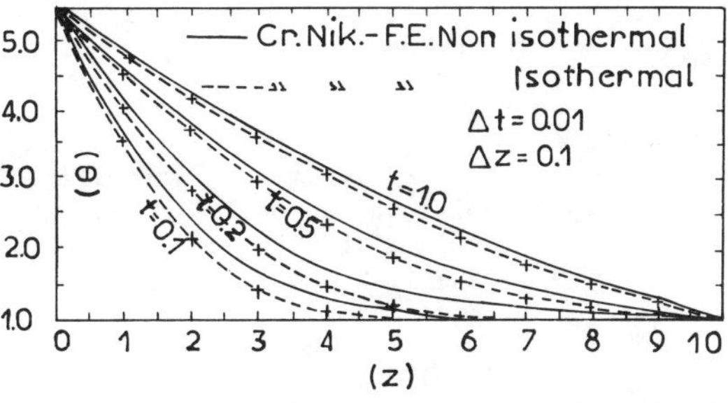

Fig.1.Moisture profiles for non-isothermal and isothermal conditions.

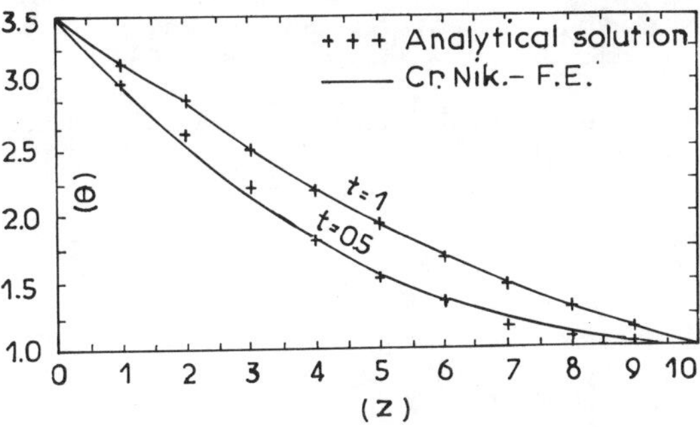

Fig.2.Comparison between analytical solution and Crank-Nicholson scheme and F.E.method with boundary condition θ=3.5 at z=o.

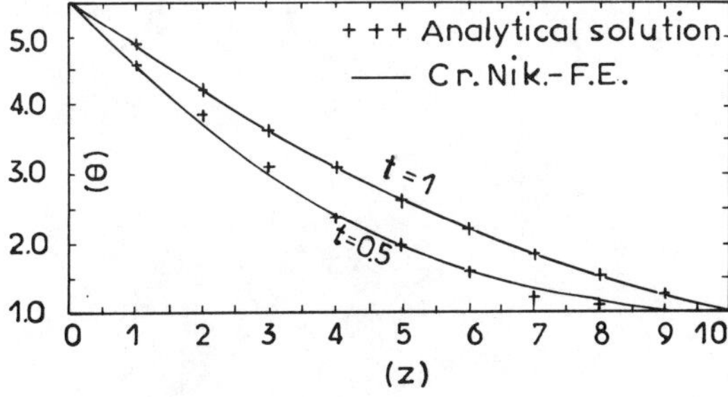

Fig.3.Comparison between analytical solution and Crank-Nicholson scheme and F.E.method with boundary condition θ=5.5 at z=o.

values of the non-dimensional time t less than 0.5 the Fourier series solution presents some oscillations and the series that comes from the Laplace transform is used.
c.The very small influence of the " phase number" ε is shown in following table.

ε	θ at t=0.5	θ at t=0.8
0.0	2.0786	2.4967
0.09	2.0702	2.4878
0.15	2.0646	2.4820
0.20	2.0600	2.4772
0.30	2.05100	2.4676

d.Linear as well as triangular finite elements were also used.Thier agreement with the finite defference solutions was excellent (up to the second decimal point).

REFERENCES

De Vries,D.A.1958,Simultaneous transfer of heat and moisture in porous media.Trans. Am.Geoph.Union 39,p.909.

Luikov,A.V.1964,Heat and Mass Transfer in Capillary-Porous Bodies.Advances in Heat Transfer p.248.Pergamon Press,Oxford.

Philip,J.R.& de Vries,D.A.1957,Moisture movement in porous materials under temperature gradients.Trans.Am.Ge.Un.38,p.222.

Rose,C.N.1968,Water transport in soil with a docily temper.wave,I,Aust.J.Soil Res.6,p.31.

Tzimopoulos C & Sidiropoulos E.1981,Flow of mass and heat in an unsaturated porous medium,Arist.Univ.,Lab.of Irrigation Report.

Proceedings of Euromech 143 / Delft / 2-4 September 1981

Flow of water treatment plant sludges through porous media

L.VIRTO & M.KHAMASHTA
Escuela Técnica Superior de Ingenieros Industriales, Terrassa, UPB, Spain

1 INTRODUCTION

Disposal of sludge from wastewater plants and surface water treatment plants for water supply is a problem not only from the economical point of view but also from the ecological one.

For the dewatering of sludge, special interest has been paid lately to drying beds and lagoons and, also, to direct land application.

It is well known that the dewatering principle is a common one in these techniques: infiltration through the porous bed, either artificial or natural, of the sludge itself. This infiltration is accompanied by several phenomena leading to the progressive separation, although a partial one, of the solid phase of the sludge: separation of the biggest size solid fraction by a similar mechanism as that of deep filtration and physical adsorption and/or physicochemical one inside the porous medium.

Broadly speaking, this problem is of a flow through a porous medium of a liquid-solid suspension, whose rheological behaviour is obviously non-newtonian and whose solid concentration decreases together with the distance through the medium.

Among the unsolved problems ocurring when simulating analytically that particular flow model -once the physical and the physicochemical characteristics of the fluid and the composition, morphology and structural characteristics of the porous medium are known- some of them demand priority and special attention, that is:

i) until what extent the different mechanisms of the separation of the solid phase: physical retention, adsorption, etc., influence the reduction of permeability of the porous medium?

ii) why the adsorption rates and likely gel formation (in the case there are polymers) are different according to the flow taking place at constant pressure or at steady flowrate.

iii) how does the intrinsic permeability of the porous medium vary with the retention rate of solids, not only versus time but also with respect to the distance up to the penetration surface of the sludge in the porous medium?

iv) which is the theoretical model defining the pressure drop in the porous medium when the retention of solids is taking place?

v) what type of kinetic equation governs the adsorption and clogging phenomena?

vi) how do the medium porosity, the average size of pores, the chemical nature of the solid matrix of the porous medium, the residence time of fluid in it and its velocity influence the retention of suspended particles smaller than the pores of the porous medium?

2 SCOPE OF THIS WORK

This is just a part of a more ample one, under development at the present time, whose scope is:

i) the mechanical and rheological characterisation of sludge originated in the decanters of surface water treatment plants for water supply

ii) the physical parametric characterisation of the artificial porous bed used in our experiments

iii) phenomenologial checking of the flow of these sludges through the porous bed in order to detect the most important mechanisms and identify the most significant variables with respect to the question pointed out in first paragraf, which will be most useful for the design of a systematic plan of experiments so as to be able to provide some information on which an analytical modelling of the flow and mass transfer and transport in the porous bed can be based.

3 CHARACTERISTICS OF THE FLUID

The sludge was collected from the discharge of the "flocons" of decanters from a surface water treatment plant, located in Llobregat river, Abrera (Barcelona), for drinking water supply.

The characterisation of the fluid is limited to: chemical composition, concentration in dry matter, distribution of particle size, sedimentation curve, rheological behaviour and specific resistance to filtration with cake formation.

3.1 Chemical composition

The sludge has a soluble fraction, approximately a 4%, whose composition is: Ca^{++} 20,8%; HCO_3^- 25,3%; Cl^- 44,15%; $SO_4^=$ 9,9%.

Of the insoluble fraction, 23,7% is constituted by calcinable matter at 650°C, its other components being SiO_2 39,5%; Al^{3+} 8,7%; Ca^{++} 5,7%; Mg^{2+} 1,1%; Na^+ 0,1%; K^+ 0,6% y $SO_4^=$ 0,7%. The rest is composed by a great variety of chemical elements in negligible percentages, although undoubtedly important as virtual pollutants (DQO is 18,67%).

3.2. Concentration in dry matter, distribution of particle size and sedimentation curve

The concentration in dry matter of the sludge varies along the time of draining of the decanter between 46,12 g/l, at 15 s, and 13,78 g/l, at 25 minutes. The average concentration is 34,44 g/l.

The determination of particle sizes and distribution of sizes was carried out by the Coulter-Counter method. Results found are shown in Fig. 1.

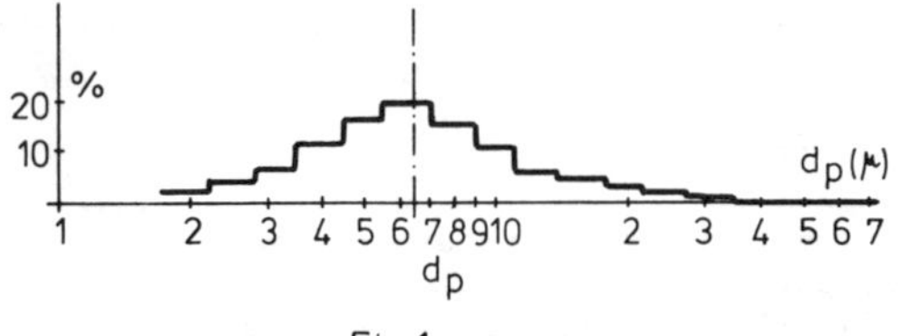

Fig.1

It has been verified that all the removable particles have sizes included in the range 1,75μ÷ 70,3μ. The average size ranges between 6,5μ and 7,3μ. 90% of the particles have sizes between 3 and 20μ.

Characteristics of thickening by column sedimentation are represented by Kynch's curve. Fig. 2.

3.3 Rheological characteristics

The flow behaviour of the sludge, defined by its rheological equation, was established from experimental results found with an extrusion rheometer. Rheograms are in Fig.3, in log-log scales.

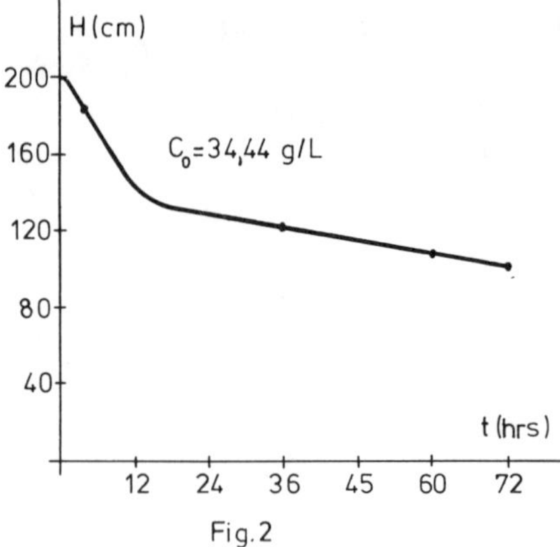

Fig.2

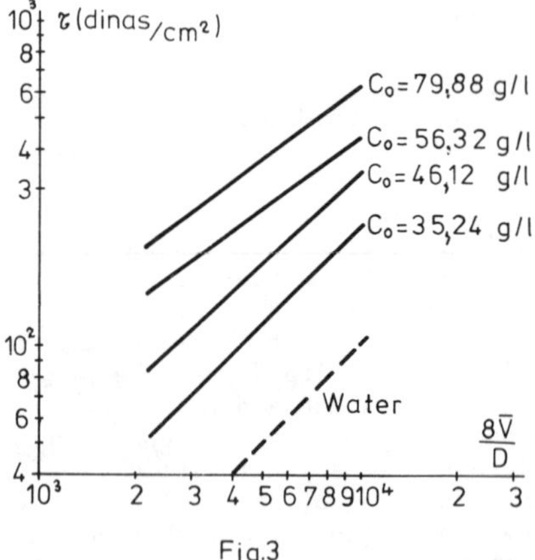

Fig.3

The rheological equation that adjusts the experimental results best, is that of Ostwald-de Waele model

$$\tau = K\,(\dot{\gamma})^n \quad (1)$$

$$\text{with } \tau = \frac{D\Delta P}{4L} \quad \dot{\gamma} = \frac{3n'+1}{4n'}\,\frac{8V}{D} \quad (2)$$

The values of K and n, in function of the concentration in dry matter of the sludge are tabulated in table 1

Table 1

C_o (g/l)	K	n
79,88	4,2	0,537
64,4	2,3	0,576
56,32	1,79	0,820
46,12	0,145	0,828
35,24	0,045	0,946

3.4 Specific resistance to filtration

This parameter is defined by this equation

$$r = 2 \frac{\Delta P}{\mu \rho_s g v_s} \cdot \frac{A^2 t}{V} \qquad (3)$$

Its assessment was made from the experimental results achieved in a pressure filtration cell. These are presented in Fig.4 as $\frac{v}{t} = f(v)$, taking the filtration pressure as a parameter.

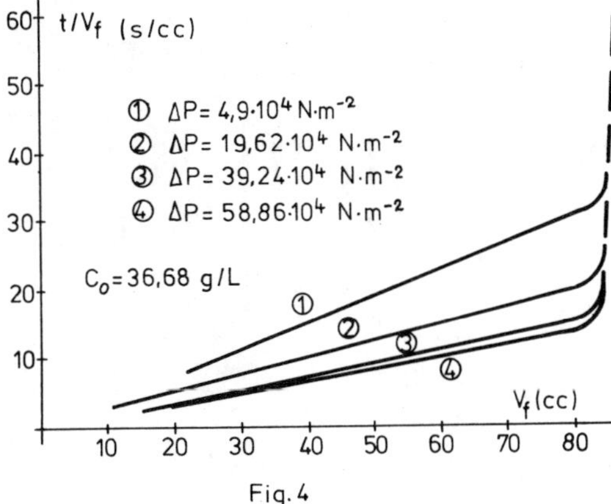

Fig. 4

The computed values of r are represented versus the filtration pressure, with C_o as a parameter, in Fig. 5

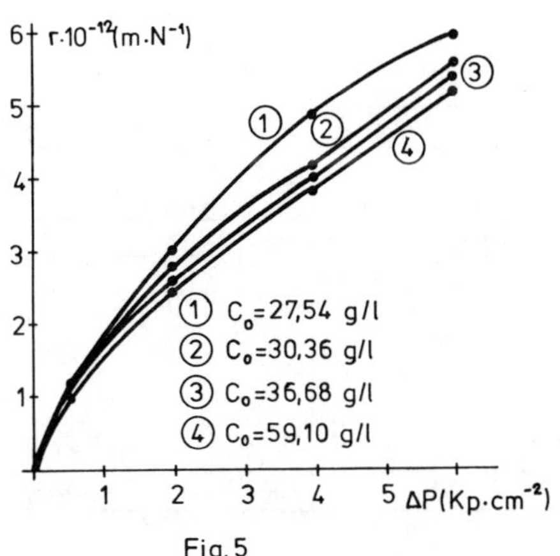

Fig.5

The specific resistance is an increasing function of the filtracion pressure and a decreasing one of the concentration in dry matter of the sludge.

For the sludges here studied, the function is expressed as

$$r = 1,1.10^9 . (\Delta P)^{0,64}$$

4 CHARACTERISTICS OF THE POROUS MEDIUM

In order to study the phenomena of convective transport and adsorption in porous media of the solid phase of the previously characterised sludges, a few porous media were built in the shape of cylindrical cartridges constituted by pilling up several rings.

In all cases, the porous medium was constituted by Styropor-P balls, packed within brass or PVC liners.

The cartridges used were differentiated from each other either by their diameter, or by their length and/or by the size of the balls. Their geometrical and structural characteristics are as follows;

Table II

Bed	Φ (mm)	L mm	nº of cartridges	d_b mm	$\bar{d}$ mm	ε	S mm⁻¹
I	80	250	5	2,5-3	2,603	0,35	149,9
II	80	200	4	2,5-3	2,603	0,35	149,9
III	80	150	3	2,5-3	2,603	0,35	149,9
IV/V	32	3/165	4/1	2,5-3	2,603	0,35	149,9
VI	80	250	5	3-4	3,038	0,349	128,6

S values were computed by this equation

$$S = \frac{6(1-\varepsilon)}{\bar{d}}$$

The intrinsic permeability of the porous medium was determined in two ways: experimentally, testing with water at 15°C and applying Darcy's law

$$j_x = \frac{K}{\mu} \rho g \quad I_x$$

and by computation from Kozeny's equation

$$K = \frac{\bar{d}_b^2}{150} \quad \frac{\varepsilon^3}{(1-\varepsilon)^2}$$

Values found were:
bed with balls of 2,603 mm of mean diameter

$$K_{experimental} = 4,64.10^{-7} \ m^2$$
$$K_{kozeny} = 6,158.10^{-7} \ m^2$$

bed with balls of 3,038 mm of mean diameter

$$K_{experimental} = 6,663.10^{-7}\ m^2$$

$$K_{kozeny} = 6,158.10^{-7}\ m^2$$

5 EXPERIMENTAL ASSEMBLY

Experiments were carried out in two experimental units (a) and (b) whose layouts are shown in Fig. 6

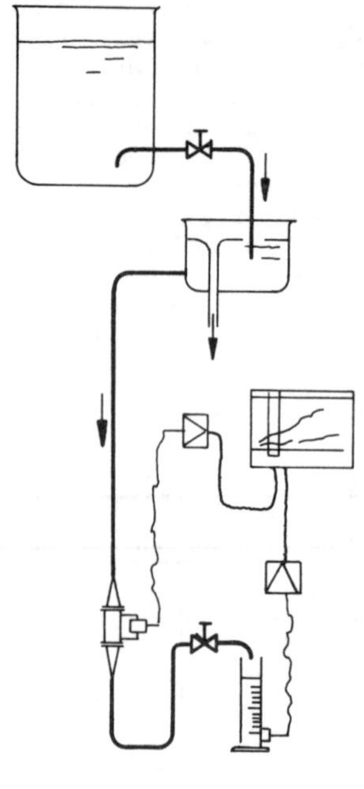

(a)

Fig.6

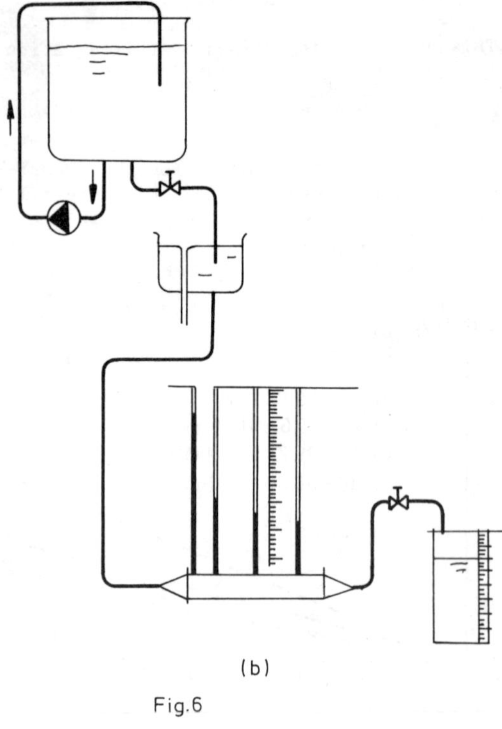

(b)

Fig.6

Unit (a), where the type IV bed-four cartridges of L = 7,5 mm and Ø 32 mm-was mounted, had some pressure transducers able to measure, at real time the values of Δ_p and v_f which were recorded by a X-Y_1-Y_2 recorder with time base.

Unit (b), where the I,II,III, and V type beds were set up -different number of cartridges of L = 50 mm and Ø = 80 mm -was provided with piezometers for the reading of the hidraulic head and with a calibrated capacity for the determination of the discharge.

Table III

Bed type	Symbol	K_w cm^2	C_o g/L	V_f cc	t s	MS_r mg	C_f g/l	L mm
IV	○	$4,64 \cdot 10^{-5}$	644	1670	170	1018		30
	△	"	"	1130	110	894		30
	□	"	"	295	82	563		30
	∨	"	"	430	89	672		30
	○	"	"	507	96	633		30
	∧	"	"	320	192	558		30
V	+	"	"	850	170			165
	×	"	"	600	70			165
I	⊠	"	50	17000		14545		250
	⌜	"	43.8	15700		11535		250
	⌟	"	64.4	6000				250
II	○	"	50	21000		14363	48.12	200
III	▽	"	50	14000		8166	48.62	150
VI	▲	$6,66 \cdot 10^{-5}$	50	17000	2160	11271	48.51	200

6 EXPERIMENTALS RESULTS

Some data about the fluid, the porous medium and the particulars on experiments are abridge in tableIII, where the symbols used to represent graphically the corresponding experimental results are also included.

Fig.7 shows v_f vs t for the same porous bed-type IV- and two different lengths of cartridge. Two remarkable influences are noticed: first, the mass of solids retained in the porous medium per unit volume of filtrate, and second, the length of the cartridge. A certain analogy with the phenomenon of cake filtration occurs. Fig. 8

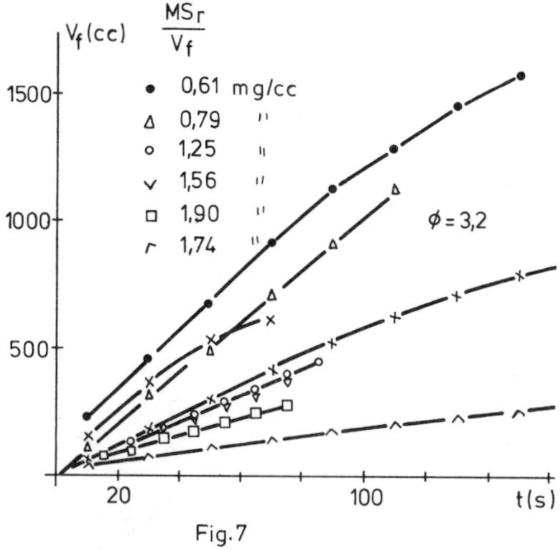

Fig.7

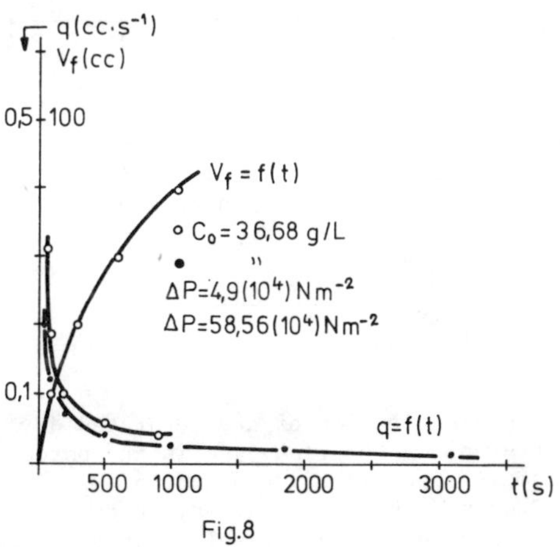

Fig.8

Again, the two above mentioned influences appear here; these, in turn, are ratified by the results shown in Fig. 10 by $\frac{q}{I} = f(MS_r)$ and in figs. 11 and 12, where $\frac{q}{I} = g(t)$ have been represented taking as a parameter the distance x from the section in wich $\frac{q}{I}$ is evaluated until the inlet section in the porous medium. All these graphs reveal a reduction of the hydraulic conductivity of the porous medium, as a function of the mass of solids retained inside the pore volume.

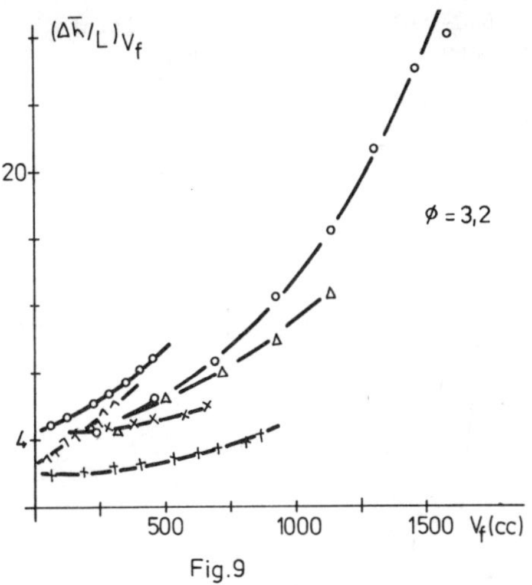

Fig.9

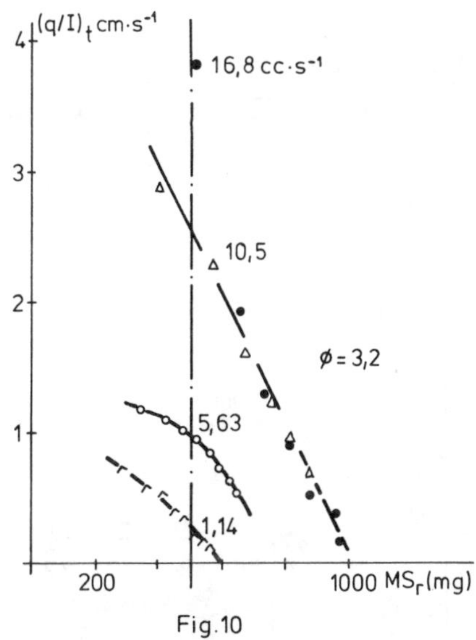

Fig.10

Assuming a cartridge length to be approximately like the path of the volumetric interception of the particles higher than 10μ, the loss of hydraulic conductivity of the porous medium seems to be a quasi-linear function of the retained solid matter. At a similar distance from the inlet porous medium, within that interception length, the change of the hydraulic conductivity along the time has a hyperbolic trend.

With respect to the retained solid matter in the pore volume of the porous medium, the variation of MS^*-relation of such a retained mass to the mass of solid matter injected in the bed- with the volume of filtrate, v_f, is represented in figs. 13 and 14.

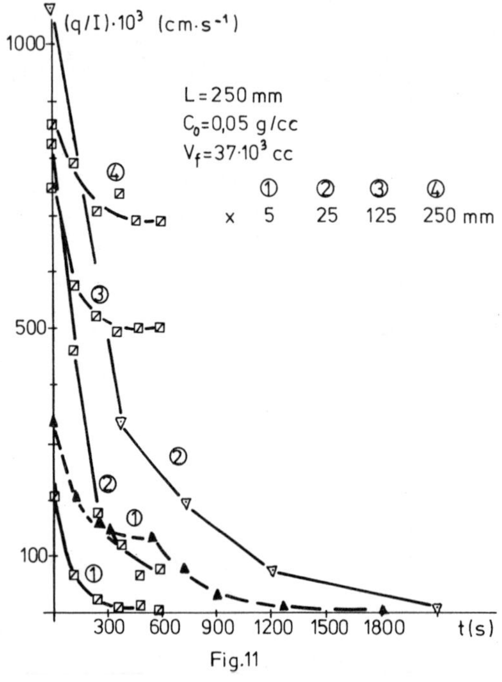

Fig.11

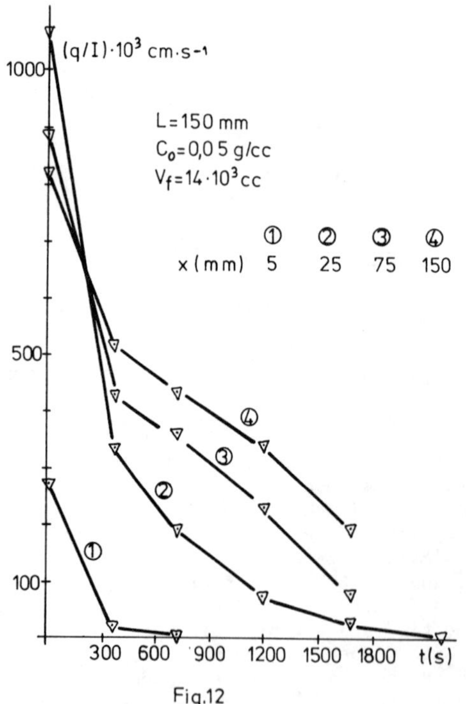

Fig.12

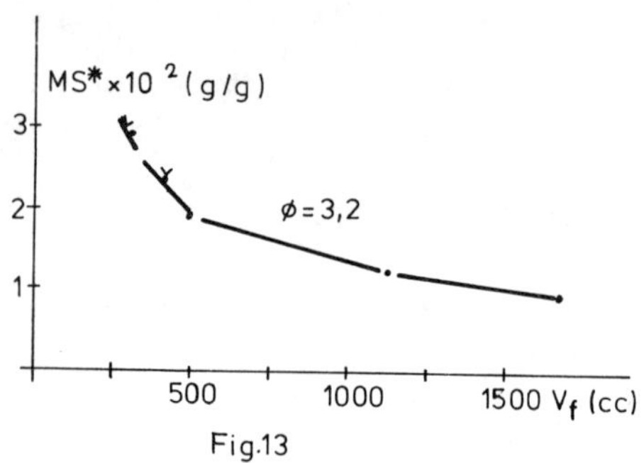

Fig.13

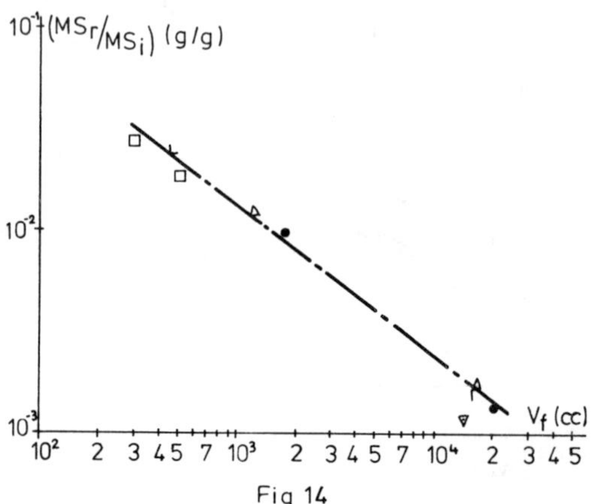

Fig 14

In view of the linear relations appearing in the representation of log (MS_r/MS_i) vs log v_f the following equation can be supposed;

$$\frac{\dfrac{MS_r}{MS_i}}{\left(\dfrac{MS_r}{MS_i}\right)_{v_f \to 0}} = \frac{1}{v_f^{\alpha}}$$

The distribution of the separated dry matter along the flow path in the porous medium is shown in figs 15 and 16.

From the results obtained, it can be seen that the separation of solid matter takes place in a very different way wheter the part of the porous medium involved is placed at

$$0 \leq \frac{X}{\bar{d}_p} < 15 \quad \text{or at} \quad 15 \leq \frac{X}{\bar{d}_p} < \infty$$

In the first case, the retention rate C or χ is high, in relative values, decreasing sharply with $\frac{x}{\bar{d}_p}$; in the latter, these rates decrease moderately with $\frac{x}{\bar{d}_p}$.

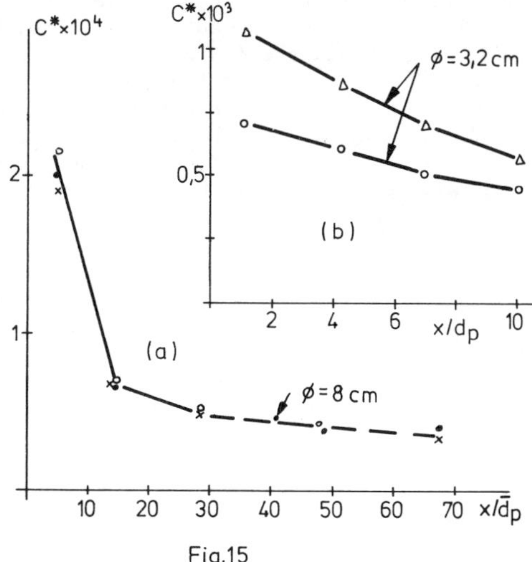

Fig.15

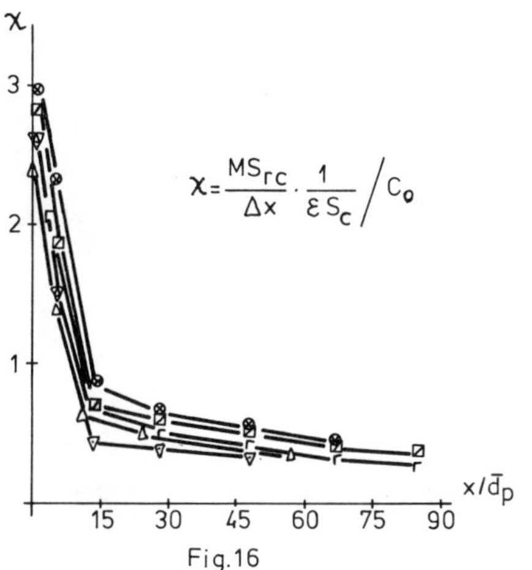

Fig.16

Experimental results seem to indicate that, in an elementary volume of the porous medium, located at the dimensionless length $\frac{x}{\bar{d}_p}$ from the boundary, the mass of solid matter retained by volume unit of the porous medium is a function of both the number of times that the pore volume has been filled up and the initial concentration in solid matter of the injected fluid.

The solid retention by volume unit of pore depends not only on the initial concentration of solids and $\frac{x}{\bar{d}_p}$ but also on v_f and the characteristic diameter of the particle, $\bar{d}_p$. It has been verified that the value of χ^p corresponding to the pores nearest to the entrance of the porous medium is extremely close to the relation between the concentration in solid matter of the fluid in the proximity of the entrance in the medium and its initial concentration, C_o.

The dry matter retained in the porous beds used, whose porosity was $\varepsilon = 0,35$, the specific surface $S \doteq 14,99$ cm^2 and the average characteristic diameters of the grains $\bar{d}_p = 2,603$ mm, was constituted by particles of a size above 15μ. The bed retains particulary all particles whose size is more than 35μ . Fig.17. The removing effect of these beds is somewhat small since, after having renovated 60 times the liquid filling the pore volume of the bed, which took 36 minutes, the concentration in dry matter of the sludge at the outlet of a 20 cm long porous cartridge was still 48,12 g/l when entering was 50 g/l.

The higher the retained dry matter the sharper will be the head loss in the layer of the bed, reaching a depth of about 5 mm, $(1,65 \div 1,42)\bar{d}_p$. In the rest of the bed, particulary from $x > 25$ mm, the hydraulic gradient for Q = cte changes very little with x. Figs. 18, 19 and 20.

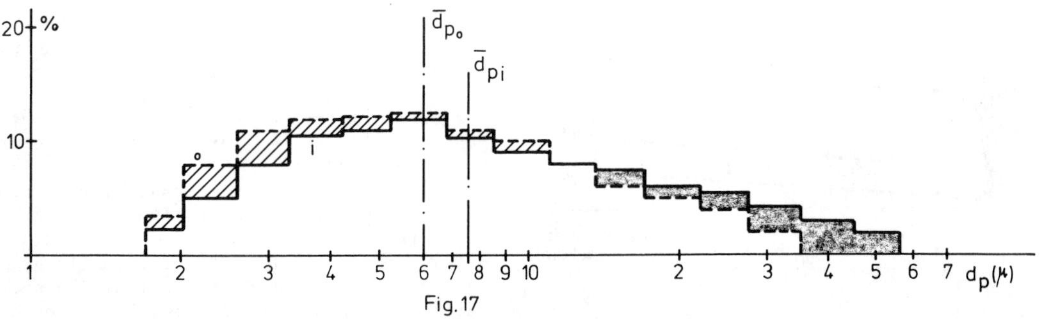

Fig.17

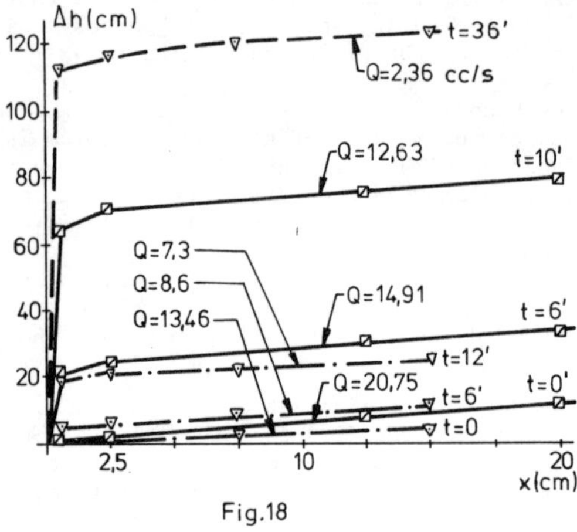

Fig.18

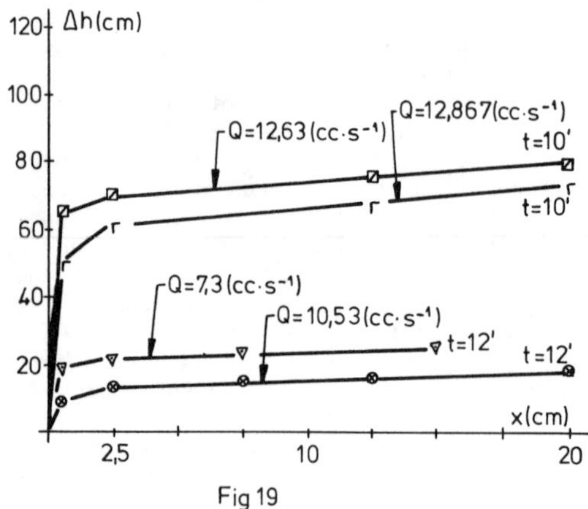

Fig 19

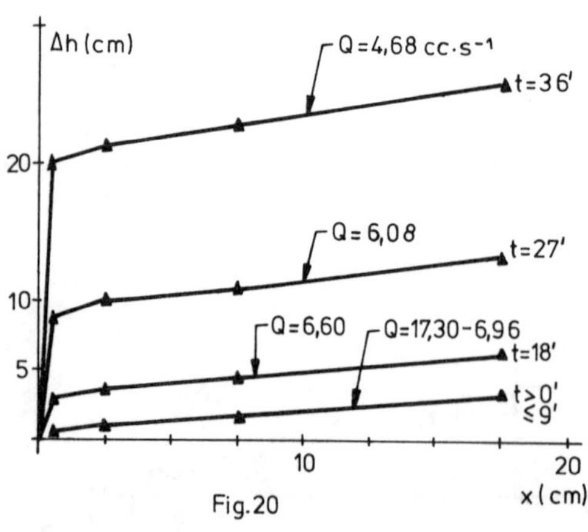

Fig.20

However, the flowrate is much lower than the one circulating by the same bed with the same head loss, should the fluid be water Figs.21 and 22. The influence of the non-newtonian behaviour of the fluid is obvious.

Fig.23 represents hydraulic gradient change with time at a distance from entrance to porous medium of about 5 m/m

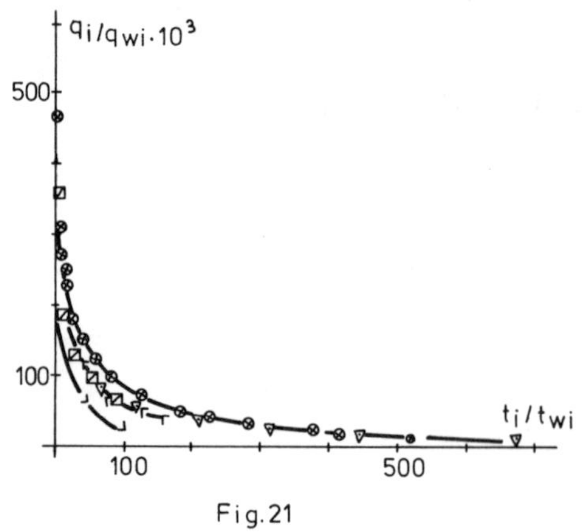

Fig.21

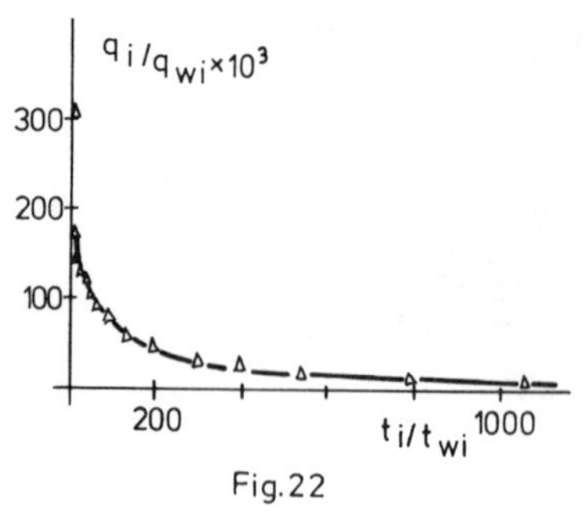

Fig.22

7 SUMMARY

Of the scope of this paper and at the present level of the experimental work, we can not draw any other conclusions than those concerning the mechanism of solid separation and the influence of the retention of suspended particles smaller than the pores of the porous medium on the pressure drop in it.

It appears crearly a mechanical retention for large particles (over 15μ), ocurring along the first layers of a porous medium. The maximum penetration depth of these large particles reaches a distance from the entry surface of about $x = (1,65 \div 1,92)\bar{d}_p$. The distribution of solid matter retained is quasi-

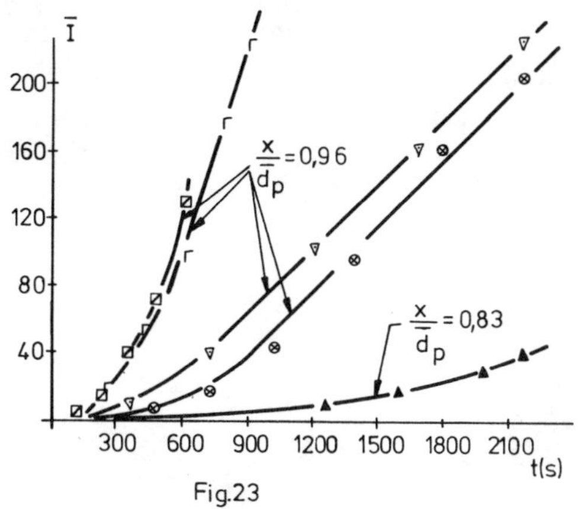

Fig.23

linear, with its maximum at the entry surface. The volume phenomena prevail and the porous media approaches to be fast clogged.

The fluid pressure drop through the porous medium increases with retention. Attempts to describe this evolution based on Kozeny's model have not been successful. Kozeny's model seems not be applicable to clogged beds, at least easily. We think more suitable the cake filtration model, with migration of smallest particles through cake.

After the layer placed at a distance from the boundary such that $x = 15\ \bar{d}_p$, the phenomena can be describe by the bulk-phase conservation equation, the mass transfer-rate mechanisms from bulk to grain surface, the kinetic adsorption-rate mechanisms at surface, the conservation relation at fluid solid surface and the modified Darcy's law for non-newtonian fluids.

8 NOMENCLATURE

C solid concentration of sludge
C_o solid concentration of sludges at the inlet of porous cartridge
C_f solid concentration of sludge at the outlet of porous cartridge
$C^* \equiv \left(\frac{MS_r}{V_c} \cdot \frac{\varepsilon\, d_p}{V_f} \right) / C_o$ separation parameter
D diameter
$\bar{d}_p$ grain mean diameter
h piezometer head
I hydraulic gradient
K intrinsic permeability
L length of the porous cartridge
$MS^* \equiv \frac{MS_r}{MS_i}$, bulk retention parameter
MS_r total solid matter retained
MS_i total solid matter injected
MS_c solid matter retained in a ring of the cartridge
P pressure
q specific flux of sludge
q_i specific flux of sludge at the time t_i
q_{wi} specific flux of water at the time t_i, for the same head loss than q_i
Q flowrate of sludge
r specific resistance of the cake
S_c cartridge cross section area
t time ; $t_{wi} = L/q_{wi}$
V_f volume of filtrate
x distance from the inlet to the cartridge along its axis
ε porosity
τ shear stress
$\emptyset$ cartridge diameter
χ dimensionless concentration of solid matter retained

9 REFERENCES

Bear, J. Dynamics of fluids in porous media. American elsevier publish.Co.Inc. 1972.

Nunge, R.J. Flow through porous media. American Chemical Society 1970.

Ison,C.R.& Ives, K.J. Removal mechanisms in deep bed filtration. Chem.Engn.Sci., 24, 717-729, 1969.

Hendricks, D.W. Convection limited sorption kinetics in packed beds. Journal of the Environmental Enginnering Division, August 1980, 728-739

Satter, A. Shum, Y.M. et all. Chemical transport in porous media with dispersion and rate - controlled adsorption. Society of Petroleum Engineers Journal, June 1980, 129-138

Litwiniszyn, J. Colmatage accompanied by diffusion. Bull. Acad.Pol.Sci.Ser.Sci.Techn. 14 (4), 295 - 381 (1966).

Maroudas, A. & Eisenklam, P. Clarification of suspensions; a study of particle deposition in granular media. Pt. I, II. Chem. Engn. Sci. (20) 867-873; 875-888 (1965).

Proceedings of Euromech 143 / Delft / 2-4 September 1981

Heat transfer in metallic screen wicks with phase change at the boundary

L.VIRTO & E.CODINA
Escuela Técnica Superior Ingenieros Industriales, Terrassa, UPB, Spain

1 INTRODUCTION

A commonly accepted hypothesis in the theoretical modelling and experimental study of heat pipe performances, particularly those with a porous wick constituted by a stainless steel screen piling-up, is that effective thermal conductivity of the wick is the same in both the condensator and the evaporator.

Experimental results recently found by the author are in disagreement with such hypothesis. In fact, the values of k'_{MP} in the evaporator can reach up to 7 times^{e}those of the condensator.

Thus, the question arises on which can be the mechanisms intervening in this possible activation of heat transfer, particularly when the possible convective effects associated to Benard's eddies cannot be necessarily the only reason explaining the heat transfer activation rate.

Despite the extensive literature on this topic, there are still many questions waiting for a definite answer.

Two are, basically, the approaches proposed to explain the mechanism of heat transfer in a wick, one of the boundaries being a free surface where a phase change takes place:

i) the predominant mechanism of heat transfer is by conduction from the wall of the pipe through the liquid saturated wick until the liquid-vapor interphase inside the wick, where vaporization takes place.

ii) the mechanism of heat transfer is by conduction through a vapor layer next to the wall, followed by conduction -convection through a wick full of liquid and, finally, vaporization in the liquid-vapor interphase inside the wick.

In our opinion, both these mechanisms are evident so that the mechanism of heat transfer is by bulk conduction and partly, by conduction in the thin film flowing towards some particular phase change sources, next to the heat surface source, followed by convective flux of the vapor phase through the partially liquid saturated wick. The preponderance of a model with respect to the other can be explained on the basis of wick material kind and its actual degree of contact with the surface through which the heat is supplied. Can be supposed that if the wick-hot wall contact is poor when applyng a determined heat flux and the space between the wall and the wick was not enough connected with the surrounding part of the wick, over heating of the liquid filling that space could take place, which could be indeed the onset of a nucleate boiling. On increasing q, boiling spreads which may give rise to a vapor continuous layer on the pipe wall. This will happen, should the screen be fine enough (large number mesh)to stop bubbles getting away and bursting at the free bundary. If the vapor bubbles get away they will simultaneously carry along a displacement of the liquid saturating the wick and this will activate the heat trasnfer (similar phenomenon to forced convection).

2 SCOPE OF THIS WORK

This paper, just a part of a major work under development, summarizes some of the experimental results achieved in a heat pipe, whose whick is constituted by several layers of stainless steel screen. A likely interpreta- of the heat transfer phenomena occurring in the evaporator is also given.

3 EXPERIMENTAL LAYOUT

Figure 1 shown the schematic arrangement for heat pipes test. Constructional details of heat pipes are summarized in Table I.

The heat pipe container was a 304 stainless steel tube, 40 mm ID and 60 mm OD, and the wick material was a package of seven layers of stainless steel screen (100 mesh). Both ends of the heat pipe were sealed off with flanges and a leak test was done.

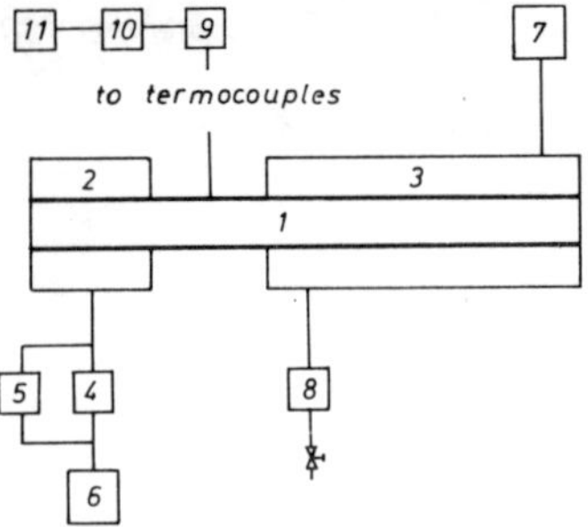

Fig. 1
1-Heat pipe, 2-Heater, 3-Calorimeter, 4-Voltmeter, 5-Ampermeter, 6-Variac, 7-Tank, 8-Flowmeter, 9-Cold juntion, 10-Scanner, 11-Recorder

Table I

Heat pipe	
Container	Stainless steel
Length	0,6 m
Evaporator length	0,15 m
Condensator length	0,30 m
Inner diameter	0,04 m
Outer diameter	0,06 m
Wick	seven layers metallic screen stainless steel-100 mesh
Fluid	distilled water

A cylindrical heater aluminium block, was used as the heat source, located at one end of the pipe and the input power to the heater was controlled by a variac. A heat-resistant agent having a good thermal conductivity filled up the space between the outside of the pipe and the inside of the heater block in order to make a good contact and the outside of the heater block was heat insulated with glass wool. Heat was picked up by the cooling water passing through a jacket calorimeter fitted to the opposite end.

Iron-konstantan thermocouples were provided along the heat pipe to monitor the axial temperature distribution along the wall heat pipe and vapor within the heat pipe.

With this heat pipe, some experiences were conducted holding the vapor temperature constant by regulating the inlet temperature and flow rate of the cooling water. At various pipe inclination angles, heat transfer rate, , temperature distribution along the length of heat pipe, and superheat have been calculated:

The heat transfer rate was calculated from heat picked up by the cooling water at the condenser

$$Q = \dot{m}\ c\ \Delta T$$

where

$\dot{m}$ = cooling water flow rate (kg/s)

c = cooling water specific heat (kcal/°C kg)

ΔT = difference between inlet and outlet temperatures

It was roughly admitted the local heat flux to be equal to the average heat flux based on the area of the evaporator inner wall. Then it was calculated from the equation

$$q = Q/S_i$$

and superheat $\Delta T|_e$ was defined as the difference between the inner heated surface and the vapor temperature.

4 RESULTS AND DISCUSSION OF RESULTS

Fig.2 shows the temperature profiles along the wall of the heat pipe, for different thermal powers. Graphs 2-a corresponds to the heat pipe vertically positioned and 2-b to the horizontally positioned one.

If the average of all temperatures of the inner wall of the pipe is taken as the characteristic temperature of the evaporator (condensator) and we calculate the effective thermal conductivity of the wick, assuming that all the heat has been carried through it by conduction, according to Fourier's law,

$$k'_{MP_e} = \frac{Q}{2\pi L_e \cdot \dfrac{\bar{T}'_e - T_v}{\ln(r_o/r_v)}}$$

we will find the results shown in figura 3

In the light of these results, this question is made: what causes the effective thermal conductivity of a liquid saturated wick to be different when the heat transfer in the liquid-vapor interphase is produced by condensation or by evaporation?.

To answer this question, new consideration must be given to this problem by examining locally the heat transfer phenomena. Thus, figs.4 and 5 show the temperature difference through the wick (superheat) in funtion of the heat flux $q = Q/S_i$ for each point of the evaporator having a thermocouple, with the exception of the one located at the end of the heat pipe where the heat transfer process can be distorted by the end effects.

In the light of above results we can verify:

1-An non-uniform distribution of the temperature around the wall of the heat pipe. This irregularity can be imputed to the own configuration of the wick. Fig.6 presents a schematic cross-section of such wick where it is observed not only that the thickne δ' is not constant but also the existence of some gaps, derived from the technique used on preparing the wick and the procedure to introduce this into the pipe.

This should give rise to a different overheating degree of the fluid in the wick.

2-The relationship between q and $\Delta T|e$ shows two differents trends, according to

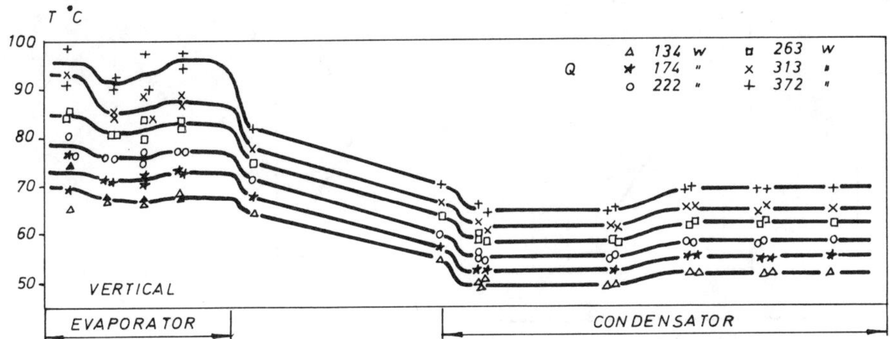

Figura 2-a

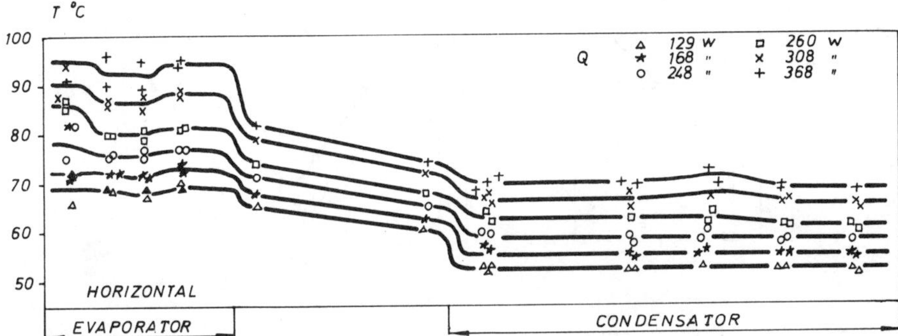

Figura 2-b

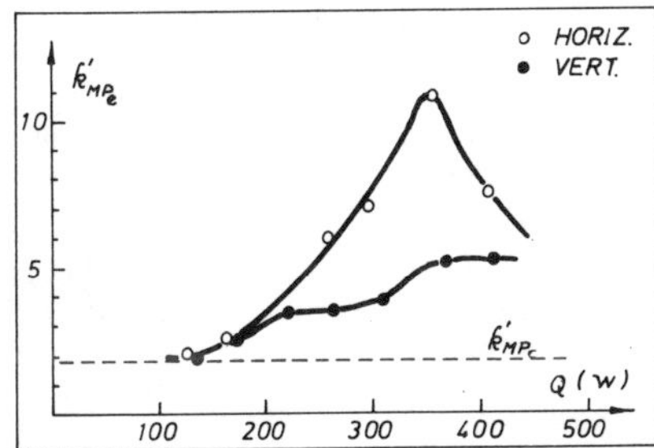

Figura 3

the level of thermal power applied to the evaporator,

A) for small values of q, the higher q, the lower $\Delta T|e$

B) for large values of q (generally q>1,5 w/cm^2): the higher q, the higher $\Delta T|e$

These two trends are fully due to the mechanisms of heat and mass transfer taking place inside the wick.

3-And, finally, a more evident outcoming of these phenomena appears as the end of the evaporator gets nearer.

An adequate interpretation of this behaviour is only possible by examining systematically all the heat transfer mechanisms taking place in the zone of the evaporator. At the beginning, let us assume these mechanisms:

i)the heat transfer mechanism from the wall of the heat pipe until the vapor region is composed by the heat transfer by pure conduction through the liquid saturated wick and evaporation in the liquid-vapor interphase.

ii)basically, the above mechanism, but this time with the addition of heat transport by convection without change of phase.

iii)initially, heat transfer is by conduction, then a nucleate boiling phenomenon occurs on increasing the thermal power. This phenomenon, apart from being a heat transport in vapor phase increases the heat transfer by convection in liquid phase. The first mechanism, that is, heat transfer by pure conduction is governed by Fourier's law

$$q = -k'_{MP_e} \frac{\partial T}{\partial r} = -k'_{MP_e} \frac{\Delta T|e}{\delta_L}$$

where δ_L is the thickness of the fully saturated liquid wick. In this case it can happen: the wick is either partially saturated $\delta_L < \delta'$

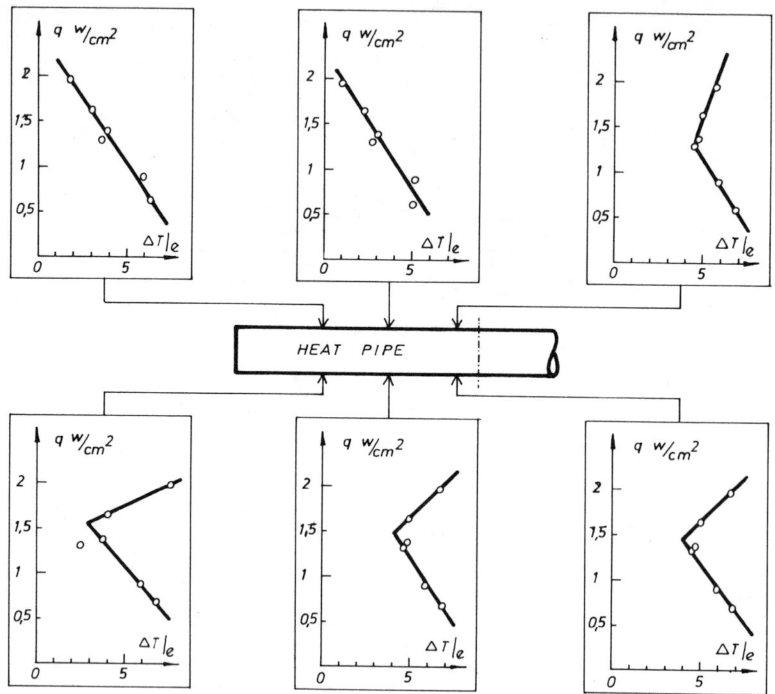

Figure 4

or totally saturated $\delta_L = \delta'$. The saturation degree of the wick will depend, in general, on the capillary suction potential.

If we examine the most unfavourable running hypothesis, we can state that $\delta_L = \delta'$ if the mass flow of liquid pumped by the action of the capillary forces is larger than the mass flow of vapor defined by:

$$\dot{m}_v = \frac{Q}{\lambda}$$

where λ is the state change latent heat of the fluid.

Considering that the wick is constituted by layers of screens, 100 mesh/inch, the curvature radius of the meniscus can be evaluated en function of the geometry of the wick by this equation:

$$R_M = \frac{t - d \text{ sen } \Psi}{2 \text{ sen } (\Psi - \theta)}$$

For $\Psi = 90°$
$\theta = 51°$
$t = 0{,}2806$ mm
$d = 0{,}1034$ mm, resulting $R_M = 137\mu$

According to Darcy's law:

$$\dot{m}_L = \rho S \, K \frac{g}{\nu} I$$

where

$$I = \frac{2 \sigma}{\rho g R_M L}$$

for $K = 1{,}4 \cdot 10^{-9}$ m^2
$R_M = 137\ \mu$
$L = 0{,}6$ m
$S = 0{,}00019$ m^2, resulting $\dot{m}_L = 0{,}9$ g/s

On the other hand, for a maximum thermal power Q=500 w it corresponds $\dot{m}_v = 0{,}2$ g/s, from where $\dot{m}_L > \dot{m}_v$ and, consequently, $\delta_L = \delta'$ for any value of the thermal power below 500 w and any orientation of the heat pipe.

In this case, according to the hypothesis formerly stated, must be $k'_{MP_e} = k'_{MP_c}$, just the opposity to whathappens in reality.

The other possible mechanism is closely connected to convection without phase-change. Since convection is characterised by Rayleigh's number

$$Ra = \frac{\beta \, g \rho^2 K c l \Delta T}{\mu k'_{MP}}$$

the convection is significant, according to STRAUS, if the value of Rayleigh's number exceedes $4\pi^2$

For $\beta = 2 \cdot 10^{-4}$ °C^{-1}
$l = \delta'$
$\Delta T = \Delta T|e$; results $Ra = 17 < 4\pi^2$

which allows us to refuse convection without change of phase as a significant mechanism of heat transfer.

Finally, we will examine the possible onset of nucleate boiling. As is known, the appearing of nucleate boiling demands a limit state of overheating ΔT^* of the fluid satura-

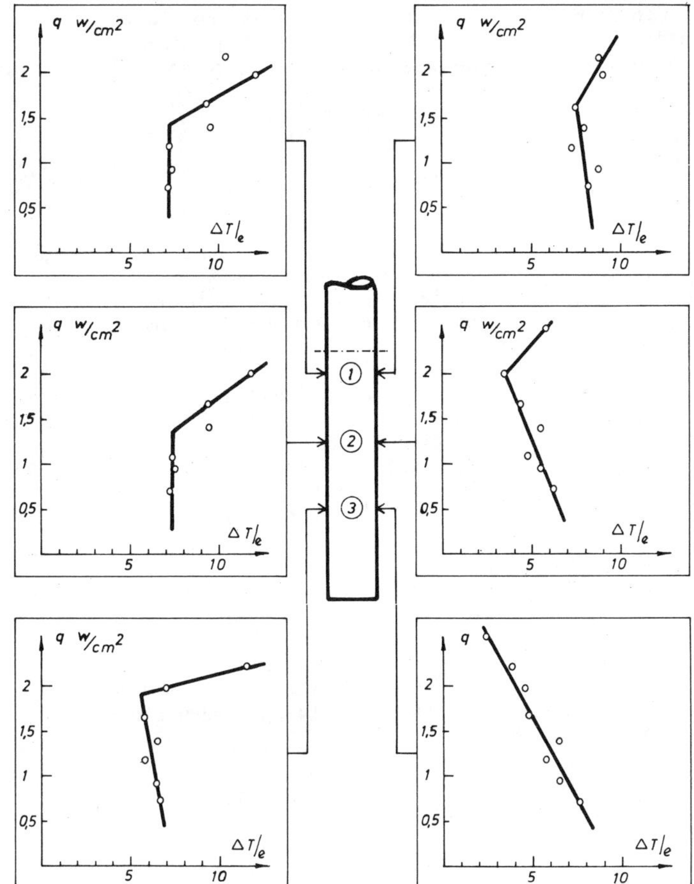

Figure 5

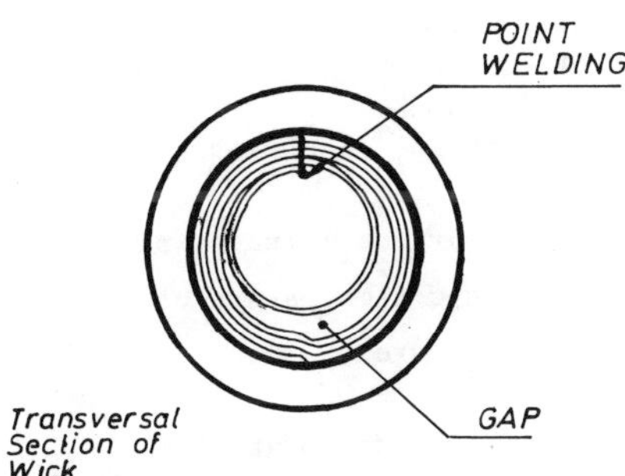

Figure 6

ting the wick in such a way that any further increase of the fluid temperature will cause the change of phase, producing vapor bubbles at a rate which depends of the thermal power.

According to HSU, the value of ΔT* is defined by the following equation:

$$\Delta T^* = \frac{3.06\ \sigma\ T}{\rho_v\ \lambda\ \delta^*}$$

where δ*, is the thickness of the thermal layer.

In a first approach, this can be equal to the mean diameter of the cavities on the heated surface.

In this particular case, for $\delta^* = 0{,}25 \cdot 10^{-3}$m it corresponds $\Delta T^* = 0{,}80$°C, which is absolutely feasible.

On the other hand, this possible mechanism justifies trend B, which is characteristical for values of the supplied thermal power flux $q > 1{,}5$ w/cm^2. So the nucleate boiling can be spreaded as the thermal flux increases, which makes the vapor on the inner wall of the pipe to move to the free boundary in the shape of discret bubbles until these get caught somewhere in the wick, acting like small heat pipes.

Similarly, this mechanism explains, too, the differences in the behaviour of the heat pipe when this acts either vertically or horizontally.

We have observed that, for the same point (i.e. point 1, representative of the nearest area to the adiabatic zone) and for $q < 1{,}5$

w/cm^2, the value of $\Delta T|e$ remains constant for a heat pipe vertically positioned while for a horizontally positioned one $\Delta T|e$ decreases as q increases. This difference is explained by taking notice of how the gravitational field is capable of restoring some liquid to the zone of the evaporator faster than the restored one by the action of capillary. This liquid restoring action prevents the increase of the overheating degree and, as a consequence, the increase of the number of active points of vapor generation. At the same time, this justifies the fact that the value of k'_{MPe} for the vertical running of the heat pipe keeps lower and more stable than the corresponding to the horizontal running of the heat pipe.Equally,it explains why the process of nucleate boiling becomes more active when the end of the evaporator gets nearer,should the orientation of the heat pipe be the same and for the same value of thermal power.In others,for trend-A the slope of graph q vs $\Delta T|e$ turns into more negative and for trend-B this be positive,however,less in absolute value.

5 SUMMARY

Of the scope of this paper we only can make some considerations with respect to the phenomena observed in our research. These are:

i) the predominant mechanism, through which the effective thermal conductivity of a fully saturated liquid wick is larger in a vaporization process than in a condensation one, is a phase change inside the wick

ii) the effectiveness of this mechanism of heat transfer depends mainly on the wick structure, particularly on the size of the screen mesh and the relative permeabilities and, also, on the contact of the wick/inner wall of the heat pipe.

iii) although the heat transfer mechanisms are complicated and rather special, three different regimes are to be distinguished:

- for $q \leqslant 0{,}5$ w/cm^2 all the heat is transferred by conduction through a fully saturated wick, the vaporization taking place in the liquid-vapor interphase. This mechanism leads to a value of the effective thermal conductivity which is similar in both the evaporator and the condensator.
- for $0{,}5 < q < 1{,}5$ w/cm^2, where discrete boiling with bubble formation is originated, the heat transfer is carried out to some extent by conduction and partially by vapor transport and simultaneous liquid convection.
- for $q > 1{,}5$ w/cm^2, where bubbles join constantly between themselves, forming great vapor cavities, the heat transfer is carried out by the same mechanisms before explained, prevailing the nucleate boiling. Because of this, the inflow of liquid into the vaporization areas becomes more and more difficult as the thermal power applied is increased, arising "hot spots".

6 REFERENCES

L.Virto and E.Codina. Effective thermal conductivity in saturated porous media 7° Symposium International de L'A.I.R.H.

E.Codina. Estudio teórico experimental de un tubo de calor capilar poroso. Thesis 1.981. Escuela Técnica Superior Ingenieros Industriales Terrassa (Spain).

J.M.Straus and G.Schubert. Modes of finite amplitude three dimensional convection in rectangular boxes of fluid-saturated porous material. J.Fluid mechanics (1.981) vol. 103 pag. 23-32.

7 NOMENCLATURE

- c- Fluid specific heat
- d- diameter
- g- intensity of gravitacional field
- K- permeability of porous medium
- k'_{MP} effective thermal conductivity
- k'_{MP_e} idem , evaporator
- k'_{MP_c} idem , condensator
- L- lenght of heat pipe
- $\dot{m}$- cooling water flow mass rate
- $\dot{m}_l$ mass flow of liquid
- $\dot{m}_v$ mass flow of vapor
- Q- thermal power
- q- heat flux
- Ra Rayleigh's number
- R_M curvature radius of meniscus
- S section
- S_i evaporator surface (inner wall pipe)
- T'_e characteristic temperature of evaporator
- T_v vapot temperature
- t pitch
- λ state change latent heat of the fluid
- δ' thickness of the wick
- δ_L thickness of liquid film
- θ contact angle
- μ fluid viscosity
- ρ density of liquid
- β coefficient of thermal expansion of the fluid
- σ surface tension

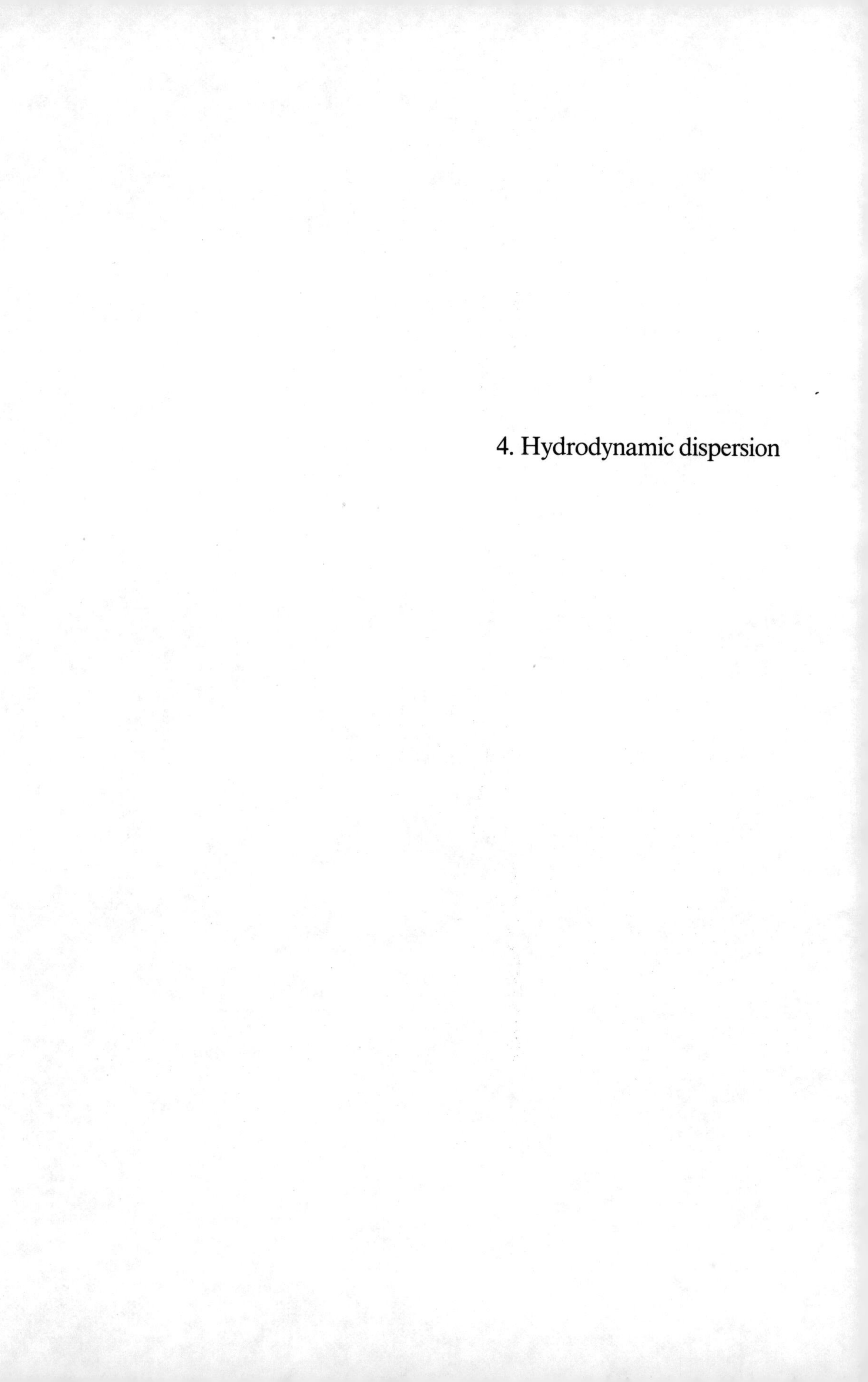

4. Hydrodynamic dispersion

Proceedings of Euromech 143 / Delft / 2-4 September 1981

The global pressure, a new concept for the modelization of compressible two-phase flows in porous media

G.CHAVENT
INRIA, Le Chesnay, France

INTRODUCTION

The aim of this paper is to present a mathematical formulation of the compressible two-phase flows in porous media, which takes into account the relative permeabilities and capillary pressure hypothesis, and the counterflows caused by gravity and heterogeneous capillary pressure.

The main feature of this formulation is the introduction of a fictitious pressure, which we call the global pressure, which has the property that the total Darcy velocity of the two fluids can be expressed in terms only of i) the gradient of the global pressure and ii) the saturation itself (but the gradient of the saturation does not enter in the formula).

The resulting equations are similar in form (with more non linear terms) to the equations governing the flow of two miscible fluids in a porous medium. This similarity is taken advantage of for the simulation of miscible displacements with immiscible simulators, which is shown to be possible, eventually with a small approximation, in rather general situations.

One other interesting features of the resulting equations is that one term is associated to every physical phenomenon (capillary diffusion, transport, gravity, heterogeneous maximum capillary pressure), allowing the discretization of each term according to its very nature (diffusion, transport), which is important for the proper numerical approximation of the solution.

1. THE IMMISCIBLE EQUATIONS

We shall establish our equations under the following hypothesis on the functional dependance of the main parameters :

. The porosity $\emptyset$ is a function of space and pressure.

. The absolute permeability tensor K is a function of space and of any dependant function (as gradient of saturation for example).

. The residual saturations of each phase are function of space only.

. The viscosities and densities are function of pressure and saturation (this latter dependance being introduced only for purpose of comparison with the miscible case).

. The relative permeability curves, when expressed in term of reduced saturation, are independant of spatial variable x.

. The capillary pressure law can be expressed as :

$$P_1 - P_2 = P_{cM}(x)\ p_c\ (S) \qquad (1)$$

where the reduced capillary pressure curve $p_c(S)$, valid all over the porous body Ω, is an increasing function of the reduced saturation , which vanishes for $S = S_c$ (usually $S_c = 1$ when S is the wetting phase saturation), and where $P_{cM}(x)$ is a spatially varying scaling factor ($P_{cM}(x)$ is the maximum capillary pressure ; it is always positive).

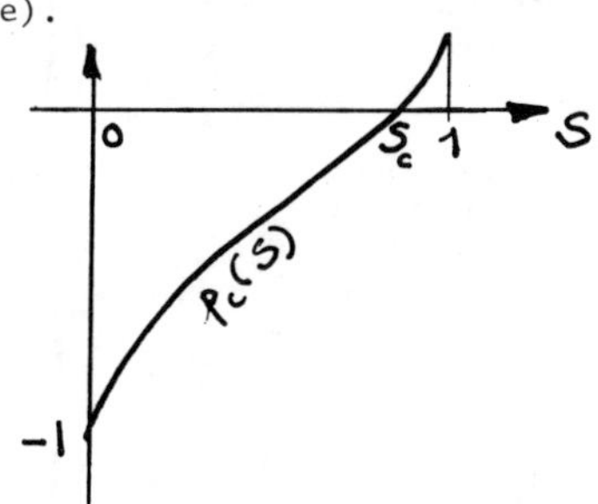

Hence the case covered by the above hypothesis is rather general and applies to different applications (oil+gas flow in reservoirs, water+air flow in dams or soil etc).

Let for a while $P(x,t)$ be any pressure field satisfying for every x in Ω and $t \geq 0$:

$$\mathrm{Min}(P_1, P_2) \leq P \leq \mathrm{Max}(P_1, P_2) \qquad (2)$$

We shall neglect the error done by evaluating the density and viscosity factors for the fluid j using P instead of P_j, and decide to evaluate the porosity $\emptyset$ using the pressure P (its here no more clear if one should physically use P_1 or P_2 !). This kind of approximation is widely used in all numerical two phase simulator.

The immiscible equations are then :

$$\left.\begin{array}{l} \vec{\phi}_j = -\psi(x) k_j(S,P)[\Delta P_j - \rho_j(S,P) g \Delta Z] \\ j = 1, 2 \quad \text{(Darcy's Law)} \end{array}\right\} \qquad (3)$$

$$\left.\begin{array}{l} \frac{\partial}{\partial t}[\phi(x,P) B_j(S,P) S_j] + \Delta \vec{\phi}_j = 0 \quad j=1,2 \\ \text{(Mass conservation Law)} \end{array}\right\} \qquad (4)$$

$$S_1 + S_2 = 1 \qquad (5)$$

$$\left.\begin{array}{l} P_1 - P_2 = P_{cM}(x)\ p_c(S) \\ \text{(Capillary Pressure Law)} \end{array}\right\} \qquad (6)$$

The purpose of the following calculations is to define a pressure field $P(x,t)$ satisfying (2) in such a manner that equations (3) to (6) be exactly equivalent to a set of two coupled equations in S and P only.

<u>Step 1</u> : The half difference of equations (4) for j=1,2 yields :

$$\left.\begin{array}{l} \frac{\partial}{\partial t}[\phi \frac{B_1 S - B_2(1-S)}{2}] + \\ + \nabla[\frac{k_2\vec{\phi}_1 - k_1\vec{\phi}_2}{k_1 + k_2} + \frac{k_1 - k_2}{k_1 + k_2} \frac{\vec{\phi}_1 + \vec{\phi}_2}{2}] = 0 \end{array}\right\}$$

which using (3) and (6) and condensed notations gives in turn :

$$\left.\begin{array}{l} \frac{\partial}{\partial t}[\phi \frac{B_1 S - B_2(1-S)}{2}] + \\ + \nabla[\vec{r} + \sum_{j=0}^{2} b_j(S,P)\vec{q}_j] = 0 \end{array}\right\} \qquad (7)$$

$$\vec{r} = -\tilde{\psi}(x, S, P)\ \nabla\ \alpha(S) \qquad (8)$$

Equation (7)(8) is a diffusion+transport equation, with a degenerated diffusion term ($\alpha'(0) = \alpha'(1) = 0$, but $\tilde{\psi}$ is bounded below by a strictly positive number) and a possibly non monotonous (if $\vec{q}_1$ or $\vec{q}_2$ are non zero) transport term.

<u>Step 2</u> : Making the half sum of equations (4) for j=1,2 yields :

$$\frac{\partial}{\partial t}[\phi \frac{B_1 S + B_2(1-S)}{2}] + \nabla \vec{q}_o = 0 \qquad (9)$$

The objective now is to express the global velocity vector $\vec{q}_o = \frac{\vec{\phi}_1 + \vec{\phi}_2}{2}$ in term of ∇P (P still to be defined !) and S only, but without ∇S. We introduce for that a function $\gamma(S,P)$ defined by :

$$\gamma(S,P) = \frac{1}{2} \int_{S_c}^{S} b_o(s,P) \frac{dp_c}{dS}(s) ds \qquad (10)$$

where $S_c \in [0,1]$ is such that $p_c(S_c)=0$. The two important properties of $\gamma(S,P)$ are :

PROPOSITION 1 :

$$|\gamma(S,P)| \leq \frac{1}{2} |p_c(S)| \quad \forall S \in [0,1],\ \forall\ P \qquad (11)$$

This follows immediately from (10) and $|b_o(S,P)| \leq 1$ for $S \in [0,1]$, as $\frac{dp_c}{dS}(s) \geq 0$ for $s \in [0,1]$.

PROPOSITION 2 :

$$\left.\begin{array}{l} |\frac{\partial \gamma}{\partial P}(S,P)| \leq \\ \frac{1}{4} \int_{S_c}^{S} |\frac{\partial}{\partial P} \mathrm{Log}(\frac{B_1}{B_2} \times \frac{M_1}{M_2})|\ \frac{dp_c}{dS} ds \end{array}\right\} \qquad (12)$$

This follows from (10 by derivation and use of inequality $4k_1k_2 \leq (k_1 + k_2)^2$.

Let now P_{min} and P_{max} be a-priori given lower and upper bounds of phase pressures P_1 and P_2. We make the hypothesis that the capillary pressure $P_{CM}(x)$ is small enough to satisfy :

$$\left.\begin{array}{l}\exists\ \theta\in]0,1[\text{ such that, }\forall P\in[P_{min}, P_{max}],\\ \forall\ S\in[0,1],\ \forall\ x\in\Omega,\ \forall j = 1,2,\text{ one has :}\\ \frac{1}{2}\left|\frac{\partial}{\partial P}\text{Log}\frac{B_j(S,P)}{M_j(S,P)}\right|\ |P_{CM}(x)| \leq \theta < 1\end{array}\right\} \quad (13)$$

This hypothesis simply means that, for each fluid, the relative variation of the density/viscosity factor $\frac{B_j}{M_j}$ for a pressure variation of the order of magnitude of the maximum capillary pressure $P_{CM}(x)$ is always smaller that two over the whole range of attaignable saturations and pressures. This is always the case in the range of pressures for which the original equations (3) to (6) are valid, thus (13) is not a practical restrictions.

Using now the proposition 2 and hypothesis (13) yields :

$$\left.\begin{array}{l}\forall S\in[0,1],\ \forall P\in[P_{min}, P_{max}],\ \forall x\in\Omega\\ |P_{CM}(x)\frac{\partial\gamma}{\partial P}(S,P)| \leq \theta < 1\end{array}\right\} \quad (14)$$

This enables us, using the contraction fixed point theorem, to define the pressure field $P(x,t)$ at every point $x \in \Omega$ and $t > 0$ by :

$$\left.\begin{array}{l}P(x,t) = \frac{1}{2}[P_1(x,t) + P_2(x,t)] +\\ P_{CM}(x)\ \gamma(S(x,t),\ P(x,t))\end{array}\right\} \quad (15)$$

Using proposition 1 one checks then easily that $P(x,t)$ defined in the above way satisfies (2), so that it can actually be used instead of P_1 or P_2 in the density or viscosity factors in the original equations (3) to (6).

One checks then easily, taking the gradient of (15) and using the capillary pressure law (1), that :

$$\left.\begin{array}{l}(k_1+k_2)\nabla P = k_1\nabla P_1 + k_2\nabla P_2\\ -\ (k_1+k_2)(1-P_{CM}\frac{\partial\gamma}{\partial P})\ \gamma_1\ \nabla P_{CM} +\\ +\ (k_1+k_2)\ P_{CM}\frac{\partial\gamma}{\partial P}\nabla P\end{array}\right\}$$

which shows that this pressure field P has the sought property concerning the global velocity vector field $\vec{q}_o = \frac{\vec{\phi}_1 + \vec{\phi}_2}{2}$:

$$\vec{q}_o = -\psi(x)d(x,S,P)[\nabla P + \sum_{j=1}^{2}\gamma_j(x,S,P)\vec{q}_j] \quad (16)$$

where $d = \frac{k_1+k_2}{2}(1 - P_{CM}\frac{\partial\gamma}{\partial S})$, γ_1 and γ_2 are given in the notation appendix, and where $\vec{q}_1$ and $\vec{q}_2$ are the same given vector fields as in the saturation equation (7) ($\vec{q}_1$ governs the effect of the heterogeneity in the maximum capillary pressure, $\vec{q}_2$ governs the gravity effects).

We call this pressure P the global pressure as its governs the global velocity $\vec{q}_o$.

Summing up the equations we get the :

Equations of Immiscible Flow in Porous Media

$$\left.\begin{array}{l}\frac{\partial}{\partial t}\{\phi(x,P)\ \frac{B_1(S,P)S - B_2(S,P)(1-S)}{2}\} +\\ +\ \nabla\{\vec{r} + \sum_{j=0}^{2} b_j(S,P)\vec{q}_j\} = 0\end{array}\right\} \quad (17)$$

$$\vec{r} = -\ \tilde{\psi}(x,S,P)\ \nabla\alpha(S) \quad (18)$$

$$\left.\begin{array}{l}\frac{\partial}{\partial t}\{\phi(x,P)\ \frac{B_1(S,P)S + B_2(S,P)(1-S)}{2}\} +\\ +\ \nabla\{\vec{q}_o\} = 0\end{array}\right\} \quad (19)$$

$$\left.\begin{array}{l}\vec{q}_o = -\ \psi(x)\ d(x,S,P)\ \nabla P +\\ +\ d(x,S,P)\ \sum_{j=1}^{2}\gamma_j(x,S,P)\vec{q}_j\end{array}\right\} \quad (20)$$

where :

S = reduced saturation of fluid # 1

P = global pressure

Once S and P have been calculated by equations (17) to (20) the filtration velocities $\vec{\phi}_1$ and $\vec{\phi}_2$ of each phase are given by :

$$\vec{\phi}_1 = \vec{q}_o + \sum_{j=0}^{2} b_j(S,P)\vec{q}_j + \vec{r} \quad (21)$$

$$\vec{\phi}_2 = \vec{q}_o - \sum_{j=0}^{2} b_j(S,P)\vec{q}_j - \vec{r} \quad (22)$$

and pressures P_1 and P_2 inside of each phase are given by :

$$P_1 = P - \{\gamma(S,P) - \frac{1}{2} p_c(S)\} P_{CM}(x) \quad (23)$$

$$P_2 = P - \{\gamma(S,P) + \frac{1}{2} p_c(S)\} P_{CM}(x) \quad (24)$$

Equations (17) to (20) are a system of two coupled evolution equations. The saturation (17) has a degenerated diffusion term $\nabla\vec{r} = -\nabla(\psi\nabla\alpha(S))$ (as $\alpha'(S) = a(S)$ vanishes for $S = 0$ and $S = 1$) and a possibly non monotonous (if $\vec{q}_1$ or $\vec{q}_2 \neq 0$) transport term $\sum_{j=0}^{2} b_j\vec{q}_j$.

The pressure equation (19) has a non degenerated diffusion term $-\nabla(\psi d\nabla P)$ (as d is always positive and bounded away from zero).

Those equations reduce, for incompressible flow, to those given in [1].

2. THE MISCIBLE EQUATIONS

We shall now recall, for comparison purposes, the equations governing the flow of miscible fluids in a porous media.

Those equations are established in [2] under the hypothesis that the gravity center is used for the definition of the Darcy law. The differential action of gravity on each component, although usually very small, is taken in account.

The notation used for the reservoir characteristics (porosity, absolute permeability, thickness) are the same as for the immiscible case ; the remaining notations are given in the appendix.

Equations of Miscible Flows in Porous Media

$$\frac{\partial}{\partial t}\{\sigma(x)\,\phi(x,P)\,B(C,P)\,\frac{2C-1}{2}\} + \nabla\{\vec{J}+(2C-1)\vec{V}-\sigma(x)DB(C,P)A(C,P)g\nabla Z(x)\}=0 \quad (25)$$

$$\vec{J} = -\sigma(x)D\,B(C,P)\,\nabla C \quad (26)$$

$$\frac{\partial}{\partial t}\{\sigma(x)\,\phi(x,P)\,\frac{B(C,P)}{2}\} + \mathrm{div}\,\{\vec{V}\} = 0 \quad (27)$$

$$\left.\begin{aligned} &\vec{V} = -\sigma(x)\frac{K(x)\,B(C,P)}{2\mu(C,P)} \times \\ &\{\nabla P - \rho(C,P)g\nabla Z(x)\} = 0 \end{aligned}\right\} \quad (28)$$

where :

C = mass concentration of one of the components

P = pressure in the fluid.

Comparing the miscible equations (25) to (28) to the immiscible equations (17) to (20), one sees that both set of equations have exactly the same structure. The natural problem which arises then is to check if those two sets of equations are equivalent, i.e. if each of them can become identifical to the other by an adequate choice of the physical coefficients.

3. COMPARISON OF MISCIBLE AND IMMISCIBLE EQUATIONS

We first list in Table 1 the physical unknown and coefficients appearing in both systems.

One sees in this table that immiscible flows depend on many more parameters than miscible ones. It seems hence hopeless to try to simulate an immiscible displacement with a miscible model.

	MISCIBLE	IMMISCIBLE
UNKNOWNS	P = pressure C = mass concentration	P = global pressure S = reduced saturation
RESERVOIR CHARACTERISTICS	$\sigma(x)$ = "thickness" $\emptyset(x,P)$ = porosity $K(x)$ = permeability	
FLUIDS CHARACTERISTICS	$\rho(C,P)$ = mass of unit volume of mixture	$\rho_1(S,P)$, $\rho_2(S,P)$ } mass of unit volume of each fluid (practically ρ_j does not depend on S)
	$\mu(C,P)$ = viscosity of mixture	$\mu_1(S,P)$, $\mu_2(S,P)$ } viscosity of each fluid (practically μ_j does not depend on S)
	D = diffusion coefficient (small and taken as constant)	$kr_1(S)$, $kr_2(S)$ } relative permeability curves
	$A(C,P)$ = function describing the differential action of gravity on the components (very small and often neglected)	$P_{CM}(x)$, $p_c(S)$ } capillary pressure $\bar{S}_m(x)$, $\bar{S}_M(x)$ = (non-reduced) residual saturations

Table 1 : The unknowns and parameters for miscible and immiscible flows.

Let us now see if the converse operation is possible, namely simulating a miscible flow with an immiscible model :

- suppose we are given a miscible flow, i.e. we know all the characteristics (reservoir, fluid) of the left column of table 1.

- we try now to choose the characteristics of the right column of the same table in such a way that the resulting immiscible equations (17) to (20) are closest to the given miscible equations (25) to (28) :

  i) Reservoir Characteristics : we keep for the immiscible flow the same σ, $\emptyset$, K as in the given miscible model.

  ii) Fluid Characteristics : we identify first the concentration C and the saturation S (i.e. C = S). We choose then for the given miscible flow reference densities and viscosities ρ_o and μ_o, and define the density and viscosity factors B(C,P) and M(C,P) of the given miscible flow by :

$$\left.\begin{aligned} \rho(C,P) &= B(C,P)\rho_o \\ \mu(C,P) &= M(C,P)\mu_o \end{aligned}\right\} \tag{29}$$

We define the, in the right column of table 1, the densities $\rho_j(S,P)=B_j(S,P)\rho_{jo}$ and the viscosities $\mu_j(S,P)=M_j(S,P)\mu_{jo}$, j = 1,2 of the fictitious immisible flow by :

$$B_j(S,P) = B(C,P), \quad M_j(S,P) = M(C,P) \quad j = 1,2 \tag{30}$$

$$\left.\begin{aligned} &\frac{1}{2}(\rho_{10} + \rho_{20}) = \rho_o \\ &K(x)S(1-S)\frac{B(S,P)}{\mu_o M(S,P)}(\rho_{10}-\rho_{20}) \simeq DA(C,P) \end{aligned}\right\} \tag{31}$$

$$\mu_{jo} = \mu_o \quad j = 1,2 \tag{32}$$

We define then the relative permeabilities and the capillary pressure of the fictitious immiscible flow by :

$$kr_1(S) = S, \; kr_2(S) = 1 - S \tag{33}$$

$$\left.\begin{array}{l} P_{CM}(x) \text{ is chosen independant of } x \text{ (in order that } \vec{q}_1 \equiv 0) \text{ and such as :} \\ K(x)\, P_{CM} \dfrac{M(S,P_o)}{M(S,P)} \simeq D \end{array}\right\} \tag{34}$$

$$\left.\begin{array}{l} p_c(S) \text{ is chosen such as :} \\ p_c(\frac{1}{2}) = 0, \; \dfrac{dp_c}{dS}(S) = \dfrac{M(S,P_o)}{S(1-S)} \end{array}\right\} \tag{35}$$

We choose finally for S_m and S_M :

$$S_m(x) \equiv 0, \; S_M(x) \equiv 1 \quad \forall x \tag{36}$$

With the above choices for the parameters of the immiscible flow the immiscible equations (17) to (20) reduce to :

$$\frac{\partial}{\partial t}\{\sigma \, \emptyset \, B \, \frac{2S-1}{2}\} + \tag{37}$$

$$+\nabla\{\vec{r}+(2S-1)\vec{q}_o-\sigma \underbrace{BKS(1-S)\frac{B}{\mu_o M}(\rho_{10}-\rho_{20})}_{\simeq DA} g\nabla Z\}=0$$

$$\vec{r} = - \sigma \underbrace{KP_{CM}\frac{M(S,P_o)}{M(S,P)}}_{\simeq D} B \cdot \nabla S \tag{38}$$

$$\frac{\partial}{\partial t}\{\sigma \, \emptyset \, \frac{B}{2}\} + \nabla\{\vec{q}_o\} = 0 \tag{39}$$

$$\left.\begin{array}{l} \vec{q}_o = - \sigma \dfrac{KB}{2\mu} \times \\ \{\nabla P - B \underbrace{(\rho_{10}S+\rho_{20}(1-S))}_{\simeq \rho_0} g\nabla Z\} = 0 \end{array}\right\} \tag{40}$$

If we compare now (37)-(40) to the miscible equations (25)-(28) we see that :

i) in the general case, the immiscible flow defined as above represents the given immiscible flow only in an approximate way ; the main approximation being located with diffusion term.

ii) if we neglect, as usual, the differential action of gravity on each of the components (i.e. if $A(C,D) \equiv 0$) then

$$\rho_{10} = \rho_{20} = \rho_0 \tag{41}$$

and the immiscible and miscible equations coincide up to the diffusion term (compare (26) and (38)), the miscible equations giving always a diffusion function of the absolute permeability of the porous medium.

iii) if we moreover suppose that :

$$\left.\begin{array}{l} \text{the absolute permeability } K \text{ is constant the miscible viscosity} \\ \mu \text{ is independant of pressure} \end{array}\right\} \tag{42}$$

then (34) reduces to :

$$K \, P_{CM} = D \tag{43}$$

so that the two system of equations fully coincide.

4. CONCLUSION

We have shown that the immiscible equations for two phase compressible flows in porous medium can be reduced, through the introduction of a fictitious "global" pressure, to a couple of two parabolic equations very similar in structure to the equations governing the miscible displacements.

As a first application, a carefull comparison of the coefficients of the two systems of equations was done, which showed that simulation of a miscible displacement with an immiscible simulator was possible, with eventually small approximations, under rather general circumstances.

Applications to the approximation of the compressible immiscible equations by finite element techniques are under way, (the incompressible case being already treated in [3], [4]).

Notations for the Immiscible Case

$\Omega \subset \mathbb{R}^n$ = porous body (n = 1, 2 or 3)
$x \in \Omega$ = geometric point
$t \geq 0$ = time
$\sigma(x)$ = "thickness" of Ω at point x =

$$= \begin{cases} \text{cross sectional area} & \text{for } n = 1 \\ \text{thickness} & \text{for } n = 2 \\ 1 & \text{for } n = 3 \end{cases}$$

$Z(x)$ = depth
g = gravity accelleration
$K(x,.)$=rock absolute permeability tensor
$\emptyset(x,p)$=rock porosity
} at point x of Ω

P_1 = pressure inside the wetting phase

P_2 = pressure inside the non-wetting phase

P = global pressure, defined in (15), $\mathrm{Min}(P_1,P_2) \leq P \leq \mathrm{Max}(P_1,P_2)$

P_o = reference pressure at which the volume balances are done

$\bar{S}(x,t)$ = wetting phase saturation at point x and time t

$\bar{S}_m(x)$ = wetting phase residual saturation at point $x \in \Omega$

$\bar{S}_M(x)$ = 1 - non-wetting phase residual saturation at point $x \in \Omega$

$S(x,t) = \dfrac{\bar{S}(x,t)-\bar{S}_m(x)}{\bar{S}_M(x)-\bar{S}_m(x)}$ = reduced saturation of the wetting phase

$S_j(x,t)$ = reduced saturation of the jth fluid $S_1 = S$, $S_2 = 1-S$

$\phi(x,P) = [\bar{S}_M(x) - \bar{S}_m(x)]\ \sigma(x)\ \emptyset(x,p)$

$\psi(x) = \sigma(x)\ K(x)$

$\tilde{\psi}(x,S,P) = \psi(x)\ P_{CM}(x)\ \dfrac{B_1(S,P)B_2(S,P)}{M_1(S,P)M_2(S,P)} \times$

$\dfrac{k_{10}(S)+k_{20}(S)}{k_1(S,P)+k_2(S,P)}$ = non degenerating part of the capillary diffusion coefficient

$\mu_j(S,P)$ = viscosity
$\rho_j(S,P)$ = mass of unit volume
$\mu_{jo}=\mu_j(\frac{1}{2},P_o)$= reference viscosity
$\rho_{jo}=\rho_j(\frac{1}{2},P_o)$=reference for mass of unit volume
$B_j(S,P)=\rho_j(S,P)/\rho_{jo}$= density factor
$M_j(S,P)=\mu_j(S,P)/\mu_{jo}$= viscosity factor
$kr_j(S)$=relative permeability, taken independant of x when expressed in term of the reduced saturation S
$k_j(S,P)=\dfrac{k_{rj}(S)}{\mu_j(S,P)} \times B_j(S,P)$=mobility curve at pressure P
$k_{jo}(S)=\dfrac{k_{rj}(S)}{\mu_{jo}}$=reference mobility curve
$k_j(S,P)=k_{jo}(S) \times \dfrac{B_j(S,P)}{M_j(S,P)}$
} of fluid j, $j=1,2$

$P_{CM}(x)$ = maximum capillary pressure at point x ; $P_{CM}(x) \geq 0$

$p_c(S)$ = reduced capillary pressure curve :

- p_c is independant of x when expressed in term of the reduced saturation S,
- $S \to p_c(S)$ is monotonous increasing, with $|p_c(S)| \leq 1$,
- p_c vanishes for $S = S_c$

$S_c \in [0,1]$ = reduced saturation ofr which the capillary pressure vanishes (usually $S_c = 1$ as S is the wetting phase saturation)

$\vec{\phi}_j$ = volumetric flow vector of jth fluid, evaluated at reference pressure P_o, $j = 1,2$

$\vec{q} = \vec{q}_o = \dfrac{\vec{\phi}_1+\vec{\phi}_2}{2}$= mean volumetric flow vector

$\vec{r} = -\tilde{\psi}(x,S,P)\nabla\alpha(S)$ = part of $\dfrac{\vec{\phi}_1+\vec{\phi}_2}{2}$ due to capillary diffusion

$\vec{q}_1 = -\psi(x)\nabla P_{CM}(x)$ = vector field governing the transport due to heterogeneous capillary pressure

$\vec{q}_2 = -\psi\nabla P_G(x)$ = vector field governing the transport due to gravity

$P_G(x) = -\frac{1}{2}(\rho_{10}+\rho_{20})\ gZ(x)$ = pseudopotential of gravity

$a(S) = \dfrac{k_{10}(S)\ k_{20}(S)}{k_{10}(S)+k_{20}(S)} \times \dfrac{dp_c(S)}{dS}$ = degenerating part of the capillary diffusion coefficient

$\alpha(S) = \int_o^S a(s)ds$

$b_0(S,P)=b(S,P) = \dfrac{k_1(S,P) - k_2(S,P)}{k_1(S,P) + k_2(S,P)} =$

= non linearity in the transport term caused by global flow

$b_1(S,P) = \dfrac{k_1(S,P)\ k_2(S,P)}{k_1(S,P) + k_2(S,P)}\ p_c(S) =$

= non linearity in the transport term caused by heterogeneous capillary pressure

$$b_2(S,P) = \frac{k_1(S,P)k_2(S,P)}{k_1(S,P)+k_2(S,P)} \times \frac{\rho_1(S,P)-\rho_2(S,P)}{(\rho_{10}+\rho_{20})/2} =$$

= non linearity in the transport term caused by gravity

Typical shapes for a, α, b_0, b_1, b_2 are

(for a given P) :

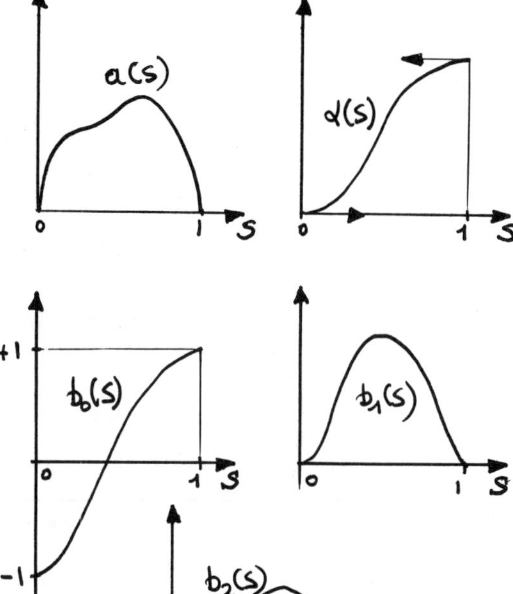

$$\gamma(S,P) = \frac{1}{2} \int_{S_c}^{S} b_0(s,p) \frac{dp_c}{dS}(s)ds =$$

= function used to define the global pressure P

$$\gamma_1(x,S,P) = \frac{1}{2(1-P_{CM}(x)\frac{\partial\gamma}{\partial P}(S,P))} \int_{S_c}^{S} \frac{\partial b_0}{\partial S}(s,P)p_c(s)ds$$

$$\gamma_2(x,S,P) = \frac{k_1(S,P)\rho_1(S,P)+k_2(S,P)\rho_2(S,P)}{k(x,S,P)} \times \frac{2}{\rho_{10}+\rho_{20}}$$

$$k(x,S,P) = \frac{1}{2}[k_1(S,P)+k_2(S,P)][1-P_{CM}(x)\frac{\partial\gamma}{\partial P}(S,P)]$$

Additionnal notations for the miscible case

$\rho(c,p)$ = mass of unit volume of mixture at concentration C and pressure P

$\rho_o = \rho(\frac{1}{2}, P_o)$ = mass of unit volume under reference conditions

$B(C,P) = \frac{\rho(C,P)}{\rho_o}$ = mass factor

$\mu(C,P)$ = viscosity of mixture at concentration C and pressure P

D = diffusion coefficient (taken independant of space)

A(C,P) = function describing the differential action of gravity over components (usually extremely small).

Indexes

1 = index of the wetting phase

2 = index of the non wetting phase

0 = index of vector fields and non linearities associated to global flow

1 = index of vector fields and non linearities associated to heterogeneity of capillary pressure

2 = index of vector fields and non linearities associated to gravity.

REFERENCES

[1] CHAVENT G., A new formulation of incompressible flow in porous media, in Lect. Notes in Mathematics, n° 503, Springer 1976.

[2] MARLE C., Notes sur la comparaison des équations des déplacements de fluids miscibles ou immiscibles en milieux poreux. Rapport IFP n° 28552, 1980, IFP, BP 311, 92506 Rueil Malmaison Cédex.

[3] CHAVENT G., SALZANO G., A finite element for the 1-D waterflooding problem with gravity. Submitted to Journal of Comp. Phys.

[4] JAFFRE J., Formulation mixte d'écoulements diphasiques incompressibles dans un milieux poreux. Rapport INRIA n°37, 1980, INRIA, Domaine de Voluceau, BP 105, 78153 Le Chesnay Cédex.

Proceedings of Euromech 143 / Delft / 2-4 September 1981

Time dependence of an "Equivalent dispersion coefficient" for transport in porous media

A.DIEULIN, G.MATHERON, G.DE MARSILY & B.BEAUDOIN
Ecole des Mines de Paris, Fontainebleau, France

ABSTRACT

A general formulation of the transport equation of solutes in porous media is developed theoretically for a slug injection, with the assumption that the distribution of the velocities in the medium is gaussian, and that molecular diffusion is negligible. The equivalent of the dispersion coefficient in the equation is then shown to be a function of time. The transport equation is also obtained from a mass balance equation.

1. INTRODUCTION

The transport of solute in non-uniform porous media is currently described by the dispersion equation, which writes:

$$\mathrm{div}\ (\overline{\overline{D}}\ \mathrm{grad}\ C - UC) = \frac{\partial C}{\partial t} \qquad (1)$$

$\overline{\overline{D}}$: dispersion tensor, with principal directions parallel and orthogonal to the velocity,
C: concentration
U: microscopic velocity.

The dispersion coefficients of the tensor $\overline{\overline{D}}$ are assumed to be functions of the velocity U only. However, a large numer of authors (e.g. ref. 1, 2, 3) have recently questioned the validity of this model, both on the basis of theoretical developments and experimental data. It is well known, for instance, that the interpretation of a tracer test gives dispersion coefficients which increase as the distance between the input well and the observation well is increased: such a deviation from the classical model is refered to as the "scale effect".

In this paper, we will first develop a different equation to represent the transport of solute in porous media, under certain simplifying assumptions, using a stochastic approach. We will then show that it corresponds to a mass balance equation.

2. A SIMPLIFIED TRANSPORT EQUATION IN A STOCHASTIC FRAMEWORK

We will develop our equation in the ordinary space $\mathbb{R}^n$ (n = 1, 2 or 3) with the three following assumptions:
(a). The transport is governed by the velocity variations of the fluid in the porous medium, molecular diffusion is negligible.
(b). The microscopic velocity field, U, which is unknown, can be regarded as a stationary random process, i.e. U is a vectorial stationary random function (SRF). We will further assume that the probability distribution function (PDF) of U is gaussian with n variables, and that U is conservative (div U = 0). This means that the flow is in steady state, with a constant porosity.
(c). A slug tracer is injected at time t = 0 at the origin $X_o = 0$ of the system. In our notations, U and X are vectors; and U^i or X^i represent one of their components:

$$X = \begin{Bmatrix} X^1 \\ \cdot \\ \cdot \\ \cdot \\ \cdot \\ \cdot \\ X^n \end{Bmatrix}$$

The lower index of a vector will denote the time: X_t.
The transport can be described by giving, as a function of time, the position X_t of a particle injected at t = 0 at the origin. As discussed in ref. 3, this method of

representing transport is equivalent to the determination, in the case of a slug injection, of the concentration C as a function of time and space, as in (1) : C(X,t) is equal to the probability density of the particle at location X and time t.
With our assumption (a) the transport equation writes:

$$\frac{dX_t}{dt} = U(X_t) \tag{2}$$

Using the assumptions (b) for the velocity U, we will develop the differential equation which is verified by the probability density $\rho(X,t)$ of the particle (or by the concentration C in the case of a slug injection).

Let:

$$V(t) = U(X_t)$$

Matheron (ref. 4) has shown that if U is a SRF with a given PDF, then the condition div U = 0 is sufficient to insure that V is also a SRF with the same PDF as U in the time domain.
We can write (2) as:

$$X_t = \int_0^t V(\tau)d\tau \tag{3}$$

We then have:

$$E(X_t) = E\int_0^t V(\tau)d\tau =$$

$$\int_0^t E\left[V(\tau)\right]d\tau = t\,E\left[V\right] = t\,E\left[U\right] = t\,\bar{U} \tag{4}$$

where $\bar{U} = E\left[U\right]$

$$K_t = \mathrm{var}\left[X_t\right] = E\left[(X_t - E(X_t))^T(X_t - E(X_t))\right]$$

$$= E\,(X_t^T X_t) - E\,(X_t)\,E\,({}^T X_t)$$

Where ${}^T X = (X^1,\ldots,X^n)$ is the transpose of X. Note that this variance is a n × n matrix.

$$K_t = E\left[\int_0^t V(\tau)d\tau \int_0^t {}^T V(\tau)d\tau\right] - t^2\bar{U}^T\bar{U}$$

$$= \int_0^t\int_0^t \{E\left[V(\tau)\,{}^T V(\tau')\right] - \bar{U}^T\bar{U}\}\,d\tau d\tau'$$

$$= \int_0^t\int_0^t E\{\left[V(\tau)-\bar{U}\right]^T\left[V(\tau')-\bar{U}\right]\}\,d\tau d\tau' ;$$

$$K_t = \int_0^t\int_0^t C(\tau-\tau')d\tau d\tau' = 2\int_0^t (t-\tau)C(\tau)d\tau \tag{5}$$

where C(t) is the n × n covariance matrix of the components of the velocity V, taken with a time lag t.

As we assume the SRF, U, to be a gaussian function with n variables, V is also gaussian and the integral X_t of V in (3) is also gaussian. Then, the PDF of the particle at location X and time t is given by:

$$\rho(X,t) = \frac{1}{(2\Pi)^{n/2}\sqrt{d}}\,e^{-P/2} \tag{6}$$

where d is the determinant of the variance matrix K_t given in (5),

$$P = {}^T(X - t\bar{U})\,K_t^{-1}\,(X - t\bar{U})$$

P is a positive quadratic form, the matrix of which is the inverse of the variance matrix K_t.
We will establish the partial differential equation which $\rho(X,t)$ verifies using the Fourier analysis of (6). The Fourier transform over space of the PDF $\rho(X,t)$ is given by:

$$\mathcal{F}_S\left[\rho(X,t)\right] = \int_{\mathbb{R}^n} \rho(X,t)\,e^{iSX}dX = \int_{-\infty}^{+\infty}\cdots\int_{-\infty}^{+\infty} \rho$$

$$(X^1,\ldots,X^n,t)\,e^{i(S_1X^1+\ldots S_nX^n)}\,dX^1,\ldots,dX^n$$

where $S = (S_1,\ldots,S_n)$ is the Fourier variable vector.
One can show that:

$$\mathcal{F}_S\left[\rho(X,t)\right] = e^{itS\bar{U} - 1/2\,SK_t{}^TS} \tag{7}$$

The expected value of X_t has been taken here as in (4), and its variance matrix K_t is given in (5).
We can use the following properties of the Fourier transform. In the case of a regular function f(X,t), we have:

$$\frac{\partial}{\partial t}\,\mathcal{F}_S\left[f(X,t)\right] = \mathcal{F}_S\left[\frac{\partial}{\partial t}\,f(X,t)\right] \tag{8}$$

$$\mathcal{F}_S\left(\frac{\partial f}{\partial x^j}\right) = -iS_j\mathcal{F}_S(f) \tag{9}$$

Here, we can write, according to (7) and (8):

$$\mathcal{F}_S\left(\frac{\partial\rho}{\partial t}\right) = \left(iS\bar{U} - \frac{1}{2}\,S\,\frac{\partial K_t}{\partial t}\,{}^TS\right)\mathcal{F}_S(\rho) \tag{10}$$

Applying (9) twice to (7):

$$\mathcal{F}_S\left(\frac{\partial\rho}{\partial x^j}\right) = -iS_f\mathcal{F}(\rho) \tag{11}$$

$$\mathcal{F}_S\left(\frac{\partial^2\rho}{\partial x^j\partial x^k}\right) = -S_jS_k\mathcal{F}(\rho) \tag{12}$$

Combining equations (10) to (12), we obtain:

$$F_S(\frac{\partial\rho}{\partial t}) = -F_S\ (\text{grad}\ \rho\bar{U}) + \frac{1}{2} F_S \left[\text{div}\ (\text{grad}\ \rho\ \frac{\partial K_t}{\partial t})\right]$$

where grad ρ is the vector $(\frac{\partial\rho}{\partial x^1} \ldots \frac{\partial\rho}{\partial x^n})$

and div is the operator $\sum_j \frac{\partial}{\partial x^j}$.

By applying F_S^{-1}, we obtain:

$$\frac{\partial\rho}{\partial t} = -\ \text{grad}\ \rho\bar{U} + \frac{1}{2}\ \text{div}\ (\text{grad}\ \rho\ \frac{\partial K_t}{\partial t}) \quad (13)$$

which can also be written as:

$$\frac{\partial\rho}{\partial t} = -\sum \bar{U}^j \frac{\partial\rho}{\partial x^j} + \frac{1}{2}\sum_j\sum_k \frac{\partial K^{jk}}{\partial t} \frac{\partial^2\rho}{\partial x^j \partial x^k} \quad (14)$$

where K^{jk} are the coefficients of the covariance matrix given in (5):

$$K^{jk} = \int_0^t\int_0^t E\{[V^j(\tau)-\bar{U}^j][V^k(\tau')-\bar{U}^k]\}d\tau d\tau' = 2\int_0^t (t-\tau)\ C^{jk}(\tau)\ d\tau$$

C^{jk} is a function of time, not of space, and:

$$\frac{1}{2}\frac{\partial K^{jk}}{\partial t} = \int_0^t C^{jk}(\tau)d\tau \quad (15)$$

$C^{jk}(\tau)$ is the covariance function of the components j and k of the velocity V, with a time lag τ.

3. MASS BALANCE EQUATION

Matheron has shown (ref. 5) that even in the case of a non stationary RF, V, a mass balance equation may be written:

$$\frac{\partial\rho}{\partial t} + \sum_i \frac{\partial}{\partial x^i}(\rho W^i) = 0 \quad (16)$$

where $\rho(X,t)$ is the PDF of X_t,

$W(t) = E\ (V(t)/X_t = x)$

the expected value of V(t) when the particle is known to be at location x. In the case of a gaussian SRF, V(t), the expected value W(t) is a linear function of the components x^i:

$$W^k(t) = \bar{U}^k + \sum_j \lambda_j^k y^j \quad (17)$$

where $y^j = x^j - \bar{U}^j_k t_k$

The coefficients λ_j^k are the solution of a linear system which, using the classical properties of gaussian variable, writes:

$$\sum_j K^{ij} \cdot \lambda_j^k = \Omega^{ik} \quad (18)$$

where:

$$\Omega^{ik} = E[(X^i - \bar{U}^i t)\ (V^k - \bar{U}^k)]$$

Using the inverse matrix G_t of the variance matrix K_t, equation (18) may be written:

$$\lambda_j^k = g_{ji}\ \Omega^{ik} \quad (19)$$

where $G_t = \{g_{ji}\}$

In equation (16), we will first calculate the term: $\rho_t\ W^i(t)$. Using equation (17) and (18), it writes:

$$\rho\ W^k = \rho\ \bar{U}^k + \sum_{ji}\sum \rho\ \Omega^{ik} g_{ji} y^j$$

The PDF, $\rho(X,t)$, is a gaussian distribution, and its deritatives may be calculated using equation (6):

$$\frac{\partial\rho}{\partial x^i} = \rho \times -\frac{1}{2}\frac{\partial P}{\partial x^i} = -\rho\ g_{ij}\ y^j$$

since $P = \sum_i\sum_j g_{ij}\ y^i\ y^j$ and

$$\frac{\partial P}{\partial x^i} = \sum_j 2\ g_{ij}\ y^j$$

Finally, equation (16) writes:

$$\frac{\partial\rho}{\partial t} + \sum \frac{\partial}{\partial x^k}(\rho\bar{U}^k - \sum_i \Omega^{ik}\frac{\partial\rho}{\partial x^i}) = 0 \quad \text{or}$$

$$\frac{\partial\rho}{\partial t} = -\sum_k \bar{U}^k \frac{\partial\rho}{\partial x^k} + \sum_k\sum_i \Omega^{ik}\frac{\partial^2\rho}{\partial x^k\partial x^i} \quad (20)$$

The PDF ρ_t being a gaussian distribution, we have:

$$\frac{\partial^2\rho}{\partial x^k\partial x^i} = \frac{\partial^2\rho}{\partial x^i\partial x^k}$$

So, the second term of equation (20) becomes:

$$\sum_i\sum_k \Omega^{ik}\frac{\partial^2\rho}{\partial x^k\partial x^i} = \frac{1}{2}\sum_i\sum_k\left[(\Omega^{ik} + \Omega^{ik})\ \frac{\partial^2\rho}{\partial x^k\partial x^i}\right]$$

Since we have

UNIVERSITY COLLEGE LIBRARY CARDIFF

$$\frac{\partial K^{ik}}{\partial t} = \Omega^{ik} + \Omega^{ki}$$

The final expression of equation (20) is:

$$\boxed{\frac{\partial \rho}{\partial t} = -\sum_k \bar{U}^k \frac{\partial \rho}{\partial x^k} + \frac{1}{2} \sum_i \sum_k \frac{\partial K^{ik}}{\partial t} \frac{\partial^2 \rho}{\partial x^i \partial x^k}}$$

which is identical to equation (14).

4. DISCUSSION

In the transport equation (14) developed here for the transport of solute in porous media, the equivalents of the dispersion coefficients are a function of time. This equation is only valid for a slug injection at t=o and X=o, and not for any other initial conditions.
It is reasonable to assume that, in general:

$$\int_o^t C^{jk}(\tau)d\tau \to A^{jk} \text{ as } t \to \infty$$

Transport will therefore be, in general, asymptotically diffusive, with a constant dispersion tensor, but only for a transient time sufficient for this integral to converge. This will normally be a function of the "correlation length" of the medium, i.e. the distance over which the velocities are correlated.
For early times, equations (13), (14) and (15) show that the "dispersion tensor" $\frac{\partial K_t}{\partial t}$ will vary with time, both in magnitude and in principal directions, as there is a priori no reason why these directions should remain constant when the coefficients vary. A constant dispersion tensor could only occur if the coveriance functions $C^{jk}(\tau)$ of the velocity components were Dirac functions, i.e. a medium displaying no correlations of the velocity components even for small time lags.
Finally, under the same simplifying assumptions, it has been shown that a mass balance equation leads to the same transport equation (14). This remark indicates the physical significance of the new transport equation.
A number of limiting assumptions have been necessary to develop the theory (e.g. gaussian distribution of velocity, no molecular diffusion, etc....). However, a few experimental results seem to support our conclusions, namely that for a slug injection of tracer, the equivalent of the dispersion coefficient, what we called the "temporal fuction of dispersion", is indeed a function of time, or of the mean travelled distance of the tracer (ref. 6 and 7).

5. CONCLUSION

Under certain simplifying assumptions, we have obtained a new transport equation in the case of a slug injection. This equation may be derived from a mass balance equation. This theoretical tool has been used in practice for the description of tracer movements: equivalent dispersion coefficients seem indeed to be a function of time.
We intend to continue the research along this line by considering the following questions:
- validity of time convolution,
- introduction of molecular diffusion,
- determination of the covariance functions of each component of the velocity, and of the principal directions of the covariance matrix.

Work should continue both theoretically and experimentally.

6. BIBLIOGRAPHY

1. Mercado, A. 1967, The spreading pattern of injected water in a permeability stratified aquifer, Int. Ass. of Sc. Hydrology, Proc. Symp. of Haīfa, Publ. 72, p. 23-26.
2. Gelhar, L.W. Gutjahr, A.L., Naff, R.L. 1979, Stochastic analysis of macrodispersion in a stratified aquifer, WRR, 15-6 p. 1387-1397.
3. Matheron, G., Marsily, G. de, 1980, Is transport in porous media always diffusive? A counter example, WRR, 16-5, p. 901-917.
4. Matheron, G. 1979, Quelques exemples simples d'émergence d'un demi-groupe de dispersion, Note interne Ecole des Mines, Centre de Morphologie Mathématique, Fontainebleau.
5. Matheron, G. 1981, Unpublished results.
6. Dieulin, A. 1980, Propagation de pollution dans un aquifère alluvial. L'effet de parcours. Thèse, Ecole des Mines de Paris-Université Paris VI.
7. Dieulin, A., Marsily, G. de, Beaudoin, B. 1981, Sur l'existence d'un effet de parcours dans le transfert d'éléments en solution en milieu poreux, C.R. Acad. Sc., 292, série II, 121.

Proceedings of Euromech 143 / Delft / 2-4 September 1981

An alternative formulation for hydrodynamic dispersion in porous media

ALLEN M. SHAPIRO
Royal Institute of Technology, Stockholm, Sweden

1 INTRODUCTION

An investigation of hydrodynamic dispersion in porous media from thermodynamic interpretations is the purpose of this report. The Coleman and Noll (1) method of extracting general restrictions and equilibrium constraints on constitutive functions from the Second Law of Thermodynamics is employed. The non-equilibrium parts of the phenomenological relationships are developed using Taylor series expansions based on the outcome of the Coleman and Noll (1) method. This procedure has been extensively applied in the development of constitutive functions for multi-component continua, i.e. mixtures of solids and fluid (2,3), and only recently has it been directed towards subsurface hydrologic phenomena (4,5). However, contaminant transport has yet to be considered in these cases. The concept of diffusion has only been briefly touched upon in topics which concern the displacement of immiscible phases (6,7). Although these investigations provide useful insight into diffusion phenomena, their applicability to dispersion in porous media is not entirely consistent with the conceptualization of a fluid phase having a known velocity and considering the velocities of the individual contaminants as unmeasurable.

2 EQUATIONS OF CONSERVATION FOR CONTAMINANT TRANSPORT IN POROUS MEDIA

Using statements of conservation for the thermodynamic properties mass, momenta, energy and entropy, the behaviour of the medium to external influences can be identified. The scope of this analysis, however, shall be limited by several restrictions. It will be assumed that one contaminant exists in the fluid phase and a chemically inert rock matrix prohibits its existence in the solid phase. The fluid and rock are also assumed to be in thermal equilibrium and surface tension characteristics are neglected. In addition, the rock phase is considered to be microscopically incompressible. The necessity of stating the latter assumption arises from the thermodynamics description of porous media, a discussion of which is found in Hassanizadeh and Gray (5) or Shapiro and Pinder (8).

The balance laws for a continuum idealization of a porous medium based on these assumptions are given by the following:

Conservation of Mass-phase

$$\frac{\partial(\rho^f\phi^f)}{\partial t} + (\rho^f\phi^f v_k^f)_{,k} = 0 \qquad (2.1)$$

$$\frac{\partial\phi^r}{\partial t} + (\phi^r v_k^r)_{,k} = 0 \qquad (2.2)$$

Conservation of Mass- constituent

$$\frac{\partial(\rho^f\phi^f\sigma)}{\partial t} + (\rho^f\phi^f\sigma v_k^f)_{,k} - J_{k,k} = 0 \qquad (2.3)$$

Conservation of Linear Momentum

$$\rho^\alpha\phi^\alpha\frac{D^\alpha}{Dt}v_k^\alpha - \tau_{lk,l}^\alpha - \rho^\alpha\phi^\alpha g_k^\alpha = \rho^\alpha\phi^\alpha\hat{F}_k^\alpha \qquad \alpha=f,r \qquad (2.4)$$

Conservation of Angular Momentum

$$\tau_{lk}^\alpha = \tau_{kl}^\alpha \qquad \alpha=f,r \qquad (2.5)$$

Conservation of Energy

$$\rho\frac{De}{Dt} - \sum_\alpha\{\tau_{kl}^\alpha d_{lk}^\alpha + q_{k,k}^\alpha - (\rho^\alpha\phi^\alpha e^\alpha v_k^\alpha)_{,k}\} - \rho h = \rho^f\phi^f\hat{F}_k^f v_k^{fr} \qquad (2.6)$$

Conservation of Entropy

$$\rho^\alpha\phi^\alpha\frac{D^\alpha\eta^\alpha}{Dt} - S^\alpha_{k,k} - \rho^\alpha\phi^\alpha b^\alpha - \rho^\alpha\phi^\alpha\hat{N}^\alpha =$$
$$= \rho^\alpha\phi^\alpha\Gamma^\alpha \qquad \alpha=f,r \tag{2.7}$$

A definition of the nomenclature is provided at the conclusion of the text.

The previous relationships are subject to restrictions which insure a global conservation of the thermodynamic properties noted above. These are stated by the following:

$$\sum_\alpha \phi^\alpha = 1 \tag{2.8}$$

$$\Sigma_\alpha\ \rho^\alpha\phi^\alpha\hat{F}^\alpha = 0 \tag{2.9}$$

$$\sum_\alpha \rho^\alpha\phi^\alpha\hat{N}^\alpha = 0 \tag{2.10}$$

Lacking in the equations given above, however, are constitutive relationships which will explicitly define the behaviour of the porous medium in terms of a set of measurable quantities, and further provide an equivalent number of equations and unknowns. The constitutive relationships are restricted from violating the balance laws or additional thermodynamic principles. The Second Law of Thermodynamics acts as a further constraint on the physical system and is expressed as a restriction on the net entropy production, i.e. the

Second Law of Thermodynamics

$$\rho\Gamma = \sum_\alpha \rho^\alpha\phi^\alpha\Gamma^\alpha \geq 0 \tag{2.11}$$

Introducing equation (2.7) in the above expression and applying the following definitions,

$$S^\alpha_k = \frac{q^\alpha_k}{\theta} \tag{2.12}$$

$$b^\alpha = \frac{h^\alpha}{\theta} \tag{2.13}$$

$$\psi^\alpha = e^\alpha - \theta\eta^\alpha \tag{2.14}$$

the ensuing form of the Second Law of Thermodynamics results:

$$\theta\Gamma = \sum_\alpha\{-\rho^\alpha\phi^\alpha(\frac{D^\alpha\psi^\alpha}{Dt} + \eta^\alpha\frac{D\theta}{Dt}) + \tau^\alpha_{lk}d^\alpha_{kl}$$
$$+ (\frac{q^\alpha_k}{\theta} - \rho^\alpha\phi^\alpha\eta^\alpha v^\alpha_k)\theta_{,k}\} - \rho^\alpha\phi^\alpha\hat{F}^f_k v^{fr}_k \geq 0 \tag{2.15}$$

Equations (2.12) and (2.13) are, in fact the specification of constitutive relationships which have often been used in the description of equilibrium thermodynamic processes for discrete continua. They are employed here because they have been shown to yield satisfactory results in the description of multi-component continua (2) including porous media (5).

3 CONSTITUTIVE FUNCTIONS

The quantities in the balance laws which require definition are the following:

$$J_k,\ \tau^\alpha_{kl},\ \hat{F}^f_k,\ q^\alpha_k,\ \psi^\alpha,\ \eta^\alpha \quad k,l=1,2,3\ ;\ \alpha=f,r \tag{3.1}$$

Other variables appearing in equations (2.1)-(2.7) can either be defined in terms of those listed above or are considered as variables upon which the quantities in (3.1) will depend, i.e.

$$\rho^\alpha\ ,\ \phi^\alpha,\ v^\alpha_k,\ \theta,\ \sigma \quad k=1,2,3;\ \alpha=f,r \tag{3.2}$$

Initially, the explicit dependence of all constitutive functions is defined as follows:

$$L = L(\rho^f,\ \sigma,\ \sigma_{,k},\ E^r_{KL},\ v^{fr}_k,\ d^f_{kl},\ \phi^f,\ \phi^f_{,k},$$
$$\theta,\ \theta_{,k},\ X^r_K) \quad k,l,K,L=1,2,3 \tag{3.3}$$

where L is any of the functions listed in (3.1). Including the material coordinates of the rock phase, X^r_K, in the list of independent variables will account for the inhomogeneities in the porous medium. The method of arriving at the dependence noted above is discussed in general by Eringen (9), and by Hassanizadeh (10) and Shapiro (11) for specific subsurface hydrologic phenomena. Only through restrictions imposed by the Second Law Thermodynamics will the assumed dependence given in (3.3) be altered.

Based on the functional dependence noted above, the quantity $D^\alpha\psi^\alpha/Dt$ in the Second Law of Thermodynamics (eq. (2.15)) can be expanded using chain rule differentiation. Restrictions on the constitutive functions are provided by introducing such an expansion into (2.15) and recognizing that the coefficients of those quantities which appear linearly must vanish to insure that the Second Law of Thermodynamics will not be violated for arbitrary thermodynamic states (1). Thermodynamic equilibrium restrictions can also be defined by investigating the mathematical implications of (2.15) when entropy production is minimized, i.e. $\rho\Gamma = 0$. As a result of these considerations the following definitions and restrictions are obtained:

$$\psi = \frac{1}{\rho}\sum_{\alpha}\rho^{\alpha}\phi^{\alpha}\psi^{\alpha} = \psi(\rho^{f}, E^{r}_{KL}, \theta,\sigma) \qquad (3.4)$$

$$\psi^{f}_{E} = \psi^{f}(\rho^{f}, \theta, \sigma) \qquad (3.5)$$

$$\psi^{r}_{E} = \psi^{r}(E^{r}_{KL}, \theta, \sigma) \qquad (3.6)$$

$$\eta = \frac{1}{\rho}\sum_{\alpha}\rho^{\alpha}\phi^{\alpha}\eta^{\alpha} = -\frac{\partial\psi}{\partial\theta} \qquad (3.7)$$

$$\rho^{f}\phi^{f}\hat{F}^{f}_{k} = P^{f}\phi^{f}_{,k} + \bar{F}^{f}_{k} \qquad (3.8)$$

$$\tau^{f}_{kl} = -P^{f}\phi^{f}\delta_{kl} + \bar{\tau}_{kl} \qquad (3.9)$$

$$\tau^{r}_{kl} = -P^{f}\phi^{r}\delta_{kl} + \rho^{r}\phi^{r}\frac{\partial\psi^{r}}{\partial E^{r}_{KL}}x_{(k,K}x_{l),L} \qquad (3.10)$$

$$P^{f} = (\rho^{f})^{2}\frac{\partial\psi^{f}}{\partial\rho^{f}} \qquad (3.11)$$

The over-barred quantities denote the non-equilibrium parts of the associated functions. These variables retain the functional dependence initially hypothesized in (3.3). The non-advective species flux and heat flux vectors vanish at equilibrium, and thus, posses only dissipative parts, i.e. $J_k = \bar{J}_k$ and $\sum_{\alpha} q^{\alpha}_{k} = \bar{q}_k$.

At this point further information cannot be extracted from theoretical interpretations. Ideally, experimental methods should be employed in the definitions of the non-equilibrium parts of the constitutive functions. However, due to the number of functions which require definition, this would prove to be an impractical task. As a result, Taylor series expansions with regard to the independent variables about the equilibrium state are offered as approximations to the forms of the constitutive functions. In most cases, linear terms are adequate in defining phenomenological relationships which correspond to accepted experimental results, i.e. Darcy's Law, Hooke's Law for linearly elastic solids, etc. The derivation of the dissipative parts of ψ^f, ψ^r, τ^{f}_{kl} and $\hat{F}_k$ are given by Hassanizadeh and Gray (5) and Shapiro and Pinder (8), and will not be further investigated here. The form of the non-advective species flux shall be considered in the following paragraphs.

A linear expansion about an equilibrium state for $\bar{J}_k$ takes the following form:

$$\bar{J}_k = \bar{J}_k|_E + \frac{\partial\bar{J}_k}{\partial d^{f}_{mn}}\Big|_E d^{f}_{mn} + \frac{\partial\bar{J}_k}{\partial\theta_{,l}}\Big|_E\theta_{,l} + \frac{\partial\bar{J}_k}{\partial v^{fr}_{l}}\Big|_E v^{fr}_{l} + \frac{\partial\bar{J}_k}{\partial\sigma_{,l}}\Big|_E\sigma_{,l} + \frac{\partial\bar{J}_k}{\partial\rho^{f}}\Big|_E(\rho^{f}-\rho_E) + \frac{\partial\bar{J}_k}{\partial\theta}\Big|_E(\theta-\theta_E) + \frac{\partial\bar{J}_k}{\partial\sigma}\Big|_E(\sigma-\sigma_E) + \frac{\partial\bar{J}_k}{\partial\phi^{f}}\Big|_E(\phi^{f}-\phi^{f}_{E}) \qquad (3.12)$$

This expression can be shown to be independent of terms which include E^{r}_{kl} and $\phi^{f}_{,k}$, since restrictions on these quantities are not required in the definition of the equilibrium state (5,8). By definition, $\bar{J}_k|_E$ must vanish, and if the medium is assumed to be isotropic the odd order tensorial quantities must also equal zero. This reduces (3.12) to the following:

$$\bar{J}_k = \lambda\theta_{,k} + A v^{fr}_{k} + \pi\sigma_{,k} \qquad (3.13)$$

where

$$\lambda = \frac{\partial\bar{J}_k}{\partial\theta_{,l}}\Big|_E\delta_{kl}; \quad A = \frac{\partial\bar{J}_k}{\partial v^{fr}_{l}}\Big|_E\delta_{kl}; \quad \pi = \frac{\partial\bar{J}_k}{\partial\sigma_{,l}}\Big|_E\delta_{kl} \qquad (3.14)$$

with λ, A and π being functionally dependent upon X^{r}_{K} only.

Since it is assumed that the quantities v^{fr}_{k}, $\theta_{,k}$ and $\sigma_{,k}$ can be chosen independently to represent unique thermodynamic states, one can consider the conditions where there are no gradients in temperature or species concentration. The above expression would then predict the unreasonable result of a non-advective species flux for non-zero values of v^{fr}_{k}. The coefficient of v^{fr}_{k} must then be set equal to zero. However, physical experiments have indicated that the non-advective species flux is highly dependent upon fluid velocity (12,13). Yet, velocity will not appear in the linear expression above, as either a distinct quantity or as part of the material coefficients. Consequently higher order terms need to be considered in the definition of $\bar{J}_k$.

The non-advective species flux can be defined by the ensuing relationship:

$$\bar{J}_k = \lambda\theta_{,k} + \pi\sigma_{,k} + a^1_{klmn}\sigma_{,l}\sigma_{,m}\sigma_{,n} + a^2_{klmn}\sigma_{,l}\sigma_{,m}v^{fr}_n + a^3_{klmn}\sigma_{,l}v^{fr}_m v^{fr}_n \quad (3.15)$$

where a^i_{klmn} are isotropic tensors specified by the following:

$$a^i_{klmn} = \gamma^{2i}\delta_{kl}\delta_{mn} + \frac{(\gamma^{2i-1} - \gamma^{2i})}{2}(\delta_{km}\delta_{ln} + \delta_{kn}\delta_{lm}) \quad (3.16)$$

and dependent upon X^r_K. Here a cubic expansion has been provided where only the relative velocity and the gradient in the species mass fraction are considered in the higher order terms. The quadratic terms in the expansion have been set equal to zero due to isotropic considerations, and the argument concerning fluid velocity that was noted above has also been invoked. An expression similar to equation (3.15) has also been intuitively developed by Whitaker (15).

4 CONCLUSIONS

In analyzing the above definitions of the non-advective species flux, one can associate the first three terms with molecular diffusion. If gradients in the species mass fraction are small the linear term will dominate. The final term on the right hand side of (3.15) closely resembles the definition of hydrodynamic dispersion which is most often employed in predictive models. Thus, one could consider the definition of hydrodynamic dispersion as given in Bear (12), as an incomplete third order expansion.

The significance of the additional term in (3.15) is unknown at this time. However, it may have some importance, since the existing phenomenological relationship for the non-advective species flux has often failed to provide adequate correlations with measured results. The magnitude of the dispersivities needed to reproduce field information has been known to vary with the distance from a given source. One could attribute this to the spatial variations in dispersivity which is often not considered in predictive models. Even so, this may account for the small values of dispersivity needed in the vicinity of the source and larger ones required at further distances, irregardless of where the source is located in the aquifer.

The functional form of the heat flux vector can be developed in a similar manner to that given above, except that the higher order terms would be defined with respect to the temperature gradient and the relative fluid velocity.

5 NOMENCLATURE

a^i_{klmn}	material coefficient in the definition of $\bar{J}_k$.
b^α	external supply of entropy for the α-phase
d^α_{kl}	deformation rate tensor, $d^\alpha_{kl} = \frac{1}{2}(v^\alpha_{k,l} + v^\alpha_{l,k})$
$\frac{D(\ )}{Dt}$	material time derivative with respect to the mixture velocity.
$\frac{D^\alpha(\ )}{Dt}$	material time derivative with respect to the velocity of the α-phase
e^α	internal energy density of the α-phase
e	internal energy density of the mixture $e = (1/\rho)\sum_\alpha \rho^\alpha\phi^\alpha e^\alpha$
E^r_{KL}	Lagrangian strain tensor
$\hat{F}_k$	momentum exchange between the α-phase and all other phases due to mechanical interaction
g^α_k	external supply of momentum for the α-phase
h^α	external supply of energy for the α- phase
h	external supply of energy of the mixture, $h = (1/\rho)\sum_\alpha \rho^\alpha\phi^\alpha h^\alpha$
J_k	non-advective species flux
$\hat{N}^\alpha$	exchange of entropy between the α-phase and all other phases due to mechanical interaction
p^f	fluid pressure
q^α_k	heat flux vector for the α-phase
s^α_k	entropy flux of the α-phase
v^α_k	velocity of the α-phase
v^{fr}_k	velocity of fluid relative to the rock phase
x_k	coordinates in the spatial reference frame
X^α_K	coordinates in the material reference frame
$x_{k,K}$	deformation gradient

Greek Notation

γ^i material coefficient in the definition of $\bar{J}_k$

Γ^α entropy production for the α-phase

Γ entropy production for the mixture

δ_{kl} Kronecker delta

η^α entropy density of the α-phase

η entropy density for the mixture, $\eta=(1/\rho)\sum_\alpha \rho^\alpha\phi^\alpha\eta^\alpha$

θ thermodynamic temperature

λ material coefficient in the definition of $\bar{J}_k$

ν_k^α velocity of the α-phase relative to the mixture velocity, $\nu_k^\alpha = v_k^\alpha-(1/\rho)\sum_\alpha \rho^\alpha\phi^\alpha v_k^\alpha$

π material coefficient in the definition of $\bar{J}_k$

ρ^α intrinsic mass density of the α-phase

ρ mass density of the mixture, $\rho=\sum_\alpha \rho^\alpha\phi^\alpha$

σ mass fraction of the chemical species

τ_{kl}^α stress tensor for the α-phase

ϕ^α volume fraction of the α-phase

ψ^α Helmholtz free energy of the α-phase

ψ Helmholtz free energy of the mixture

Subscripts

k (lower case indices) spatial reference frame

K (upper case indices) material reference frame

E evaluation at thermodynamic equilibrium

Superscripts

α denotes a given phase in the continuum, (r-rock, f-fluid)

^ denotes the exchange of a thermodynamic property between phases

- non-equilibrium part of a constitutive function

6 REFERENCES

1. Coleman, B.D. and W. Noll, Thermodynamic of elastic materials with heat conduction and viscosity, Arch. for Rat. Mech. and Anal., Vol. 13, 167-178, 1963.
2. Ingram, J.D. and A.C. Eringen, 1967, A continuum theory of chemically reacting media II. Constitutive equations of reacting fluid mixtures, Int.J. of Engr.Sci., Vol. 5, 289-322.
3. Muller, I., Thermodynamics of mixtures of fluids, J.de Mech., Vol. 14, 267-303, 1975.
4. Kenyon, D.E., The theory of an incompressible solid fluid mixture, Arch. for Rat. Mech. and Anal. Vol. 62, 131-147, 1976.
5. Hassanizadeh, M. and W.G. Gray, General conservation equations for multi-phase systems: 3. Constitutive theory for porous media flow, Adv. in Wat.Res., Vol.3, 25-40, 1980.
6. Muller, I., A thermodynamic theory of mixtures of fluids, Arch. for Rat. Mech. and Anal., Vol. 28, 1-39, 1968.
7. Drumheller, D.S. and A. Bedford, A thermo-mechanical theory for reacting immiscible mixtures, Arch. for Rat. Mech. and Anal., Vol.73, 257-284, 1980
8. Shapiro, A.M. and G.F. Pinder, Fractured porous media: 1. Formulation of the governing equations, (in preparation) 1981.
9. Eringen, A.C., Mechanics of Continua, John Wiley, New York, 1967.
10. Hassanizadeh, M., Macroscopic description of multiphase systems - A thermodynamic theory of flow in porous media, Ph.D.Thesis, Princeton Univ., 1980.
11. Shapiro, A.M., Fractured porous media: Equation development and parameter identification, Ph.D.Thesis, Princeton Univ. 1981.
12. Bear, J., Dynamics of Fluids in Porous Media, Elsevier, New York, 1972.
13. Scheidegger, A.E,, The Physics of Flow through Porous Media, Univ. of Toronto, Toronto, 1974.
14. Scheidegger, A.E., General theory of dispersion in porous media, J. of Geophys. Res., Vol.66, 3273-3278, 1961.
15. Whitaker, S., Diffusion and dispersion in porous media, Amer.Inst. of Chem. Engr. J., Vol. 13, 420-427, 1967.

Proceedings of Euromech 143 / Delft / 2-4 September 1981

Degeneration of second order dispersion equation

R.SOERJADI
Delft University of Technology, Netherlands

SUMMARY

A second order dispersion equation with constant coefficients degenerates, when the determinant of the coefficient matrix of the second order term equals zero. Such a degenerated problem, with inital and boundary conditions of a point injection, without additional sources or sinks, is treated, and illustrated by an example.

1 INTRODUCTION

Dispersion is defined, in this paper, as the scattering of a cloud of particles caused by some stochastic spreading mechanism. Such a process can be described approximately by a differential equation. This equation turns out to be a partial differential equation of infinite order (ref. 2,5). The spreading mechanism is supposed to be place and time independent. In that case, the governing equation possesses constant coefficients. If, moreover, these coefficients converge rapidly, the problem can be approximated by a low order, preferably second order, equation. Such a second order equation, with initial and boundary conditions of the simplest form, namely a point injection, will be treated.

2 SECOND ORDER DISPERSION EQUATION

Let in the n-dimensional space (with Cartesian coordinates, which can be considered as the components of the place vector $\mathbf{r}$) and time (t) continuum, the particle density be denoted by $u(\mathbf{r},t)$. Dispersion of particles in this space can be assumed to fulfil the following second order equation (ref.2,5,6),

$$u_t = -\mathbf{a}' \operatorname{grad} u + \operatorname{div}(B \operatorname{grad} u) \qquad (1)$$

with constant coefficients $\mathbf{a}$ and B . The vector $\mathbf{a}$ denotes the average particle displacement per unit time, or average particle velocity, and the matrix B is the covariance matrix of the particle displacement per unit time. In equation (1) the subscript t means partial differentiation with respect to that variable, and the accent denotes transposition showing a scalar multiplication of two vectors.

By translating the coordinates, according to

$$\mathbf{r}_1 = \mathbf{r} - \mathbf{a}t \qquad (2)$$

equation (1) becomes (ref.6)

$$u_t = \operatorname{div}(B \operatorname{grad} u) \qquad (3)$$

Physically, transformation (2) means that the coordinate system is moving with the average velocity $\mathbf{a}$.

By rotating the coordinate system, according to

$$\mathbf{r}_2 = R\,\mathbf{r}_1 \qquad (4)$$

in which R is a matrix, and such that

$$RBR' = \bar{B} = \text{diagonal matrix} \qquad (5)$$

equation (3) becomes (ref.6)

$$u_t = \operatorname{div}(\bar{B} \operatorname{grad} u) \qquad (6)$$

The elements of the transformation matrix R in (4) are the cosine of the angles between the new axes ($\mathbf{r}_2$) and the corresponding old axes ($\mathbf{r}_1$). It is an orthogonal matrix. This part is, in fact, an eigenvalue-problem, which can be characterized as finding the principal axes of an n-dimensional ellipsoid.

Another scale factor along the coordinate axes $\mathbf{r}_3$,

$$\mathbf{r}_3 = (\bar{B})^{-\frac{1}{2}} \mathbf{r}_2 \tag{7}$$

brings the diagonal matrix $\bar{B}$ to the unity matrix, reducing equation (6) to

$$u_t = \text{div grad}\, u = \Delta u \tag{8}$$

in which Δ denotes the n-dimensional Laplacian. Equation (8) is also known as the heat conduction or diffusion equation. From a particular point of view, diffusion can be regarded as a special case of dispersion, i.e. molecular dispersion.

With initial and boundary conditions of a point injection in the origin $\mathbf{r}_3 = \mathbf{O}$ at time $t = 0$, and without additional sources or sinks, the solution of (8) will be (ref.1,4.6), in normalized form,

$$u(\mathbf{r}_3, t) = \frac{1}{(2\sqrt{\pi t})^n} \exp\left\{-\frac{\mathbf{r}_3'\mathbf{r}_3}{4t}\right\} \tag{9}$$

Solution (9) can be transformed back in the original coordinate system $(\mathbf{r}, t)$,

$$u(\mathbf{r}, t) = \frac{1}{(2\sqrt{\pi t})^n\sqrt{|B|}} \times \exp\left\{-\frac{(\mathbf{r}-\mathbf{a}t)'B^{-1}(\mathbf{r}-\mathbf{a}t)}{4t}\right\} \tag{10}$$

in which $|B|$ denotes the determinant of B.

For any arbitrary time $t > 0$, solution (10) denotes an n-dimensional Gaussian distribution with mean value $\mathbf{a}t$ and covariance matrix $2Bt$.

Looking at $\sqrt{|B|}$ and B^{-1} in (10), two questions arise. What happens when: (1) B is a singular matrix, or (2) the determinant $|B|$ is less than zero. The last question will not be discussed at all in this study. The first one is dealing with a degenerated equation, which will be treated in the next section.

3 DEGENERATION

For the sake of simplicity, only the two-dimensional case, $n = 2$, will be considered. More-dimensional cases, $n > 2$, can be treated similarly. Equation (1), in which

$$\mathbf{a} = \begin{bmatrix} a_1 \\ a_2 \end{bmatrix} \quad \mathbf{r} = \begin{bmatrix} x \\ y \end{bmatrix} \quad B = \begin{bmatrix} b_{11} & b_{12} \\ b_{12} & b_{22} \end{bmatrix} \tag{11}$$

with $|B| = 0$

can be written as

$$u_t = -a_1u_x - a_2u_y + b_{11}u_{xx} + 2b_{12}u_{xy} + b_{22}u_{yy} \tag{12}$$

The coordinate translation (2),

$$\left.\begin{aligned} x_1 &= x - a_1t \\ y_1 &= y - a_2t \end{aligned}\right\} \tag{13}$$

transforms equation (12) into

$$u_t = b_{11}u_{x_1x_1} + 2b_{12}u_{x_1y_1} + b_{22}u_{y_1y_1} \tag{14}$$

which is the same as equation (3) for $n = 2$.

The coordinate rotation can be denoted by

$$\begin{bmatrix} x_2 \\ y_2 \end{bmatrix} = \begin{bmatrix} \cos\phi & \sin\phi \\ -\sin\phi & \cos\phi \end{bmatrix} \begin{bmatrix} x_1 \\ y_1 \end{bmatrix} \tag{15}$$

in which

$$R = \begin{bmatrix} \cos\phi & \sin\phi \\ -\sin\phi & \cos\phi \end{bmatrix} \tag{16}$$

where ϕ is the angle between the x_1 and x_2 axes. This transformation, however, gives some difficulties. For, since $|B| = 0$, the transformation (5) gives only one non-zero eigen-value. We obtain

$$RBR' = \bar{B} = \begin{bmatrix} B_{11} & 0 \\ 0 & 0 \end{bmatrix} \tag{17}$$

Then, instead of (3), we obtain

$$u_t = B_{11}u_{x_2x_2} \tag{18}$$

which is, in fact, a one-dimensional dispersion equation. It can be said, that the equation has degenerated. Equation (18), with initial and boundary conditions of a point injection at the origin ($x_2 = 0$, $y_2 = 0$) at time $t = 0$, has the normalized solution

$$u(x_2, y_2, t) = \frac{1}{2\sqrt{\pi B_{11}t}} \exp\left\{-\frac{x_2^2}{4B_{11}t}\right\} \tag{19}$$

(ref.6), which does not contain y_2 anymore, showing its one-dimensional character. It denotes, for any arbitrary time $t>0$, a Gaussian distribution along the x_2 axis, with $\mu(x_2)=0$ and $\sigma^2(x_2)=2B_{11}t$.

Solution (19) can be transformed back in the original coordinate system x,y,t,

$$u(x,y,t) = \frac{1}{2\sqrt{\{\pi(b_{11}+b_{22})t\}}} \times$$

$$\exp\left\{-\frac{\{\sqrt{b_{11}}(x-a_1t)+\sqrt{b_{22}}(y-a_2t)\}^2}{4(b_{11}+b_{22})t}\right\} \tag{20}$$

It should be noted, that (20) is not proportional to $1/t$, as would be expected for a two-dimensional problem, but proportional to $1/\sqrt{t}$, showing the degeneration to a one-dimensional problem.

A similar procedure can be applied to more-dimensional problems of this kind.

4 EXAMPLE

4.1. A non-degenerated equation

A two-dimensional problem, governed approximately by a second order equation with constant coefficients, has been studied by Soerjadi (ref.6.). An example from that study is dealing with the equation

$$u_t = -\frac{8}{7}u_x - \frac{8}{7}u_y + \frac{148}{441}u_{xx} + \frac{220}{441}u_{xy} + \frac{100}{441}u_{yy} \tag{21}$$

The determinant of the coefficient matrix B,

$$B = \begin{bmatrix} \frac{148}{441} & \frac{110}{441} \\ \frac{110}{441} & \frac{100}{441} \end{bmatrix} \tag{22}$$

turns out to be not equal to zero. Hence, equation (21) is non-degenerated. With initial and boundary conditions of a point injection at the origin ($x=0$, $y=0$) at time $t=0$, it has, according to (10), the solution

$$u(x,y,t) = \frac{147}{40\sqrt{3}\pi t}\exp\left\{-\frac{1}{300\,t}\times \{23(7x-8t)^2-55(7x-8t)(7y-8t)+37(7y-8t)^2\}\right\} \tag{23}$$

It should be noted, that (23) is proportional to $1/t$, being an appropriate solution of a two-dimensional problem. The solution denotes, at any time $t>0$, a two-dimensional Gaussian distribution. It has the mean values

$$\mu(x) = \frac{8}{7}t \qquad \mu(y) = \frac{8}{7}t \tag{24}$$

and the principal standard deviations

$$\sigma_1 = 1.03584\sqrt{t} \qquad \sigma_2 = 0.22750\sqrt{t} \tag{25}$$

The angle ϕ, between the x axis and the direction of σ_1 can be computed.

$$\phi \simeq 39^{\circ} \tag{26}$$

The principal standard deviations, σ_1 and σ_2, are the principal axes of the dispersion ellipse (see also figure).

4.2. A degenerated equation

De Josselin de Jong and Shao-Chih Way (ref.3) have studied the dispersion of particles transported by fluid flowing through a system of fissures in rock. In that study, it has been assumed, that the fissure planes are parallel to one coordinate axis, z axis say, and that the conditions in all planes perpendicular to that axis are identical. Thus the problem to be treated is two-dimensional, and only the x,y coordinates are essential (ref.3, p.1). A worked example from that study (see also ref.6) has been leading to the equation

$$u_t = -(\tfrac{3}{8}\sqrt{3}\,v_o)u_x - (\tfrac{1}{8}v_o)u_y + (\tfrac{3}{256}v_o l)u_{xx} + (\tfrac{6\sqrt{3}}{256}v_o l)u_{xy} + (\tfrac{9}{256}v_o l)u_{yy} \tag{27}$$

in which l and v_o are constants, denoting unit fissure length and unit particle velocity, respectively. The determinant of the matrix B,

$$B = \begin{bmatrix} \frac{3}{256}v_o l & \frac{3\sqrt{3}}{256}v_o l \\ \frac{3\sqrt{3}}{256}v_o l & \frac{9}{256}v_o l \end{bmatrix} \tag{28}$$

turns out to be zero. Hence, equation (27) has degenerated. With the initial and boundary conditions of a point injection at the origin ($x=0$, $y=0$) at time $t=0$, it has, according to (19) the normalized solution

$$u(x,y,t) = \frac{4}{\sqrt{3\pi t v_o l}} \times \exp\left\{-\frac{4(x+y\sqrt{3}-\frac{1}{2}\sqrt{3}\,v_o t)^2}{3v_o l\,t}\right\} \tag{29}$$

Expression (29) denotes, for any arbitrary time $t>0$, a one-dimensional Gaussian distribution along the x_2 axis defined by (15). Its mean values and principal standard deviation are, respectively,

$$\mu(x) = \frac{3\sqrt{3}}{8} v_o t \tag{30}$$

$$\mu(y) = \frac{1}{8} v_o t \tag{31}$$

$$\sigma(x_2) = \frac{\sqrt{6}}{8}\sqrt{v_o l t} \tag{32}$$

Strictly speaking, these values should be completed by $\sigma(y_2)=0$. The latter means, physically, that there is no spreading of particles perpendicular to the x_2 axis.

The angle ϕ between the x_2 and x axes can be computed.

$$\phi = 60^{o}$$

All these values are in accordance with the results found by De Josselin de Jong and Shao-Chih Way (ref.3,6).

5 CONCLUSIONS

On solving a second order dispersion equation with constant coefficients, one might be concerned with degeneration. This depends on the coefficient matrix of the second order term. The determinant of this matrix can be (1) positive, (2) zero, or (3) negative. The first case is regular and can be treated normally (section 2). The third case is not discussed at all in this study. The second case, in which the determinant equals zero, has turned out to give a degenerated equation (section 3). It has turned out that such a problem degenerates to a lower dimensional problem than seemed to be. Physically, it means that particles are not spreaded in some particular directions. Solutions of degenerated and non-degenerated equations, with initial and boundary conditions of a point injection, have been illustrated by some examples (section 4).

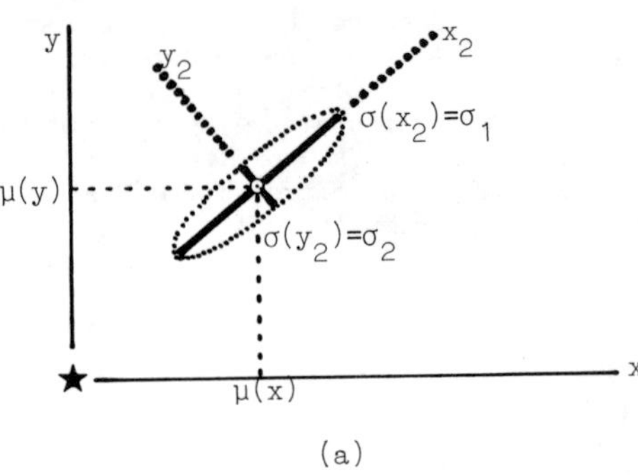

(a)

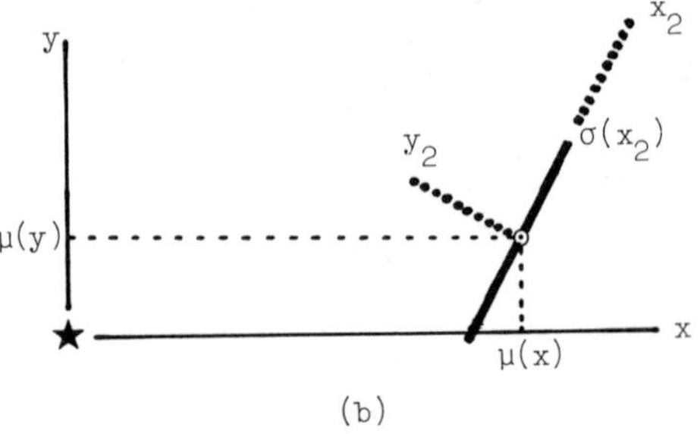

(b)

Dispersion ellips of the particle distribution:
(a) non-degenerated (example 4.1),
(b) degenerated (example 4.2).
The star denotes the point of injection.

6 REFERENCES

1. Chandrasekhar, S.,Stochastic Problems in Physics and Astronomy, Reviews of Modern Physics, Vol.15, No.1, 1943.
2. De Josselin de Jong, G. 1972, Dispersion described by differential equation developed with Lagrangian Correlation Functions, Delft University of Technology, Geotechn. Lab.
3. De Josselin de Jong, G.& Shao-Chih Way, 1972, Dispersion in fissured rock, New Mexico Institute of Mining and Techn., Geoscience Department, Socorro, New Mex.
4. Lax, Peter, Samuel Burstein, Anneli Lax, 1976,Calculus with Applications and Computing, Vol.1, Springer, New York.
5. Soerjadi, R. 1980, An alternative development of the dispersion equation. Internal Report of Dept.of Civ.Eng., Delft University of Techn. To be published.
6. Soerjadi, R., Two-dimensional dispersion governed by second order equation. Internal Report. Delft University of Techn. To be published.

Proceedings of Euromech 143 / Delft / 2-4 September 1981

Physico-chemical exchanges and transfer of pollution in aquifers

J.ROCHON & D.THIERY
Bureau de Recherches Géologiques et Minières, Orléans, France

1 INTRODUCTION

This paper is divided into two parts. The first one is a mathematical analysis of different laws for physic and chemical exchanges occuring during a natural flow in an aquifer. The second part shows, with field experiments at Bonnaud (Jura), how it is possible to determine the parameters of an exchange law in order to forecast the long term evolution of the concentration of a solute which has reached the aquifer.

2 INTRODUCTION TO THE FIRST PART

This part will deal with the study of the concentrations in the water of an aquifer after an injection of tracer. The hypothesis relative to the aquifer are :

- homogeneous and isotropic aquifer,
- uniform and permanent flow.

The evolution of the concentration will result of three phenomenous :

- advection : it is displacement of the tracer with the flow,
- dispersion,
- physic and chemical exchanges between phase.

It is assumed that there are no exchanges with the lower and upper boundaries of the aquifer. The general equation is :

$$\frac{\partial C}{\partial t} = D\frac{\partial^2 C}{\partial x^2} - u\frac{\partial C}{\partial x} - \frac{1-n}{n}\cdot\frac{\partial S}{\partial t}$$

with the relation

$$\frac{\partial S}{\partial t} = f(C,S)$$

C = concentration in the liquid phase (water)
S = concentration in the solid phase (aquifer)
D = dispersion coefficient
n = kinematic porosity
u = effective velocity
t = time
x = distance

Many exchanges laws may be studied : an exchange law may be reversible or not, its equilibrium equation (or "isotherm") may be linear or not (e.g. Langmuir scheme), see J. Rochon 1978.

We choose to study some of these laws which seem to be the most useful.

2.1. Linear exchanges with linear exchange kinetic

$$\frac{\partial C}{\partial t} = D\frac{\partial^2 C}{\partial x^2} - u\frac{\partial C}{\partial x} - \frac{1-n}{n}k_c(k_e C - S)$$

$$\frac{\partial S}{\partial t} = k_c(k_e C - S)$$

which may be writen with dimensionless variables :

$$\frac{\partial C_R}{\partial t_R} = D\frac{\partial^2 C_R}{\partial x_R^2} - \frac{\partial C_R}{\partial x_R} - C_a . N_e (C_R - S_R)$$

$$\frac{\partial S_R}{\partial t_R} = N_e (C_R - S_R)$$

with

$$x_R = x/L \qquad t_R = t/\frac{L}{u}$$

$$C_R = C.L.n/m \qquad S_R = S.k_e.L.n/m$$

Three dimensionless numbers appears :

$P_e = \frac{uL}{D}$ it is the Peclet number

$N_e = \frac{L.k_c}{u}$ it is the number of Echanges during the advection along the distance L

$C_a = \frac{1-n}{n}k_e$ it is the Exchange Capacity of the aquifer

This system of two dimensionless differential equations has been solved by finite differences with B.R.G.M. numerical code M.O.D.E.L. (J. Ausseur, B. Kanehiro, J.P. Sauty, 1979) and made it possible to design dimensionless type curves describing the

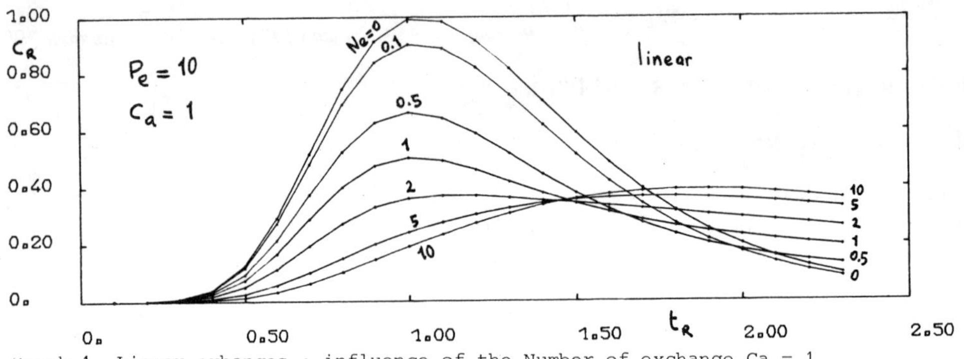

Graph 1. Linear exhanges : influence of the Number of exchange Ca = 1

Graph 2. Linear exchanges : influence of the Number of exchange Ca = 2

evolution of concentrations following a slug injection of tracer. The calculations have been performed for a Peclet number equal to 10, a number of exchange ranging from 0.1 to 10 and a Retention Capacity from 0.1 to 5. Graphs 1 and 2 shows the type curves for Ca = 1 and Ca = 2.

2.2. Exchanges according to Langmuir isotherm

The equilibrium equation is :

$$S_e = \frac{k_e C}{1 + C/C_d}$$

with

C_d = half saturation concentration in the liquid phase (water)

S_e = equilibrium concentration in the solid phase (aquifer)

The exchange kinetic is of second order :

$$\frac{\partial S}{\partial t} = k_c \left(k_e C \cdot \frac{S_e - S}{S_e} \right)$$

The transformation of the variable is the same as in the previous chapter. The dimensionless equation may be written :

$$\frac{\partial C_R}{\partial t_R} = \frac{1}{P_e} \cdot \frac{\partial^2 C_R}{\partial x_R^2} - \frac{\partial C_R}{\partial x_R} - C_a \cdot N_e \left(C_R - S_R - C_R S_R / S_a \right)$$

$$\frac{\partial S_R}{\partial t_R} = N_e \left(C_R - S_R - C_R S_R / S_a \right)$$

with a fourth dimensionless number

$$S_a = \frac{C_d \cdot L \cdot n}{m}$$ it is the Saturation Constant.

if Sa is equal to infinity which means no saturation effect, the system of equations is similar to the system describing the linear exchanges.

As the solution depends on four dimensionless numbers, it is not possible to draw a set of type curves.

We choose to draw type curves for instantaneous exchanges corresponding to N_e equal to infinity.

If the Peclet number is determined, only two dimensionless number are left : Ca and Sa. Type curves have been drawn to study the influence of each of these numbers. See graphs 3 and 4.

2.3. Ionian exchanges

The hypothesis are the following :

- the sum of the cations in the liquid phase is T,
- the sum of the cations in the solid phase is Qo,
- the anions are not adsorbed by the solid phase,
- the total quantity of injected cations is m.

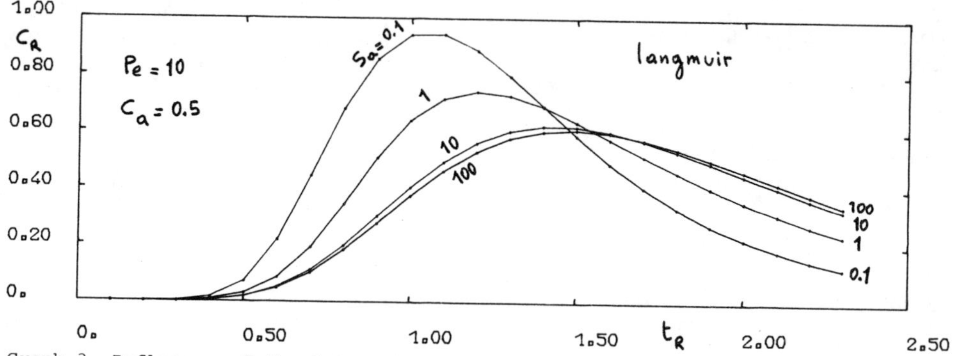

Graph 3. Influence of the Saturation Constant

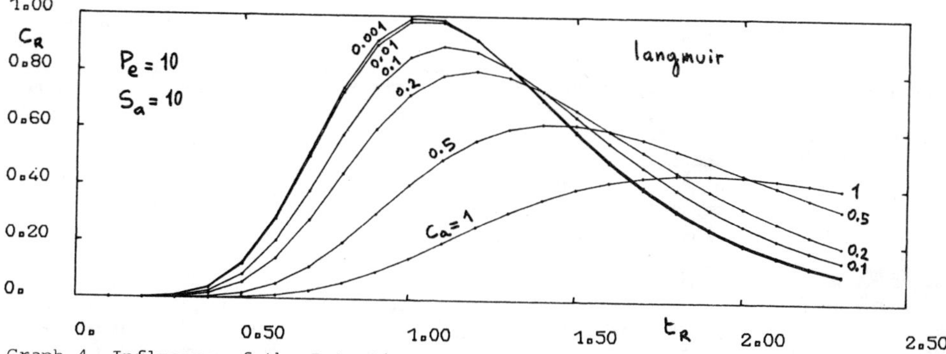

Graph 4. Influence of the Retention Capacity

The relation between phases is the following :

$$S_e = \frac{k\, Q_0 C}{T + (k-1)C}$$

when the equilibrium is reached or at any time, if the exchange kinetic is instantaneous. k is the constant of the mass action law.

The sum of all the cations in solution was equal to To. At the time of injection, it is increased at the injection point and then is described by the law of dispersion without exchanges :

$$\frac{\partial T}{\partial t} = D \frac{\partial^2 T}{\partial x^2} - u \frac{\partial T}{\partial x}$$

The reason is that the anions are not adsorbed which is an hypothesis, so the sum of the anions in solution is not involved in exchanges. The electrical neutrality tells that the sum T of the cation too will not be involved in exchanges (see J. Rochon, 1979).

Assuming that the exchange kinetic is instantaneous, it is possible to derive the following equilibrium relation which holds at any time

$$S = \frac{k_e C}{T_R + C/C_d}$$

with the following transformations :

$$T_R = T/T_0$$

$$k_e = k\, Q_0/T_0$$

$$C_d = T_0/(k-1)$$

The equilibrium relation has the same expression than Langmuir one, when T_R is equal to unity, i.e. for low concentrations of injection.

The dimensionless system of differential equations is :

$$\frac{\partial C_R}{\partial t_R} = \frac{1}{P_e}\frac{\partial^2 C_R}{\partial x_R^2} - \frac{\partial C_R}{\partial x_R} - \frac{C_a}{(T_R + C_R/S_a)^2}\left(T_R \frac{\partial C_R}{\partial t_R} - C_R \frac{\partial T_R}{\partial t_R}\right)$$

$$\frac{\partial T_R}{\partial t_R} = \frac{1}{P_e} \cdot \frac{\partial^2 T_R}{\partial x_R^2} - \frac{\partial T_R}{\partial x_R}$$

when $C_o \ll T_o$, $T_R \simeq 1$ and is constant, and these equations are similar to Langmuir scheme. However the second equation which has no exchanges term shows that the variation of T is much faster than the variation of C. After a time equal to about twice the advection time the evolution of concentrations is identical to Langmuir scheme.

3 INTRODUCTION TO THE SECOND PART

All the tracer experiments in uniform flow which are analysed in this part took place in 1974 at Bonnaud, near Lons-le-Saunier (Jura - France). The experiments were designed together by two teams : the Bureau de Recherches Géologiques et Minières (BRGM) and the Section d'Application des Radio-éléments du Commissariat à l'Energie Atomique (SAR-CEA). The aim of the research was to study extensively with field experiments the underground migration of various matters soluble in water. All the results are described in B. Gaillard, J. Guizerix, J. Margat, J. Molinari, P. Peaudecerf, 1976.

3.1 Description of the experiment field

The aquifer has the following characteristics :

- the water table is of limited thickness (3 to 4 m) which helps for the measurements,
- the aquifer is confined which implies no influences of the fluctuation of the water table and no exchanges with the non saturated zone,
- the aquifer is not thick and is well defined,
- the naturel variations of level are small,
- the hydraulic conductivity is high which helps the advection of the injected tracers (transmissivity : 10^{-3} m^2/s)

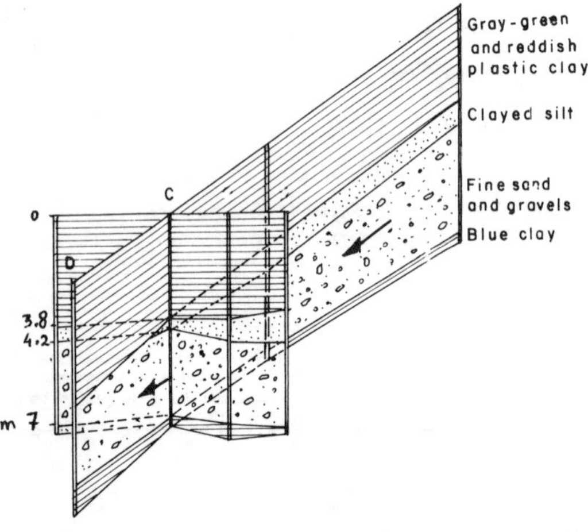

Graph 5. Longitudinal and transversal geological cross-sections of aquifer

In order to get a uniform gradient and increase the velocities to keep the experiment duration reasonable, a uniform flow has been imposed and controlled by a set of injection wells (M) and pumping wells (N) with constant rates (see graph 6).

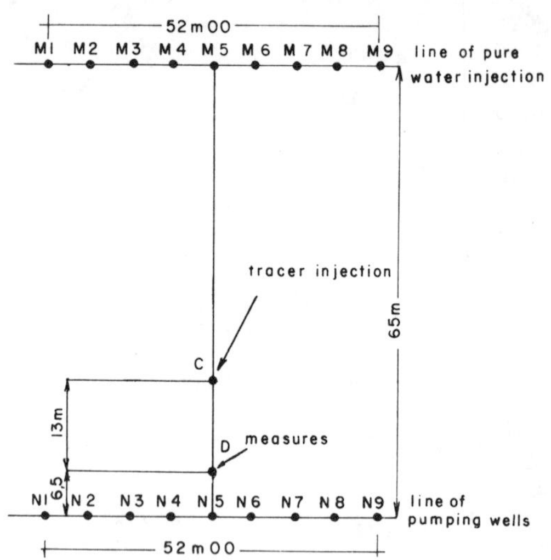

Graph 6. Situation of wells

3.2 The field experiment to be analysed

Among all the experiments three of them have been selected and are analysed in the following lines. Each injection took place in well C and measurements were in well D, 13 meters further. The main characteristics of these experiments are gathered in table 1.

Table 1. Main characteristics of the field experiments at Bonnaud

Injected matter	I-Na	Rhodamine WT	Uranine
Mass injected (g)	20	4	40
Date of injection	23/10/74	18/10/74	25/10/74
Uniform flow 10^{-3} m^3/s	3.39	2.42	2.36
Incidents		36 hours stop	
Maximum concentration measured µg/l	295	17	105
Estimation of residual concentration (from previous experiments µg/l)	40	4	6

3.3 Method of interpretation

The three experiments have been analysed separately by ajustment of the parameter with a mathematical model. The number of parameter is 4 : D, u, k_c, k_e and also the cross-section area of the part of the aquifer in which the migration takes place.

Other experiments with "perfect" tracers like tritium for which there are no exchanges have been performed on the same experiment field. The analysis of these experiments (P. Peaudecerf, J.P. Sauty, 1978) made it possible to determine the parameters describing advection and diffusion :

- effective velocity $u = 3.8\ 10^{-5}$ m/s (3,3 m/day),
- (kinematic porosity n = 33 %),
- diffusion coefficient $D = 2,5\ 10^{-5}$ m^2/s (2.2 m^2/day).

These parameters lead to a dispersivity $\alpha = D/u = 0.66$ m and a Peclet number : $Pe = uL/D = 19.8$ (L = distance = 13 m).

This high value of Peclet number shows that the advection term is much larger than the diffusion one.

The experiments with the other matters (I-Na, Rhodamine WT, Uranine) have been analysed with these same values for D, u and n. The analysis have been done only by ajusting the parameters of exchanges between phases.

3.3.1. Choice of the model of exchange

Various injections have been realised on the same location with the same tracers but with different injected mass (ratio of 10). None of these test show any mass effect ; in other words, for a given tracer, and after a given duration the concentration is proportionnal to the injected mass of tracer. Moreover, as the capacity of cationic exchange Q_o of the aquifer and the chemical nature of the water To are not known, it has not been possible to try any interpretation with ionian exchange scheme or Langmuir scheme. The model which has been chosen is then the reversible and linear exchanges. It must be noted that, as the duration of the measurements is sometimes short and as some incidents occur during the experiments, it should have been illusory to use a more complicated model with more parameters.

3.3.2. Determination of the parameter

For each experiment two parameters have been determined :

- the kinetic constant k_c (corresponding to the half exchange time : $t_d = Ln2/k_c$),
- the equilibrium constant k_e.

The corresponding dimensionless parameters are :

- number of exchanges : $Ne = L.kc/u$
- retention capacity : $Ca = k_e\ (1-n)/n$

The interpretation scheme is one dimensional so the transverse difusion is neglected The flow is assumed to take place in a prism of aquifer of section A (thickness h and width l) in which the concentration is uniform. The thickness of the aquifer ranges between 2.9 m at injection point to 2.5 m at measurement point. The width l is not known. It is possible to compute it by the ratio of the injected mass to the initial injected concentration which is a parameter of the model.

The parameters obtained after calibration are gathered in table 2.

Table 2. Parameters of exchange

	I-Na	Rhodanine WT	Uranine
kinetic constant kc (10^{-6} s^{-1})	3.0	8.0	7.5
half exchange time td (days)	2.7	1.0	1.1
equilibrium constant ke	0.22	1.12	1.49
number of exchanges Ne	1.03	2.74	2.57
Retention capacity Ca	0.45	2.27	3.03
equivalent width of prism l (m)	6.7	8.4	6.7

The comparison of the measured and computed concentration appear on graph 7 to 9. These graphs show that the calibration is reasonable, keeping in mind the incidents which occur during the experiments and also the fact that the concentration measurements seems sometimes affected by erratic variations.

3.4. Interpretation of the exchange parameters

The experiment with Sodium Iodide (I-Na) shows little exchange because the Retention Capacity is small (Ca = 0,45) and the kinetic of exchange is slow (Ne = 1.0).

The experiments with Rhodamine and Uramine give results which are quite similar (see graph graph 8 and 9) with much more exchanges : the Retention capacity ranges from 2 to 3, the Number of Exchanges is equal to about 2.5. The computed width of the equivalent prism of aquifer is equal to 7 or 8 meters for the 3 experiments. This width does not describe the lateral extension of the plume of tracer but rather the width of aquifer influenced by the pumping at the measurement point D.

4 CONCLUSION

The use of models makes it possible to predict the migration of a pollution in the water of an aquifer, taking into account the physic and chemical interactions of the aquifer. To determine the exchange parameters it is necessary to do the following :

- perform a first experiment on the field

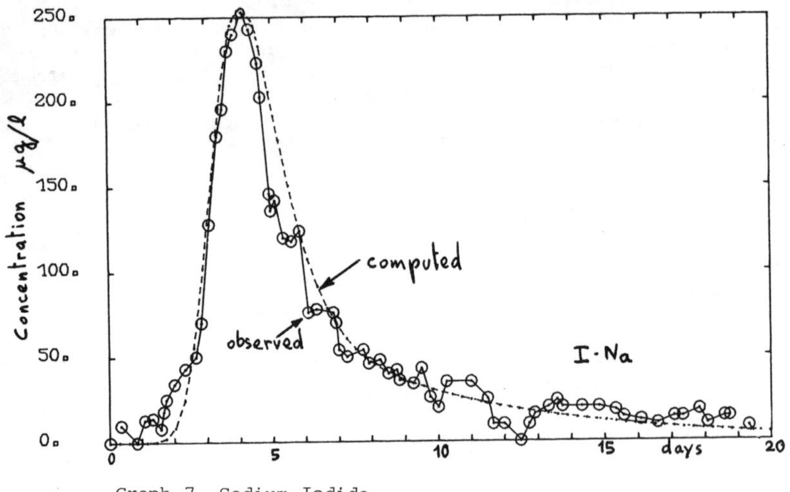

Graph 7. Sodium Iodide

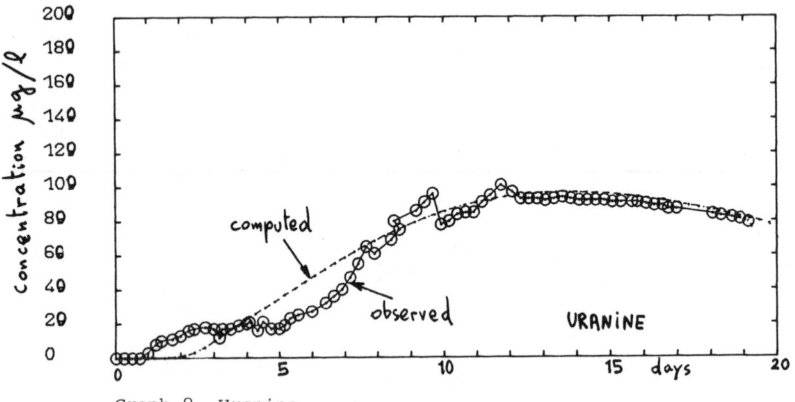

Graph 8. Uranine

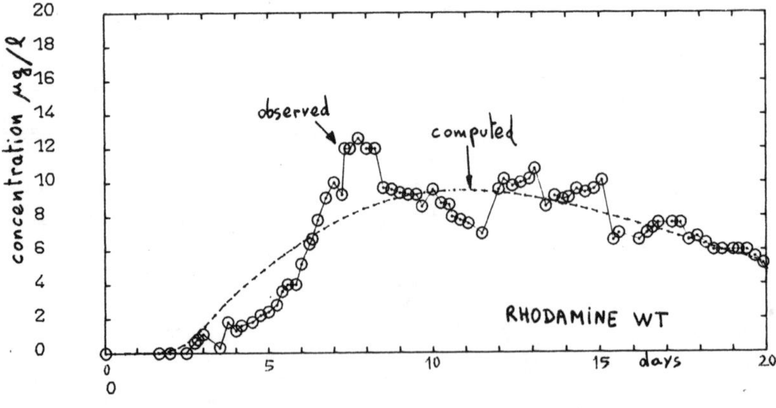

Graph 9. Rhodamine WT - Tracer injection at Bonnaud (Jura)

with a "good" tracer (i.e. with no or little exchanges) in order to determine the parameters describing advection and dispersion,

perform a second test -preferably a field experiment- with a solute similar to the product to study but which is not a pollutant. When such an experiment may not be performed (for economical reasons or because there is no solute with properties similar to the polluant to study) a dynamic test may be realised in laboratory with a sample of the ground and with the pollutant. From this second experiment, the exchange parameters will be deducted. However a laboratory test will only give a rough estimation of the parameters by not taking into account the heterogeneity of the real field.

NOTATIONS

C = Concentration in the liquid phase

C_a = Retention Capacity

C_e = Equilibrium Concentration

C_R = Dimensionless Concentration in the liquid phase

C_d = Half-saturation Concentration (kg/m^3)

D = Dispersion Coefficient

k = Equilibrium Constant

k_c = Kinetic of exchange constant

k_e = Equilibrium Constant

L = Distance between injection point and measurement point

m = Injected mass of solute or quantity of injected cations

n = Kinematic porosity of the aquifer

N_e = Number of exchanges

P_e = Peclet Number = uL/D

Q_o = Sum of the cations in the solid phase

S = Concentration in the solid phase

S_a = Saturation Constant

S_R = Dimensionless Concentration in the solid phase

T_o = Sum of the cations initially in the liquid phase

T = Sum of the cations in the solid phase

t = Time

t_R = Dimensionless time

u = Effective velocity or poral velocity

x = Distance (m)

x_R = Dimensionless Distance

BIBLIOGRAPHY

Ausseur, J.Y., Kanehiro, B., Sauty, J.P. Mass transfer through fractured rocks "M.O.D.E.L." Code. Convection Dispersion Exchange Decay, Bureau de Recherches Géologiques et Minières, rapport n° 79 SGN 748 HYD (to be printed)

Gaillard, B., Guizerix, J., Margat, J., Molinari, J., Peaudecerf, P., 1976, Action thématique programmée "hydrogéologie", Etude des caractéristiques de transfert de substances chimiques dans les nappes (site de Bonnaud - Jura), Bureau de Recherches Géologiques et Minières, rapports 75 SGN 056 AME, 76 SGN 235 AME, 76 SGN 236 AME, 76 SGN 240 AME, 76 SGN 226 AME.

Peaudecerf, P., Sauty, J.P. 1978, Application of a mathematical model to the characterisation of dispersion effects on groundwater quality, Prog. Wat. Tech., vol. 10 n° 5/6, pp 443-454, Pergamon Press

Rochon, J. 1978, Propagation de substances miscibles en interaction physico-chimique avec le substrat, Approche simplifiée pour l'utilisation en hydrogéologie, Thèse de Doc. Ing., Institut National Polytechnique de Grenoble, 25 janvier 1978.

Rochon, J. 1979, Echange cationique au cours d'un écoulement unidirectionnel en milieu poreux saturé, in : "workshop OCDE/NEA", L'utilisation des matériaux argileux pour l'isolation des déchets radioactifs, Paris, 10-12 septembre 1979.

Villermaux, J., Van Swaaij, W.P.M. 1969, Modèle représentatif de la distribution des temps de séjour dans un réacteur semi-infini à dispersion axiale avec zones stagnantes, Application à l'écoulement ruisselant dans des colonnes d'anneaux Raschig, Chemical Engineering Science, Vol. 24.

Proceedings of Euromech 143 / Delft / 2-4 September 198

Transport in structured porous media

P.A.C.RAATS
Institute for Soil Fertility, Haren, Netherlands

1 INTRODUCTION

Aggregates, cracks, channels left behind by decayed roots, and by animals, and layering often have a large influence on the transport of water and of solutes in soils. Physico-mathematical models for transport in soils with such structures usually are based on a distribution of water and/or solutes among a mobile and a stagnant phase, roughly corresponding to networks of large and of small pores. For any constituent the combined balances of mass for the mobile and stagnant phases can be written as

$$\frac{\partial \rho_m}{\partial t} + \frac{\partial \rho_s}{\partial t} = - \nabla \cdot \underset{\sim}{F} , \qquad (1)$$

where t is the time, ∇ is the vector differential operator, ρ_m is the mass of the constituent in the mobile phase per unit bulk volume (including the volume occupied by the solid phase of the porous material), ρ_s is the mass of the constituent in the stagnant phase per unit bulk volume, and $\underset{\sim}{F}$ is the flux of the constituent in the mobile phase. Important further ingredients in these models are the mechanisms of transport in the mobile phase and the nature of the storage capacities of the phases and the associated mechanisms of exchange between the phases.

2 MECHANISMS OF TRANSPORT IN THE MOBILE PHASE

In the literature the transport is often assumed to be either purely convective or purely diffusive, but sometimes it is assumed to be simultaneously convective and diffusive. Purely convective transport has been assumed to describe mass transport in packed beds (Klinkenberg 1948) and soils (Gardner and Brooks 1957; Raats 1973); it also appears in mathematically related theories for heat exchangers (Nusselt 1911; Anzelius 1926) and for transport of sediment (Einstein 1936; Polya 1937). Purely diffusive transport occurs in theories describing the influence of tides along coasts and in rivers on water levels in adjoining aquifer systems (Steggewentz 1933; van der Kamp 1973), transport of water and oil in structured porous media (Barenblatt and Zheltov 1960; Barenblatt et al. 1960; Raats 1969; Streltsova 1976), secondary consolidation in soil mechanics (Taylor and Merchant 1940; Gibson and Lo 1961) and in mathematically related theories describing heat transfer in media with two temperatures (Maxwell 1867; Chen and Gurtin 1968). Simultaneous convective and diffusive transport has been used to describe transport of solutes in packed beds (Lapidus and Amundson 1952) and soils (see Bolt 1979 and Van Genuchten and Cleary 1979 for reviews). In the context of this paper, no special complications arise from assuming simultaneous convective and diffusive transport. For definiteness, all equations will be written for a solute dissolved in water, but the results are easily adaptable to any entity, including solutes, water and heat, and to either purely convective or purely diffusive transport. Thus the flux $\underset{\sim}{F}$ is assumed to be given by

$$\underset{\sim}{F} = \theta_m \underset{\sim}{v} c - D_m \nabla c, \qquad (2)$$

where θ_m is the volumetric water content of the mobile phase, $\underset{\sim}{v}$ is the velocity of the water in the mobile phase, c is the concentration of the solute in the mobile phase, and D_m is the dispersion coefficient associated with the mobile phase.

3 SOME SIMPLE CAPACITY RELATIONSHIPS

The bulk density ρ_m and the concentration c are assumed to be related by

$$\rho_m = (\theta_m + k)\ c, \qquad (3)$$

where k describes instantaneous, linear adsorption. A vanishing storage capacity in the mobile phase ($\theta_m + k \to 0$) represents an important special case (e.g., Anzelius 1926; Steggewentz 1933; Polya 1937; Raats 1969, 1973). Introducing equations (2) and (3) into equation (1) and using the mass balance for water in the mobile phase (assuming there is no exchange of water between the mobile and stagnant phases),

$$\frac{\partial \theta_m}{\partial t} = - \nabla \cdot (\theta_m \underset{\sim}{v}), \qquad (4)$$

gives

$$M = - \frac{\partial \rho_s}{\partial t}, \qquad (5)$$

where the mobile phase operator M is given by

$$M = (\theta_m + k) \frac{\partial c}{\partial t} + \theta_m \underset{\sim}{v} \cdot \nabla c - D_m \nabla^2 c. \qquad (6)$$

The main concern of this paper is the exchange between the mobile and stagnant phases, accounted for by the term $\partial \rho_s / \partial t$ in equations (1) and (5). In most papers cited above the rate of exchange is assumed to be given by an expression of the form

$$\frac{\partial \rho_s}{\partial t} = - \alpha\ (\rho_s - \kappa\ \rho_m), \qquad (7)$$

where α is an exchange constant, and κ is the capacity ratio of the stagnant and mobile phases at equilibrium. Equation (7) is tantamount to assuming that the two phases are separated by a membrane with conductance α and that the phases themselves are perfectly mixed. If the time rates of change are very slow, then the equivalent conductance model given by equation (7) reduces to the equilibrium model given by

$$\rho_s = \kappa\ \rho_m,$$

$$\text{or} \quad \frac{\partial \rho_s}{\partial t} = \kappa \frac{\partial \rho_m}{\partial t} = \kappa\ (\theta_m + k) \frac{\partial c}{\partial t}. \qquad (8)$$

Some studies suggest that the effect of structure upon transport of solute can be represented by an equivalent dispersion coefficient D_e proportional to v^2 (Passioura 1971):

$$\frac{\partial \rho_s}{\partial t} = \kappa\ (\theta_m + k) \frac{\partial c}{\partial t} - D_e \frac{\partial^2 c}{\partial x^2}. \qquad (9)$$

In this paper the exchange among the phases will be treated as a full-fledged diffusion process. As a result, equation (5) will become a linear, partial integro-differential equation. It will be shown that, by using Laplace transforms, this equation can be converted to a partial differential equation in which derivatives with respect to time of all orders occur. The equilibrium, equivalent conductance, and equivalent dispersion models, given by equations (8), (7), and (9) respectively, will be shown to be approximations of such partial differential equations.

4 DIFFUSIVE EXCHANGE BETWEEN PHASES

I will sketch the theory for the simplest type of structured medium, namely one in which the mobile, stagnant and inert phases occur in layers (Fig. 1).

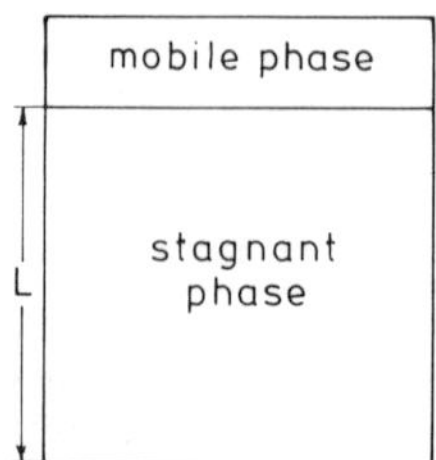

Fig. 1. Layered structure.

The exchange among the phases can then be treated as a one-dimensional diffusion process over distance L, with an effective diffusion coefficient D within the stagnant phase. The ratio L^2/D is the characteristic time of the exchange process and can be used to define a dimensionless time t^* by

$$t^* = \frac{D}{L^2} t. \quad (10)$$

Using Duhamel's theorem, the concentration c in the mobile phase can be shown to satisfy (Skopp and Warrick 1974):

$$M = -(\theta_m + k)\,\kappa \frac{D}{L^2} \int_0^{t^*} \frac{\partial c}{\partial t^*}(\underset{\sim}{x}, \tau^*)$$

$$\sum_{n=0}^{\infty} 2 \exp - \beta_n^2 (t^* - \tau^*)\, d\tau^*; \quad (11)$$

where

$$\tau^* = \frac{D}{L^2} \tau \quad (12)$$

denotes instants in the past, and

$$\beta_n = \frac{(2n + 1)\pi}{2} \quad (13)$$

denotes a geometry factor. The combination $L^2/(D\beta_n^2)$ represent a discrete spectrum of relaxation times. The sum

$$\sum_{n=0}^{\infty} 2 \exp - \beta_n^2 (t^* - \tau^*) \quad (14)$$

describes the memory of the stagnant phase for changes of concentration at the boundary between the mobile and stagnant phases. Fig. 2 shows successive terms of this sum.

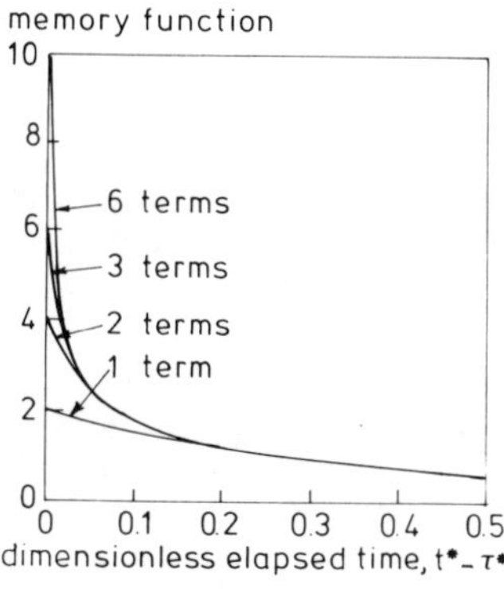

Fig. 2. The memory function.

Taking the Laplace transform of the relaxation integral form (11) of the transport equation gives

$$\overline{M} = -\sum_{n=0}^{\infty} \frac{2}{s^* + \beta_n} (\theta_m + k)\,\kappa\, s\overline{c}, \quad (15)$$

where $\overline{M}$ is the Laplace transform of the mobile phase operator,

$$\overline{M} = (\theta_m + k)\, s\overline{c} + \theta_m \underset{\sim}{v}\cdot\nabla\overline{c} - D_m \nabla^2 \overline{c}, \quad (16)$$

and

$$s^* = (L^2/D)s. \quad (17)$$

The infinite series in equation (15) can be expressed in terms of a tanh function:

$$\overline{M} = -(s^*)^{-\frac{1}{2}} \tanh (s^*)^{\frac{1}{2}} (\theta_m + k)\kappa s\overline{c} \quad (18)$$

The tanh function can in turn be expanded in another infinite series:

$$\overline{M} = -\{1 - \tfrac{1}{3} s^* + \overline{R}_1\} (\theta_m + k)\kappa s\overline{c} \quad (19)$$

where the remainder $\overline{R}_1$ is given by

$$\overline{R}_1 = \frac{2}{15}(s^*)^2 - \frac{7}{315}(s^*)^3 + \ldots . \quad (20)$$

Inversion of the Laplace transforms in equations (19) and (20) gives the first differential form of the transport equation:

$$M = -\{1 - \frac{1}{3}\frac{L^2}{D}\frac{\partial}{\partial t} + R_1\}$$

$$(\theta_m + k)\kappa \frac{\partial c}{\partial t}, \quad (21)$$

where the remainder R_1 is given by

$$R_1 = \frac{2}{15}\{\frac{L^2}{D}\}^2 \frac{\partial^2}{\partial t^2}$$

$$- \frac{7}{315}\{\frac{L^2}{D}\}^3 \frac{\partial^3}{\partial t^3} + \ldots . \quad (22)$$

An alternative form can be derived by using in equation (18) the reciprocal relationship between the tanh and the ctanh functions:

$$(s^*)^{\frac{1}{2}} \text{ ctanh } (s^*)^{\frac{1}{2}} \bar{M} = -(\theta_m + k)\kappa s\bar{c}. \quad (23)$$

Expanding the ctanh function in an infinite series gives:

$$\{1 + \tfrac{1}{3} s^* + \bar{R}_2\}\bar{M} = -(\theta_m + k)\kappa s\bar{c}, \quad (24)$$

where the remainder $\bar{R}_2$ is given by

$$\bar{R}_2 = -\tfrac{1}{45}(s^*)^2 + \tfrac{2}{915}(s^*)^3 + \ldots \quad (25)$$

Inversion of the Laplace transforms in equations (24) and (25) gives the second differential form of the transport equation:

$$\{1 + \tfrac{1}{3}\frac{L^2}{D}\frac{\partial}{\partial t} + R_2\} M = - (\theta_m + k)\, \kappa \frac{\partial c}{\partial t}, \quad (26)$$

where the remainder R_2 is given by

$$R_2 = -\tfrac{1}{45}\{\frac{L^2}{D}\}^2 \frac{\partial^2}{\partial t^2} + \tfrac{2}{915}\{\frac{L^2}{D}\}^3 \frac{\partial^3}{\partial t^3} + \ldots. \quad (27)$$

5 FOUR APPROXIMATIONS

Substituting (6) into (21), neglecting the remainder R_1, and rearranging gives the first approximate differential form of the transport equation:

$$(1 + \kappa)(\theta_m + k)\frac{\partial c}{\partial t} + \theta_m \underset{\sim}{v} \cdot \nabla c - D_m \nabla^2 c = \tfrac{1}{3}\frac{L^2}{D}(\theta_m + k)R\frac{\partial^2 c}{\partial t^2}. \quad (28)$$

Neglecting the right hand side of equation (28) amounts to assuming that the two phases are in equilibrium with each other:

$$(1 + \kappa)(\theta_m + k)\frac{\partial c}{\partial t} + \theta_m \underset{\sim}{v} \cdot \nabla c - D_m \nabla^2 c = 0 \quad (29)$$

The second order time derivative on the right hand side of equation (28) describes the negative "inertia" of the structured medium with regard to changes of concentration.

Substituting (6) into (26), neglecting the remainder R_2, and rearranging, gives the second approximate differential form of the transport equation:

$$(1 + \kappa)(\theta_m + k)\frac{\partial c}{\partial t} + \theta_m \underset{\sim}{v} \cdot \nabla c - D_m \nabla^2 c = -\tfrac{1}{3}\frac{L^2}{D}(\theta_m + k)\frac{\partial^2 c}{\partial t^2} - \tfrac{1}{3}\frac{L^2}{D}\frac{\partial}{\partial t}\{\theta_m \underset{\sim}{v} \cdot \nabla c - D_m \nabla^2 c\}. \quad (30)$$

This form of the transport equation also results if equations (1), (2), and (3) are combined with the equivalent conductance model given by equation (7), provided the conductance α is chosen to be

$$\alpha = \frac{3D}{L^2}. \quad (31)$$

This result is a nice surprise!

The equilibrium equation (29) represents a zeroth-order approximation, while equations (28) and (30) are two alternative first-order approximations. Solving the zeroth-order equation for $\theta_m \underset{\sim}{v} \cdot \nabla c - D_m \nabla^2 c$ and substituting the result in the last term of equation (30) gives equation (28). Thus the right hand sides of equation (28) and (30) are roughly equivalent, as one would hope for two approximations of the same order.

The zeroth-order equation (29) can be used in yet another way to modify the right hand sides of the first order equations (28) and (30). Assuming the flow is one-dimensional and v is constant and neglecting the term accounting for the dispersion, equation (29) reduces to

$$\frac{\partial c}{\partial t} = -\frac{\theta_m v}{(1 + \kappa)(\theta_m + k)}\frac{\partial c}{\partial x}. \quad (32)$$

Substituting the proportional relationship between time and space derivatives implied in equation (32) into the right hand side of equation (28) or of equation (30) gives:

$$(1 + \kappa)(\theta_m + k)\frac{\partial c}{\partial t} + \theta_m v \frac{\partial c}{\partial x} - (D_m + D_e)\frac{\partial^2 c}{\partial x^2} = 0, \quad (33)$$

where the equivalent dispersion coefficient D_e is given by:

$$D_e = \frac{1}{3}\frac{L^2}{D} \frac{\theta_m^2}{(\theta_m + k)^2} \frac{\kappa}{(1 + \kappa)^2} v^2 \qquad (34)$$

This expression for the equivalent dispersion coefficient was also obtained by Passioura (1971; see also Bolt 1979). This result is another nice surprise!

6 CONCLUDING REMARKS

In this paper a variety of linear theories for transport in structured porous media were all shown to have a common basis. The pattern of links between the various theories is similar to that long familiar in linear viscoelasticity (cf., Freudenthal and Geiringer 1958; Leitman and Fisher 1973). The type of analysis presented here for a layered structure can in principle be carried out for any geometry. Results for media with spherical stagnant regions will be presented at the colloquium Euromech 143.

7 REFERENCES

Anzelius, A. 1926, Über Erwärmung vermittels durchströmender Medien. Z. Angew. Math. Mech. 6: 291-294.

Barenblatt, G.I. & Yu.P. Zheltov 1960, Fundamental equations of filtration of homogeneous liquids in fissured rocks. Dokl. Akad. Nauk SSSR. 132: 545-548 (Russian)/522-525 (English translation).

Barenblatt, G.I., Yu.P. Zheltov & I.N. Kochina 1960, Basic concepts in the theory of seepage of homogeneous liquids in fissured rocks. Prikl. Mat. Mekh. 24: 852-864 (Russian)/1286-1303 (English translation).

Bolt, G.H. 1979, Movement of solutes in soil: principles of adsorption/exchange chromatography. In G.H. Bolt (ed.), Soil Chemistry. B. Physico-chemical Models. p. 285-348. Amsterdam, Elsevier.

Chen, P.J. & M.E. Gurtin 1968, On a theory of heat conduction involving two temperatures. Z. Angew. Math. Phys. 19: 614-627.

Einstein, H.A. 1936, Der Geschiebetrieb als Wahrscheinlichkeitsproblem. Dissertation, Eidgenössischen Technischen Hochschule, Zürich. 112 p. (Also: Mitt. Versuchsanstalt für Wasserbau, Eidg. Techn. Hochschule, Zürich).

Freudenthal, A.M. & H. Geiringer 1958, The mathematical theories of the inelastic continuum. In S. Flügge (ed.), Handbuch der Physik, Band VI, Elastizität und Plastizität, p. 229-433. Berlin, Springer Verlag.

Gardner, W.R. & R.H. Brooks 1957, A descriptive theory of leaching. Soil Science 83: 295-304.

Gibson, R.E. & K.Y. Lo 1961, A theory of consolidation for soils exhibiting secondary compression. Norwegian Geotechnical Inst. Publ. No. 41.

Kamp, G.S.J.P. van der 1973, Periodic flow of groundwater. Dissertation Free University, Amsterdam. Amsterdam, Editions Rodopi.

Klinkenberg, A. 1948, Numerical evaluation of equations describing transient heat and mass transfer in packed beds. Ind. Eng. Chem. 40: 1992-1994.

Lapidus, L. & N.R. Amundson 1952, Mathematics of adsorption in beds. VI. The effect of longitudinal diffusion in ion exchange and chromatographic columns. J. Phys. Chem. 56: 984-988.

Leitman, M.J. & G.M.C. Fisher 1973, The linear theory of viscoelasticity. In S. Flügge & C. Treusdell (eds.) Handbuch der Physik. Band VIa/3, Festkorpermechanik III, p. 1-123. Berlin, Springer Verlag.

Maxwell, J.C. 1867, On the dynamical theory of gases. Philos. Trans. Roy. Soc. London. 157: 49-88.

Nusselt, W. 1911, Der Wärmeübertragung im Kreutzstrom. Z. Ver. Deutscher Ingenieure. 55: 2021-2024.

Passioura, J.B. 1971, Hydrodynamic dispersion in aggregated media. Soil Sci. 111: 339-344.

Polya, G. 1937, Zur Kinematik der Geschiebebewegung. Mitteilung der Versuchsanstalt für Wasserbau an der E.T.H. Zürich, Verlag Rascher.

Raats, P.A.C. 1969, The effect of a finite response time upon the propagation of sinusoidal oscillations of fluids in porous media. Z. Angew. Math. Phys. 20: 936-946.

Raats, P.A.C. 1973, Propagation of sinusoidal solute density oscillations in the mobile and stagnant phases of a soil. Soil Sci. Soc. Amer. Proc. 37: 676-680.

Skopp, J., and A.W. Warrick 1974, A two-phase model for the miscible displacement of reactive solutes in soils. Soil Sci. Soc. Amer. Proc. 38: 545-550.

Steggewentz, J.H. 1933, De invloed van de getijbeweging van zeeën en getijrivieren op de stijghoogte van het grondwater. Proefschrift Delft, 138 p.

Streltsova, T.D. 1976, Hydrodynamics of groundwater flow in a fractured formation. Water Resources Res. 12: 405-422.

Taylor, D.W. and W. Merchant 1940, A theory of clay consolidation accounting for secondary compression. J. Math. Phys.

19: 167-185
Van Genuchten, M.Th., and R.W. Cleary 1979, Movement of solutes in soil: computer-simulated and laboratory results. In G.H. Bolt (ed.) Soil Chemistry, B. Physico-Chemical Models. Developments in Soil Science 5B, p. 349-386. Amsterdam, Elsevier.

Proceedings of Euromech 143 / Delft / 2-4 September 1981

Salt movement in unsaturated soil profiles

J.W.VAN HOORN
Agricultural University, Wageningen, Netherlands

1 INTRODUCTION

Models describing the movement of salt in the soil can be continuous column models that use either dispersion, diffusion, and pore water velocity, or theoretical plate thickness as parameters, but also discontinuous column models, in which the column is split up into a series of layers.

All models offer a fair, although not fully satisfactory, description of the leaching process. From laboratory experiments in soil columns it appears that tailing occurs at the end of the leaching process, indicating incomplete mixing of applied water and the original soil solution. This may be ascribed to several factors such as unsaturated flow, the presence of micropores within soil aggregates, and the effect of pore water velocity. By introducing the concept of mobile and immobile water and a retardation factor, however, the results obtained with models can be improved.

Field experiments, too, show that only part of the water is efficient in removing salt. Moreover they reveal a large variation in the amount of leaching water required. This can be attributed to great differences in soil profiles and in methods of water application. Since the volume of soil in field tests is much larger than in laboratory tests, spatial variabilities in soil structure in the field will undoubtedly have a greater influence on water and salt movement.

To simulate salt movement under field conditions, experiments covering a period of several years were conducted in large tanks. The results were analysed according to the plate model theory and by introducing a leaching efficiency coefficient to express the incomplete mixing of leaching water and soil water. The values obtained for the theoretical plate thickness and the leaching efficiency coefficient were compared with those derived from field experiments.

2 EXPERIMENTAL PROCEDURE

The experiments were done in tanks with a surface area of 1.20 m x 1.20 m and 2.40 x 2.40 m and a depth of 1.25 m. In them, a layer of coarse sand and gravel, 0.15 m thick, was overlain by a repacked soil profile of 1 m. At the bottom of each tank, a pipe serving as drainage outlet connected the tank with a drainage reservoir that kept the watertable at a depth of 1 m below the surface. The drainage outlet pipe was provided with a tap for taking samples of the drainage water.

Three tanks, one of 2.40 x 2.40 m and two of 1.20 x 1.20 m were filled with silty clay loam and three other tanks of the same dimensions were filled with sandy loam. The soils were salinized by applying water containing between 250 and 300 meq/liter chlorides (NaCl and $CaCl_2$). Afterwards they were desalinized with water containing approximately 20 meq $CaCl_2$/liter.

Soil water samplers consisting of porous ceramic cups were installed at depths of 12.5, 37.5, 62.5 and 87.5 cm in the silty clay loam and at depths of 17.5, 42.5, 67.5 and 92.5 cm in the sandy loam, in 4 replicates in the smaller tanks and in 8 replicates in the larger ones.

Soil water samples were taken once a week and analyzed for electrical conductivity and chloride concentration. Since chloride salts do not precipitate or dissolve at concentrations used in this experiment and the chloride ion is neither adsorbed nor released by the adsorbtion complex of the soil, the chloride ion can be used as a tracer for water movement.

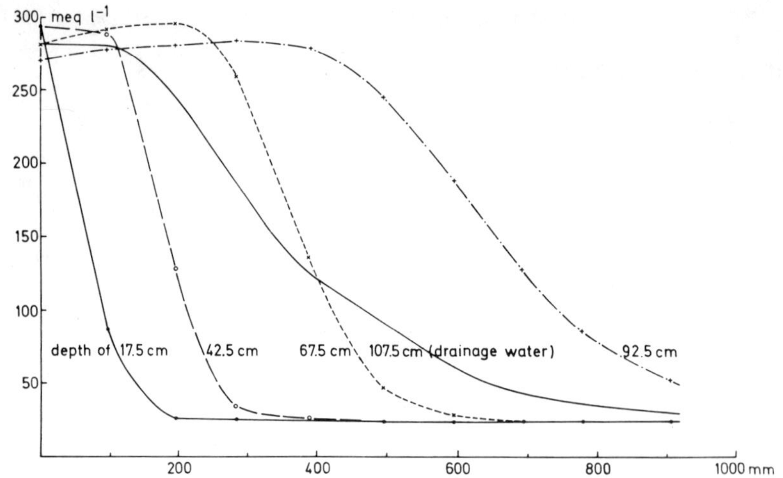

Fig.1 Chloride concentration of soil water versus amount of drainage water. Sandy loam- Desalinization 1979

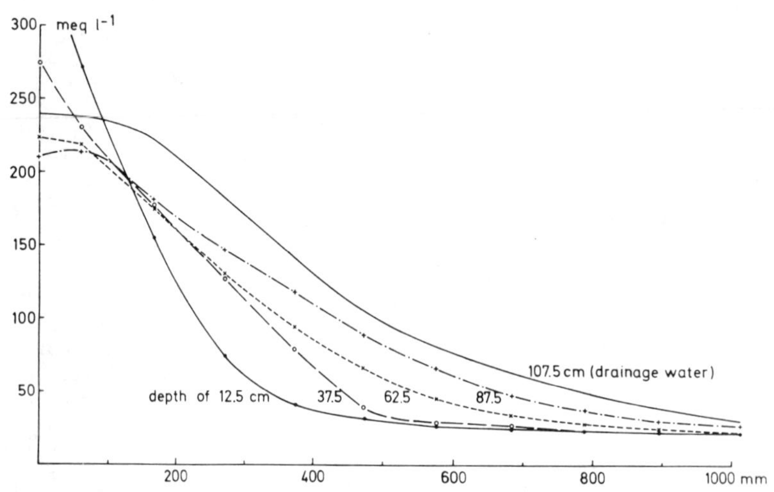

Fig. 2 Chloride concentration of soil water versus amount of drainage water. Silty clay loam- Desalinization 1979

3 RESULTS AND DISCUSSION

3.1 Theoretical plate thickness obtained from leaching curves

Figures 1 and 2 show the chloride concentration of the soil water versus the amount of drainage water of sandy loam and silty clay loam respectively during desalinization in 1979. From these figures it appears that:

- The slope of the curves presenting the relation between the chloride concentration and the amount of drainage water gradually decreases with depth, which is in agreement with the theoretical plate concept of the leaching process;
- The slope of the curves is steeper for sandy loam than for silty clay loam,

which can be ascribed to a much greater electrolyte dispersion in silty clay loam than in sandy loam;
In sandy loam the chloride concentration of the drainage water - being the same as that of the soil water at a depth of 107.5 cm in the underlying sand layer - soon reduces to values less than those at depths of 67.5 and 92.5 cm. After about 100 mm of drainage water, part of the percolation water moves apparently directly from the upper layers through large, already desalinized pores in the lower layers towards the sand layer without mixing with the soil water present in the lower part of the profile. This phenomenon appears more clearly in sandy loam than in silty clay loam. The differnce between the two soil types may be explained by the presence of the continuous large pores in the sandy loam, whereas in the silty clay loam the cracks and large pores are not continuous, leading to mixing over a greater depth in the soil profile.

Results as presented in Figs. 1 and 2 are used to calculate the theoretical plate thickness. For this purpose a continuous column model is used, in which dispersion, diffusion and pore water velocity are taken into account by selecting the thickness of the theoretical plate as a parameter. Van der Molen (1956), using Glueckauf's concept (1955), arrived for desalinization of a homogeneous soil profile at the following equation:

$$C_t - C_i = \frac{C_o - C_i}{2}\left[\operatorname{erfc}\frac{p-1}{(2p)^{\frac{1}{2}}}N^{\frac{1}{2}} - \exp(2N)\operatorname{erfc}\frac{p+1}{(2p)^{\frac{1}{2}}}N^{\frac{1}{2}}\right] \quad (1)$$

where erfc: complementary error function
- C_t : salt concentration at time t
- C_o : original salt concentration
- C_i : salt concentration of the irrigation water
- N : number of theoretical plates above depth d
- $p = q_t/\theta d$: ratio between the amount of water passed through the soil profile and the amount of water contained in the soil profile
- q_t : amount of water passed through the profile at time t, in mm
- d : depth below soil surface, in mm
- θ : water content of the soil, in cm^3/cm^3

By differentiating Eq. (1) and neglecting its second term for large values of N, we obtain:

$$N = 2\pi\left[\frac{q_t}{C_o - C_i} \cdot \frac{dC}{dq}\right]^2_{C_t=(C_o+C_i)/2} \quad (2)$$

where q_t: amount of water passed through the profile that corresponds with the half concentration value $C_t=(C_o+C_i)/2$ on the leaching curves (Figs.1 and 2)

$\frac{dC}{dq}$: slope of the leaching curve at $C_t = (C_o+C_i)/2$.

By dividing the depth by the number of layers N we obtain the theoretical plate thickness or effective mixing length L.

The leaching process can also be described by the following equation:

$$C_t - C_i = \frac{C_o - C_i}{2}\operatorname{erfc}\frac{vt-d}{(4Dt)^{\frac{1}{2}}} \quad (3)$$

where D : dispersion or apparent diffusion coefficient, in cm^2/day
- $v = v'/\theta$: pore water velocity, in cm/day
- v' : filter or Darcy velocity, in cm/day
- t : time, in days

When the second term of Eq.(1) is neglected, it is identical with Eq.(3), provided the ratio between the dispersion coefficient and the pore water velocity is put equal to half the theoretical plate thickness L:

$$\frac{D}{v} = \frac{L}{2} = \frac{d}{2N} \quad (4)$$

Table 1 Summary of experiments on plate thickness

Origin	L=2D/v in cm
Laboratory soil columns	
-clay loam (Dutt, 1964)	2
-sand (Biggar et al, 1966)	1-3
Field experiments	
-sandy loam (van der Molen, 1956)	
desalinization by rainfall	5
salinization by inundation	10-15
-sandy loam (van Hoorn, 1969)	10-20
silty clay loam (the same)	20-40
-loam (Biggar et al, 1976)	16
Tank experiments	
-silty clay loam (Wierenga, 1976)	15
-sandy loam (van Hoorn, 1981)	11.5
silty clay loam (the same)	23

Table 1 summarizes the results of the tank experiments described in this paper as well as those of other experiments, carried out in laboratory soil columns, in tanks or in the field. The values of the plate thickness obtained in tank and field experiments are in general more than 10 cm, whereas those obtained in laboratory soil columns do not exceed a few centimeters. The use of large diameter tanks instead of laboratory soil columns with a small diameter probably creates a variability in pore structure comparable with that found under field conditions.

3.2 Changes in the standard deviation of the chloride concentration obtained from soil water samplers

Since the soil water samplers were installed in 4 replicates in the small tanks and in 8 replicates in the large tanks and all tanks received the same treatment, 16 replicates were available for calculating the mean value and the standard deviation.

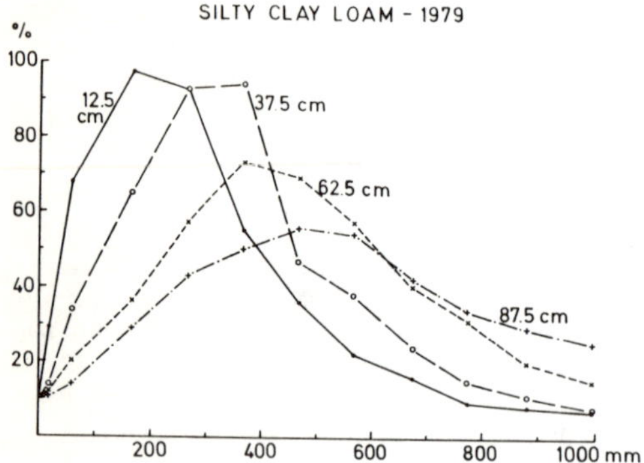

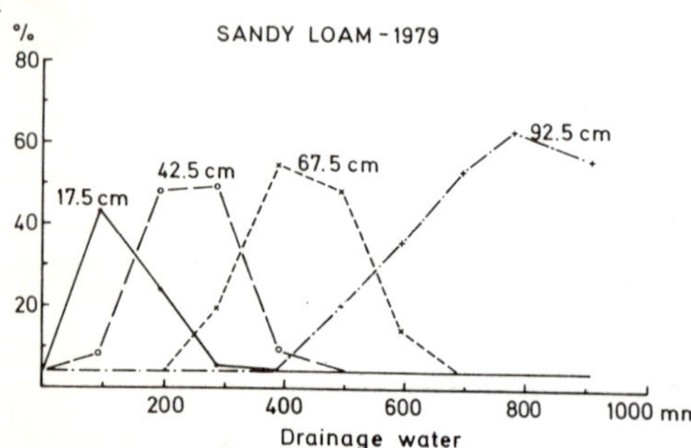

Fig.3 Standard deviation in % of the mean chloride concentration of soil water at different depths during desalinization.

Figure 3 shows the standard deviation in % of the mean value during desalinization. The standard deviation changed strongly during the leaching process and, by the time interval during which it was changing, also shows the difference in dispersion between sandy loam and silty clay loam. In sandy loam its maximum values for the successive layers were clearly separated in time, whereas for silty clay loam they overlapped.

So the sampler observations show a large variation in the chloride concentration of the soil water during the period when the salinity level is rapidly decreasing of increasing. This points towards a highly heterogeneous water movement through the soil profile, which is made visible by the chloride ion as a tracer. At a stable chloride concentration at the end of the leaching process, the variation in the chloride concentration is small. This does not mean that the water movement has become less heterogeneous, but that as soon as the chloride concentration in the soil profile has become more or less the same, the chloride ion has lost its role as a tracer for the water movement.

3.3 The leaching efficiency coefficient calculated from the leaching process.

According to the column theory, the leaching process can be calculated by different methods. The first uses Eq. (1), for which the theoretical plate thickness must be estimated as a parameter for the operation as a continuous model. The second method uses a discontinuous model by splitting up the column into a series of layers or reservoirs. This approach results in:

$$C_t^N = C_i + (C_o - C_i)\ e^{t/T} \sum_{n=o}^{n=N-1} \frac{t^n}{n!\ T^n} \qquad (5)$$

where N : number of the layer or reservoir

t : time in days

$T = \theta L / v'$: time of residence in days, i.e. the ratio between the amount of water contained in a layer and the flux passed through the layer.

If the depth of a reservoir is taken equal to the plate thickness, the results of Eq. (5) are in close agreement with those of Eq. (1). The numerical dispersion in the discontinuous series then represents, with fair accuracy, the physical dispersion in a continuous column.

If both the salt content and the water content vary with depth, the equations become quite complex. It is then better to apply a numerical method based on the following principle. Irrigation water at concentration C_1 mixes with the soil water of the first layer at concentration C_s. The concentration of the soil water in the first layer

Table 2 Summary of experiments on leaching efficiency

Origin	Result	
Fields experiments		
Holland (van der Molen, 1956)	- f coarse texture	> f fine texture
Iraq (Boumans, 1963)	- f coarse texture	> f fine texture
Tunisia (van Hoorn, 1969)	- f coarse texture	> f fine texture
(UNESCO, 1970)	f good structure	> f bad structure
	f small waterappl.	> f large waterappl.
	f increasing with depth	
Spain (Beltran, 1978)	- f good structure	> f bad structure
	f increasing with depth	
Tank experiments (van Hoorn, 1981)	- f decreasing with time	

after mixing ($C_{x_1}^{1}$) can be calculated from:

$$a \text{ mm of irrigation water} \times C_i + b \text{ mm of soil moisture} \times C_{s_1} = (a+b)C_{x_1} \quad (6.1)$$

If the first layer retains an amount of application water equal to c mm, an amount (a-c) mm with a concentration C_{x_1} will percolate and mix with the water of the second layer. The concentration of the soil water in the second layer after mixing (C_{x_2}) can be calculated from:

$$(a-c)\ C_{x_1} + d\ C_{s_2} = (a-c+d)\ C_{x_2} \quad (6.2)$$

By taking small increments in the amount of irrigation water added, the differences between the results obtained by the numerical method and those obtained with Eqs. (1) and (5) are practically negligible, as was shown by van der Molen and van Hoorn (1973).

From Figure 1 it is obvious that soon after the start of the leaching process part of the irrigation water was moving through the soil profile without mixing with the soil water. The degree to which mixing takes place can be expressed by a leaching efficiency coefficient, which can be defined as the percentage of irrigation water mixing with soil water, the remainder of the water flowing through a bypass formed by large pores. The introduction of a leaching efficiency coefficient means that the full amount of water percolated through the soil profile is replaced by the efficient amount of water. This can be obtained in Eq. (1) by using fp instead of p, in Eq. (5) by using ft/T instead of t/T, and in the numerical method by multiplying the water application with the coefficient f.

By comparing the observed chloride concentrations of soil water at different depths with chloride concentrations calculated for a range of f-values, it is possible to obtain an estimation of the leaching efficiency coefficient. In experiments carried out in the field as well as in tanks it was found that the leaching efficiency coefficient is in general less than 1 and that its value depends upon soil texture, structure, depth, time and water application, as is shown by the summary presented in table 2.

4 CONCLUSION

From the previous paragraphs we may conclude that under field conditions the theoretical plate thickness or effective mixing length of a soil profile is much greater than in laboratory columns because of a large variability in pore structure and that irrigation water does not mix completely with the soil water, part of the water moving through a bypass formed by large pores and cracks. The mixing process is affected by the physical properties of the soil and the water application method.

The very heterogeneous water movement through a soil profile, which appears from the large variation in the chloride concentration, explains that chemical constituents transported by soil water move to a much greater depth than would be expected from calculations that are based upon values for theoretical plate thickness obtained in laboratory soil columns and assume complete mixing.

5 REFERENCES

Biggar, J.W., D.R.Nielsen and K.K.Tanji 1966, Comparison of computed and experimentally measured ion concentrations in soil column effluents. Trans. ASAE 9:784-787.

Biggar, J.W., and D.R.Nielsen 1976, Spatial variability of the leaching characteristics of a field soil. Water Res.Res. 12:78-84.

Boumans, J.H. 1963, In reclamation of salt-

affected soils in Iraq. Publ. 11, ILRI, Wageningen.
Dutt, G.R. 1964, Effect of small amounts of gypsum in soils on the solutes in effluents. Soil Sc.Soc.Am.Proc.28:754-757.
Glueckauf,E. 1955, The theoretical plate concept in column separations. Trans. Faraday Soc. 51:34-44.
Hoorn, J.W. van 1969, Leaching efficiency. Int.Conf.on arid lands in a changing world. University of Arizona, Tucson.
Hoorn, J.W. van 1981, Salt movement, leaching efficiency and leaching requirement.Agr. Water Management (in press).
Martinez Beltran, J. 1978, Drainage and reclamation of salt-affected soils in the Bardenas area, Spain. Publ. 24, ILRI, Wageningen.
Molen,W.H.van der 1956, Desalinization of saline soils as a column process. Soil Sci. 81:19-27.
Molen,W.H.van der and J.W.van Hoorn 1973, Salt balance and leaching requirement. Drainage Principles and Applications. Publ. 16, Vol.II,ILRI, Wageningen.
UNESCO 1970, Research and training on irrigation with saline water. Techn.report of UNDP project Tunisia 5.
Wierenga,P.J., M.J.Shaffer, S.P.Gomez and G.A.O'Connor 1975, Predicting ionic distributions in large soil columns. Soil Sci. Soc. Am. Proc. 39:1080-1084.